D1236980

THE NEW GUYS

THE NEW GUYS

The Historic Class of Astronauts That Broke Barriers and Changed the Face of Space Travel

MEREDITH BAGBY

wm

WILLIAM MORROW

An Imprint of HarperCollins*Publishers*

For information, address HarperCollins Publishers, 195 Broadway, New York, NY 10007.

HarperCollins books may be purchased for educational, business, or sales promotional use. For information, please email the Special Markets Department at SPsales@harpercollins.com.

FIRST EDITION

Designed by Nancy Singer

Library of Congress Cataloging-in-Publication Data has been applied for.

ISBN 978-0-06-314197-1

23 24 25 26 27 LBC 5 4 3 2 1

This work is dedicated to all those NASA astronauts, engineers, managers, and administrative staff who put their hearts and souls into the space shuttle program, some of whom made the ultimate sacrifice to further our understanding of the universe.

The spirit of this book belongs to my mom and dad, who, on nights when I could not sleep, took me outside to see the moon—assuring me the universe was a wide and wondrous place—and to my loved ones today who remind me to look up at it still.

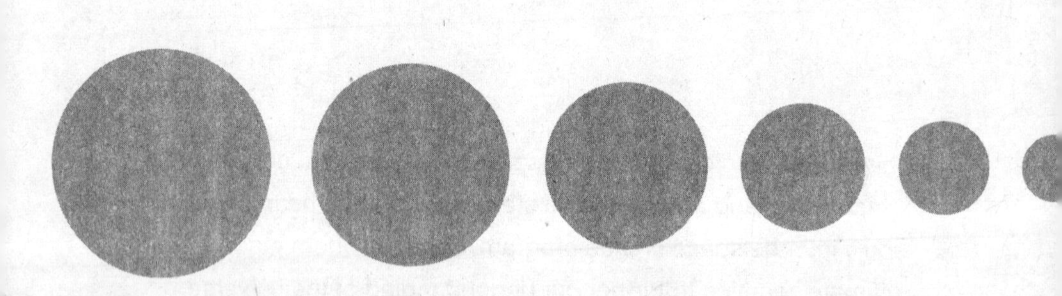

Non est ad astra mollis e terris via.

There is no easy way from the earth to the stars.

—Lucius Annaeus Seneca

Contents

Author's Note

This narrative pulls extensively from interviews conducted for Johnson Space Center's oral history project, as well as congressional hearings, the Rogers Commission and *Columbia* investigation reports, and the expansive wealth of written and audiovisual NASA histories available through the Johnson Space Center archives.

In addition, much of this book's content is sourced from nearly one hundred interviews conducted from 2017 to 2022 with the members of Astronaut Group 8, the eponymous New Guys—as well as their family, friends, and former colleagues—in Texas, Florida, and California, and of course, over Zoom.

The book draws heavily on accounts written by and about Astronaut Group 8, including autobiographies of the New Guys: *Go for Orbit: One of American's First Women Astronauts Finds Her Space* by Rhea Seddon, *Handprints on Hubble: An Astronaut's Story of Invention* by Kathryn D. Sullivan, *Riding Rockets: The Outrageous Tales of a Space Shuttle Astronaut* by Mike Mullane, and *Tumbleweed: Six Months Living on Mir* by Shannon Lucid. I relied on two biographies to provide background on Ron McNair and Sally Ride, respectively: *In the Spirit of Ronald E. McNair—Astronaut: An American Hero* by Carl S. McNair and *Sally Ride: America's First Woman in Space* by Lynn Sherr. For the history of shuttle astronaut selection, I leaned on *NASA's First Space Shuttle Astronaut Selection: Redefining the Right Stuff* by David J. Shayler and Colin Burgess, and for background on George Abbey, *The Astronaut Maker: How One Mysterious Engineer Ran Human Spaceflight for a Generation* by

Michael Cassutt. For their firsthand accounts of watershed moments in NASA's history I looked to: Michael D. Leinbach and Jonathan H. Ward's *Bringing Home Columbia: The Untold Story of a Lost Space Shuttle and Her Crew*; Allan McDonald's *Truth, Lies, and O-Rings: Inside the Space Shuttle Challenger Disaster*; and Joseph D. Atkinson and Jay M. Shafritz's *The Real Stuff: A History of NASA's Astronaut Recruitment Policy.*

To capture the viewpoints of those who have sadly passed on, I have relied on their own firsthand accounts recorded in television interviews, archival videos, newspaper articles, and personal papers as well as recollections of their peers who went through many of the same experiences. I have also attempted to re-create certain scenes from their perspectives, so that their voices are not lost from this story. These scenes have been carefully stitched together from accounts provided by friends and colleagues who lived and experienced these moments with them.

Throughout the book, to foster a sense of intimacy, I refer to main characters by their first names and secondary characters by their last names. Wherever possible, I have used direct quotations from interviews and additional research to illustrate and enliven the story. However, for many interior thoughts and feelings of these individuals—moments of quiet reflection and honest reaction—direct quotes were not always available. In these cases, I have interpreted how these individuals likely may have thought and felt and represented those thoughts, feelings, and reactions in italics.

These italicized thoughts and reactions are closely based on personal interviews and research and remain true to the spirit of the sentiments conveyed by those who were there. They have been included to attempt to capture the joy, fear, sorrow, confusion, and excitement felt all throughout the shuttle's storied history by those who had a direct hand in this incredible program. The hope is that these liberties help to bring these heroes to life.

Character List

THE NEW GUYS

Guion "Guy" Bluford Jr.

An aerospace engineer and US Air Force fighter pilot, he was the first African American to space, flying four space shuttle missions.

Anna Lee Fisher

A physician, she was the first mother and the fourth American woman to space. She married Group 9 astronaut Bill Fisher and served as chief of the Astronaut Office's space station branch.

Robert "Hoot" Gibson

A US Navy fighter pilot, he flew on five space shuttle missions, married fellow New Guy Rhea Seddon in 1981, and served as chief of the Astronaut Office and the deputy director of flight crew operations.

Frederick "Fred" Gregory

A US Air Force fighter and helicopter pilot, he was the first African American to pilot, and later command, the space shuttle. He became the first African American deputy administrator in the agency's history.

Steven "Steve" Hawley

A PhD in astronomy and astrophysics, he flew five space shuttle missions, married fellow New Guy Sally Ride, and served as deputy chief of the Astronaut Office.

Shannon Lucid
A PhD in biochemistry, she was the sixth American woman to space, flew five space shuttle missions, and held the record for most flight hours in orbit by any woman in the world until June 2007.

Ronald "Ron" McNair
A PhD in physics, he was the second African American to space and was tragically killed in the *Challenger* disaster.

Ellison "El" Onizuka
A US Air Force flight test engineer and test pilot, he was the first Asian American to space and was tragically killed in the *Challenger* disaster.

Judith "Judy" Resnik
A PhD in electrical engineering, she was the second American woman and the first Jewish astronaut to space and was tragically killed in the *Challenger* disaster.

Sally Ride
A PhD in physics, she was the first American woman to space, married fellow New Guy Steve Hawley, and served on the Rogers Commission, investigating the *Challenger* accident.

Margaret "Rhea" Seddon
A surgeon, she was the fifth American woman to space, flew on three space shuttle missions, and married fellow New Guy Hoot Gibson.

Kathryn "Kathy" Sullivan

A PhD in geology, she was the third American woman to space and the first American woman to perform a spacewalk.

FELLOW ASTRONAUTS

James "Jim" Bagian

A physician and Air Force flight surgeon, he became a NASA astronaut in 1980 and played a key role in *Challenger*'s recovery.

Robert "Crip" Crippen

A US Navy pilot who became a NASA astronaut in 1969, he was the first pilot of the space shuttle, led the recovery effort following the *Challenger* disaster, and was the director of the space shuttle program at NASA Headquarters.

William "Bill" Fisher

A physician and member of the 1980 astronaut class, he was married to New Guy Anna Lee Fisher.

John Young

A US Navy pilot and Apollo astronaut, he flew in space six times and was the first commander of the space shuttle and chief of the Astronaut Office from 1974 to 1987.

NASA BRASS

George Abbey

A US Air Force captain assigned to NASA in 1964 during the Apollo program, he was rejected from the astronaut corps in 1965 but went on to serve as director of flight operations at Johnson Space Center, selecting the first class of space shuttle astronauts—the New Guys.

James Beggs

A former business executive who served as the sixth administrator of NASA from 1981 to 1986.

Carolyn Huntoon

A PhD in physiology who joined NASA in 1970, she was the head of the Endocrine and Biochemistry Laboratories at Johnson Space Center, the only woman on the Group 8 astronaut selection board in 1978, and the first woman director of Johnson Space Center.

Christopher Kraft

The legendary NASA flight director who helped create Houston's Mission Control, he served as director of Johnson Space Center from 1972 to 1982.

CHAPTER 1

AD ASTRA

Washington, DC. Spring 1977.

Judy Resnik clicked her way up Independence Avenue with the Washington Monument behind her and the Capitol building in her sights. A breeze swept over the Tidal Basin, edging cherry blossoms off their branches, creating a flurry of white and pink petals. It was spring and the air carried a cool dewiness that she associated with beginnings.

A spirited, twenty-eight-year-old electrical engineer with a newly minted PhD, Judy had secured a plum job studying lasers at the Xerox Corporation in El Segundo, California. She was scheduled to start work later that fall, but today she was thinking of tossing that very well-honed future away.

A few weeks earlier, Judy had heard a story on the radio: The National Aeronautics and Space Administration (NASA) was, for the first time ever, recruiting women and minorities to become astronauts in its new space shuttle program.[1] Since the dawn of the agency, NASA had culled its astronauts exclusively from the ranks of white, male military pilots. During the Apollo era, the astronaut corps relaxed its requirements to include civilians, but there had never been a female or minority astronaut. Now NASA was creating a new role—*mission specialist*—for the shuttle. Larger than previous spacecraft, the shuttle was designed to carry seven passengers, leaving plenty of room for scientists, not just pilots, to journey to space. Citizens with strong backgrounds in scientific fields, including engineers and medical doctors, were encouraged to apply. The ad piqued Judy's interest, then slowly began to consume her thoughts.

Judy worked all angles to make herself an exceptional candidate. Standing five feet, four inches with dark wavy hair and a cherubic face, the Akron, Ohio, native had never been much of an athlete.[2] Now she ran every day. She ate a low-carb, high-protein diet to shed extra pounds—lots of steaks and salads—and signed up for pilot lessons on the weekends.[3] "I'm sort of a competitive person," Judy said.[4] "If I want something, I want it."[5]

Judy, with help from her ex-boyfriend Len Nahmi, found and read everything she could on NASA, including astronaut Michael Collins's autobiography, *Carrying the Fire*. Collins had flown the Apollo 11 command module *Columbia* around the moon in 1969 while his crewmates, Neil Armstrong and Buzz Aldrin, made their first landing on the lunar surface. His book detailed his experiences at NASA in the Gemini program through his time on Apollo and provided insight on how NASA chose its astronauts. Who better to give Judy tips on her NASA application or the selection process?

As Judy passed the verdant National Mall, the government buildings, and the war monuments, she made her way toward her destination: the National Air and Space Museum. Gerald Ford, in one of the last acts of his short presidency, cut the red ribbon on the new museum a year earlier.[6] Collins was now spending his days as the director of the museum, archiving his predecessors' and his own contributions to space exploration.

Here, a continuum of hangars housed the world's largest collection of aircraft and spacecraft. Exterior bays, surfaced with pink Tennessee marble, flanked a mighty all-glass atrium.[7] Walking up the stairs, Judy came to a towering sculpture that welcomed visitors. Three stainless steel shafts ascended one hundred feet into the air and came together in a pointed tip, exploding in a triple gold-star cluster. "Ad Astra," Latin for "to the stars," symbolized man's conquest of space.[8]

In the atrium, Judy came face-to-face with the 1903 *Wright Flyer*, the first successful powered plane to ever leave Earth's surface. On December 17, 1903, Orville Wright traveled a triumphant 120 feet, staying airborne for twelve glorious seconds on the dunes of Kitty Hawk, as his brother Wilbur ran beside him, marking the beginning of human aviation. Considered hobbyists at the time, the two young men wrote to the Smithsonian for any and all information on manned flight, obtaining technical papers on Samuel Langley's Aerodromes and Otto Lilienthal's German hang gliders. Among other documents they studied were pages from Leonardo da Vinci's *Codex on the Flight of Birds*. Four centuries earlier, the Florentine artist had sketched out aviary techniques, over five hundred drawings, and accurately identified the four fundamental forces that governed flight: lift, drag, thrust, and gravity.[9]

How far da Vinci's observations would take us, no one yet knew.

In another room hung the *Spirit of St. Louis*, the single-engine monoplane that Charles Lindbergh had flown from New York to Paris in 1927 without even a windshield or a parachute. Five years later, Amelia Earhart became the first woman to cross the Atlantic in her cherry-red *Lockheed Vega*.

World War II's flying fortresses and V-2 buzz bombs gave way to Cold War stealth aircraft and strategic nuclear missiles. The inventors of those instruments of destruction helped create the revolutionary vehicles that took us beyond Earth's bounds. The *Freedom 7* Mercury space capsule that had spirited Alan Shepard to the heavens lay cocked on the atrium floor like a genie's bottle. Its design had given way to the roomier Gemini IV capsule, from which Americans conducted the first spacewalk. Eight years earlier, while Judy attended her last year of college at Carnegie Tech, she watched Neil Armstrong take his first steps on the moon. Now, Armstrong's Apollo 11 spacesuit stood encased before her in glass, pitched slightly forward, as if whispering a secret.

With the trajectory of aviation and space travel laid out in progression, it was easy to see that humans are historical creatures. One generation conceded its hard-earned knowledge to the next. Expedition, inquiry, and exploration ran through the human heart. Through wars and famine, the rise and fall of nations, the change of landscapes and cultures, even when our humanity escaped us, science marched forward. Judy wanted to join the procession of science, the journey to the stars. *Ad Astra*. If this museum extolled the accomplishments of mostly men, with the notable exception of Amelia Earhart, then she would serve as a representative for the other half of the planet.

Judy strolled past the exhibits, the throngs of tourists, and museum workers. She took an elevator up to the museum's offices. Collins was not expecting her. She did not have an appointment, nor did she have any credentials for entry. She did, however, have a hand-drawn floor plan of the museum that she and Len had pieced together. Hopefully, that map would lead her to his office. She strode down the hallway, hoping a security guard would not spot her. The hum of office work—phones ringing, papers shuffling, and doors opening—floated through the air. Judy turned a corner and saw his office doorplate. Poking her head in, Judy noted a man with the same high forehead and thinning brown hair she had seen in pictures. Kind brown eyes looked up at her.

Can I help you?

Judy smiled. She did not mince words.

"Hi, Mike, how are you? My name's Judy Resnik, and I want to be an astronaut."

Malibu, California. Spring 1977.

"What do you mean, you're going to be an astronaut?" Carl McNair pressed the phone against his ear to make sure he heard his brother Ron correctly.

"NASA's looking to recruit for the new space shuttle program," Ron said. "So, I applied."

Applied? Carl thought. "Won't there be thousands of people applying? What makes you think NASA will choose you?"[10]

"Why not? I have the credentials," Ron said.

Carl shook his head and laughed. Only eleven months separated the two brothers. They had spent most of their lives in sync. No one knew Ron better than Carl. Even so, Ron could still surprise him.[11] Sure, Ron was one of the few Black men who had earned a PhD in physics from the Massachusetts Institute of Technology (MIT). He was a star athlete in high school and college and a black belt in karate. He had recently started working as a laser physicist at Hughes Research Laboratory in Malibu, California—the very company that had invented lasers in the 1960s.[12] *If anyone's qualified to work for NASA*, Carl thought, *it's Ron*. Still, an *astronaut*? Astronauts were "larger-than-life heroes like Neil Armstrong and John Glenn," not little ol' Ron McNair from Lake City, South Carolina.[13]

Ron had grown up in the segregated South only three generations removed from slavery. Being from the "wrong side of the tracks" was more than an expression in Lake City. Actual railroad tracks separated the Black and white neighborhoods.

"We knew our place," Carl said.[14] The Ku Klux Klan made sure of that, threatening Black citizens they deemed "uppity," running their preacher out of town for supporting the National Association for the Advancement of Colored People (NAACP), and burning a cross on their Boy Scout leader's lawn.[15,16]

Ron, the second son born to father Carl Sr., a mechanic, and mother Pearl, a schoolteacher, grew up in humble surroundings.[17] The first home he could remember did not have indoor plumbing. The roof on the old "unpainted weather-beaten frame house" leaked; an entire room was made unlivable from water damage.[18] On rainy days, the family set out pots and pans on the floors and the furniture to catch the dripping water. On stormy nights, Ron would fall asleep to the metallic *plink plink plink* of raindrops striking cookware.

To help their family make ends meet, Ron and Carl, then twelve and thirteen, headed out to the fields to harvest cotton, tobacco, beans, and cucumbers.[19] On summer break or holidays, they would wait on the side of the road at sunrise for work. "Y'all boys want to crop my 'bacca?" a passing farmer would yell out his truck window.[20] "Yes, sir," the boys would bellow back, climbing into the back of the pickup.

From sunrise to sunset, under the merciless South Carolina sun,

Ron and Carl fell into a flow with the other tobacco croppers: *Stoop, pluck, shuffle.*[21] Farmers preferred employing young folks like Ron and Carl to pick the lower leaves of their tobacco, priming the plant to grow taller, faster.[22] The veteran croppers spurned the work. After a few hours, the boys understood why. "By midday, I felt as if someone had stuck a knife blade deep into my lower back," said Carl. By day's end, the boys could barely stand.[23] The tobacco leaves were rough on the skin and harbored tiny worms that could bite. More perilous still were the rattlesnakes hiding in the loamy furrows. Picking cotton was worse. The spiny husks tore at their hands, making them bleed.[24]

"I gained qualities in that cotton field," Ron said. "I got tough. I learned to endure. I refuse to quit."[25] He observed the hard faces of the "gray-haired men and women who spent their whole lives at such labor" and knew he did not want the same fate.[26]

Education would be his way out. By age three, Ron was reading. By age four, his family said, "he was too smart to stay home."[27] Carl Sr. lied about Ron's age to get him into kindergarten.[28] By five, Ron was already wowing teachers, marching around school with a pencil behind his ear and a notebook in his hand. Ron flew through the material, skipped another grade, and joined his brother Carl's class.[29] Some siblings might have felt competitive with their genius kid brother, but instead Ron and Carl became inseparable, bonded by a shared love of learning.[30] "Ron inspired the rest of us," Carl said. "We wanted to beat him, and even when we couldn't, we wanted to close the gap."[31] On Saturday mornings, Ron and Carl paged through their family's World Book Encyclopedia, dreaming about the wide world outside Lake City.[32]

Inside Lake City, Ron and Carl studied in segregated and underfunded schools, with used textbooks, overworked teachers, and no extracurriculars.[33] In elementary school, Ron's teachers told him and his peers that they would have to "try twice as hard, work twice as diligently, and learn twice as much" as whites to succeed. Ron took that message to heart, ranking number one in his class and becoming a star on the school's football team.[34]

In 1966, Ron's interest in science was sparked with the debut of Gene Roddenberry's *Star Trek.*[35] Ron never missed an episode, racing home with Carl to warm up the TV before the stroke of the hour when the show would start.[36] They marveled that a Black woman, the beautiful and talented Nichelle Nichols, played a leading role on primetime television as Lieutenant Nyota Uhura. "As a Black woman, she had two strikes against her in our twentieth-century world," Carl said. "But in the enlightened society of the twenty-third century, she was a high-ranking and fully capable starship officer."[37,38]

Ron followed his love of science to a high school summer program at Virginia Union University in Richmond. There, a simple question changed his life's trajectory. "Have you ever considered pursuing a PhD?" a professor asked him, after observing Ron's capabilities. By the time Ron returned home, he had formed a grand plan. "I'm going to get a PhD," he declared.

After graduating as valedictorian at sixteen years of age, Ron followed Carl to North Carolina Agricultural and Technical (A&T) State University, one of the largest Historically Black Colleges and Universities (HBCUs) in the South. He earned a bachelor's degree in physics and, after completing a summer program at MIT, was accepted into their physics PhD program. He toiled for five years in "the dungeon," his nickname for MIT's basement physics labs. Upon graduation, he received his job offer from Hughes Laboratories. At twenty-seven, he and his new wife, Cheryl, moved into a sun-splashed apartment in Malibu, California.

Wearing his bucket hat, jogging shorts, and athletic socks pulled all the way up to his knees, Ron took long runs on the beach. Although he would only venture ankle deep in the water, he loved the expanse of the ocean, the way it suggested the infinite. Looking out over the cliffs of Malibu at the Pacific, he was worlds away from those cold Boston winters; the hot tobacco fields were nothing more than a distant memory. Using his intelligence and determination, he had climbed his way out of poverty.[39] He was living the dream in Malibu: a happily married, financially comfortable laser physicist.

Still, Ron felt something was missing.

Ron had grown accustomed to tackling life, not simply enjoying it. Carl noticed the restlessness: "[Ron's] work at Hughes was exciting, but the shift into a five-day workweek routine felt like a step toward the humdrum."[40] "After watching him surmount one obstacle after another, I understood he couldn't be content on the plateau. No sooner did he reach an arduous peak than he was searching the horizon for a loftier mountain to scale."[41]

That new challenge came in the form of a black-and-white brochure that landed in his work mailbox announcing NASA's search for a new astronaut class. A few months later, Nichelle Nichols, the famed Lieutenant Uhura herself, appeared on a television advertisement pleading NASA's case. Sporting a royal blue flight suit and posed in the Apollo Mission Control Center, she appealed to "the whole family of humankind, minorities and women alike. Now is the time," she said. "This is your NASA." Ron sat stunned, as if hit by a *Star Trek* phaser beam. She was talking to him.

Harbor General Hospital at the University of California, Los Angeles (UCLA). June 1977.

Anna Lee Tingle sat across from Harbor General's chairman of the department of surgery, knots forming in her stomach. Even *she* could not believe what she was saying.

"I'm forfeiting my appointment to the surgery department." Anna was turning down the most prestigious residency in the hospital, the position she had worked her heart out for these last three years. Watching the chairman's reaction was like watching a volcano erupt. This man had been her mentor, her advocate. He had gone out on a limb to recommend her—after all, she was a woman, the only woman to be accepted to the program *ever.*

Why are you turning down this once in a lifetime opportunity, might I ask? He was livid.

"I've applied to NASA to be an astronaut," Anna said.

The chairman sat in unbelieving silence, as if to say . . . *Preposterous. What do you mean you're applying to be an astronaut?*

Anna stumbled through an explanation. NASA was recruiting doctors to become astronauts, and yes, even women were encouraged to apply. The chairman balked at her decision.

You'll never get a job in surgery anywhere in this country if I have anything to do with it, he added before ending the meeting.

Anna's ears burned red; her heart beat fast as she left the room. She had set her career aflame, and for what? Maybe he was right. There were no doubt thousands of qualified scientists who had applied to NASA for the new position. Why in God's name would NASA ever pick her?

Slight with fine features and straight brown hair layered into a no-nonsense cut, Anna looked younger than her years, starkly different from those chiseled, thrill-seeking rocket jockeys who populated the first astronaut classes and were made famous by Tom Wolfe's *The Right Stuff.* Yet, like them, Anna possessed a sharp, focused mind and an unflappable disposition, both attributes honed by her medical training. And just like them, her life had been shaped by the United States military.

Anna had been a military brat, ferried around the globe by her father, a drill sergeant. They lived an untethered life, whisked away at a moment's notice from friends and family, moving to far-flung places across the United States and Europe. Anna's mother, Elfriede, tried to create a sense of normalcy by bringing a box of keepsakes with them to decorate each place they lived. As the family grew with the arrival of three brothers, the moves became harder and the housing more cramped. By the time Anna was ten, she had had thirteen different homes. Naturally shy, Anna felt displaced, always the outsider. She longed for a stable, storybook home.

Without a consistent community, Anna turned inward to her studies. She was gifted at math and science. She devoured biographies of women pioneers like Susan B. Anthony, Marie Curie, and Elizabeth Blackwell and dreamed of one day making her own mark on the world. She stole her brothers' science fiction books but tossed them away when none featured female characters. *Why couldn't women be explorers?*

On May 5, 1961, Anna's imagination sparked. That morning in Fort Campbell, Kentucky, Anna's seventh-grade physical education teacher took the class outside to listen to Alan Shepard's historic space flight. The launch, scheduled for the early hours, had been delayed as squalls swept up the coast of Florida, past the Cape Canaveral launchpad where

Shepard sat strapped into his Mercury *Freedom 7* space capsule. Anna sat in the dewy grass with her classmates, listening to their teacher's transistor radio as Shepard waited impatiently for the flight controllers to give him clearance.

"Fix your little problem and light this candle!" he barked.

NASA finally did light Shepard's candle, taking the *Freedom 7* into space for fifteen minutes and twenty-eight seconds. That short flight catapulted America into the Space Race with the Soviet Union. For Anna, it was the beginning of a fascination that would not quit, as if Shepard had passed the flame directly to her. The launch, the scope and majesty of it, ticked all the boxes for Anna—adventure, science, and being part of something big and meaningful. But Anna, age eleven, demure, waifish, barely breaking sixty pounds, an introvert with downcast eyes and a timid voice, seemed an improbable candidate to be an astronaut. After all, NASA did not even allow civilians to apply, much less girls.

Fifteen years after Shepard's launch, Anna met someone with whom she could share her passion. During her second year at UCLA medical school, she began her residency at Harbor General, where she met Bill Fisher, a visiting medical resident from University of Florida.

At twenty-six years old, Anna had already married and divorced a fellow graduate student. She was not looking for romance, but as she got to know Bill over coffee in the hospital cafeteria, she liked him more and more. Like Anna, Bill was a military brat, a nomad looking for a home. He had a quick mind, a wide grin, and the all-American look of a 1950s movie star. *Gosh,* she thought, *he'd be easy to fall in love with.*

One day after work, Anna found a note from Bill on the windshield of her car. "I have cheese and wine. Would you like to have a glass?"[42]

She shoved the note in her pocket and walked toward the medical residents' trailers where Bill lived. As promised, he served cheap red wine and smooth Camembert. They went for a walk, looking up at the dark sky.

"Do you know how to tell the difference between a planet and a star?" Bill asked.

"Of course!" she said. "Stars blink, planets don't."

Bill confessed that he had applied to be a NASA astronaut when he was twelve years old. Project Mercury was in its infancy then, and Alan Shepard's historic flight had not even taken place. The space agency kindly wrote back, advising him to get a little more schooling.

Anna could not believe her ears. Should she share her secret? What if Bill laughed at her? It was a risk, but she decided to take it.

"I want to be an astronaut, too," she confessed. "It's been a dream all my life."

Bill did not laugh. He understood. In fact, he fell in love with her because of it. In the spring, they moved in together, getting a little beach apartment in Rancho Palos Verdes with a view of the Pacific. At night, they sat on the porch and named the constellations. That December, Bill hung an engagement ring on the Christmas tree and asked her to marry him; she said yes. Six months later, Anna was finishing her rounds at Harbor General when Bill changed her future again.

"Paging Dr. Tingle," she heard over the loudspeaker.

Oh no, she thought, *another surgery*. But this was the *good* kind of emergency. Moments before, Bill had learned from a friend that NASA was recruiting men and women with medical backgrounds to become astronauts. Here was Bill and Anna's chance to make their dream a reality. They had only three weeks to finish their applications before the June 30 deadline. For two overworked, sleep-deprived medical residents, the task was daunting, made even more so by the sheer volume of information the applications required: school transcripts, letters of recommendation, and a complete medical history. But finish they did, and in the nick of time. Anna postmarked hers the day before the deadline, Bill the day of.

The next day, not wanting to deprive someone else of a spot, Anna declined her surgical residency. If she had set her career aflame, so be it. Now that she knew that someone like her could become an astronaut, there was no other choice but to apply and keep applying until NASA said yes.

New York City, New York. July 1977.

Do you know Sally Ride?

A man in a trench coat appeared outside Molly Tyson's midtown office, asking a lot of questions. Molly, a young associate editor at *womenSports* magazine, puzzled at her strange visitor.[43]

Yes, I know Sally Ride. Did Sally rob a bank?

The man flashed a badge too quickly for Molly to see what government agency he worked for, but he was clearly an investigator, and he would be the one asking all the questions.

It had been two years since Molly and Sally roomed together at Stanford University. Sally was a studious physics major interested in black holes and white dwarfs. Molly was an outgoing English major prone to quoting Shakespeare. Their relationship went back even farther. Yes, Molly knew Sally Ride, but why was this mysterious man asking?

Because she gave us your name as a personal reference for her application . . . to NASA.

Then it clicked. Sally had asked Molly to write a reference letter to go along with her bid for NASA's astronaut program.

Like a go-up-to-space astronaut? Molly had been floored when Sally had asked. *Why would you want to spend the best years of your life in space?*

I'm an astrophysicist, where better to study the stars than in space? Sally said.

It was outrageous that Sally wanted to use Molly as a personal reference. After all, Molly was the one person in the world who knew the secret that could keep Sally from being chosen.

The government agent fired questions at Molly. *Did Sally drink, smoke, or take drugs? How did she react to stress or emergencies? Did she keep her room neat and her hair combed? Was she a good leader? How was she at following directions?*[44]

Molly reported that she was a "model witness." Or maybe it was easy to speak about Sally in glowing terms. After all, Sally was an extraordinary person, equally talented intellectually and physically. Sally, who grew up a true "Valley Girl" in Encino, California, was the only girl invited to play softball with the neighborhood boys.[45] Young Sally dreamed of playing shortstop for the Dodgers until her mother, Joyce, explained to her flummoxed daughter that only men could play baseball.[46,47] To compensate, Joyce stuck a tennis racket in Sally's hand.[48] It was love at first swing.

Sally played like a natural but lacked some of the discipline required to dominate the courts. *Was it laziness or a fear of failure?* No one knew. "She made a point of being defiant about it," said Susan Okie, her childhood friend. "She'd be lying on the floor watching TV, and her father would say, 'Go run around the block.'"[49] By middle school, Sally ranked twentieth in the Southern California junior tennis circuit and won a tennis scholarship to the prestigious Westlake School for Girls in the posh Los Angeles neighborhood of Holmby Hills.[50]

Even though Sally did not always fit in with the "beautiful beach blonde California girls," she benefited from the small class sizes and engaging teachers.[51] One Westlake teacher, Dr. Elizabeth Mommaerts, challenged Sally's reluctance to apply herself. "She had an eagle eye out for any girl that was interested in science." Mommaerts gave Sally puzzles to solve, invited her and the other girls to her home for "fabulous French meals," and took Sally's class to university-level scientific lectures at UCLA. By the time Sally graduated from Westlake, she was hooked on astronomy and physics.[52]

Sally matriculated at Swarthmore, a liberal arts college outside Philadelphia, to study physics but was disappointed by the school's tennis scene. With dreams of going pro, she came back to sunny California, where she could hone her court skills. Maybe she would win Wimbledon and then a Grand Slam or two? That was when she bounded back into Molly's life. After zooming up to Palo Alto in her new red Toyota, Sally knocked on Molly's dorm room door wearing cutoff shorts and a white T-shirt and spinning a tennis racket in her hand.

Want to hit some balls? Sally lobbed an opening line.

I thought you were at Swarthmore, Molly said.

I missed the sunshine, Sally replied.

From that day forward, the two women were inseparable. They volleyed every day—sometimes until 11 PM—under the floodlights at the Stanford tennis courts, betting pomegranate seeds. The goal was to keep the ball in play, not hit a winning shot. Molly drew Sally into the English department, where Sally earned a double major. Sally made Molly her doubles partner. They traveled across the nation to tennis tournaments, delighting in working Shakespeare into their conversations on long car rides.[53] Molly counseled Sally on her decision to give up her tennis career when Sally came face-to-face with the reality that she might not be big enough to dominate the courts—a fact she learned playing women's tennis champion Billie Jean King in an exhibition doubles match.

Not interested in pursuing a career in which she could not be spectacular, Sally decided to focus her attention on academia. She applied for a PhD program in physics at Stanford. If not Wimbledon, maybe she could win a Nobel prize. In 1977, one year away from earning that PhD, Sally took another leap and applied to NASA.

"I only lied once," Molly said later of her encounter with the government agent. Sally must have held her breath.

Did you tell him our secret?

In their junior year, Sally broke up with her long-distance boyfriend. Then Molly and Sally took a risk. "There was a lot of fumbling around in the dark," Molly said. "We were clueless people coming up against a pretty formidable taboo. Kissing felt incredibly daring, sleeping in the same bed felt incredibly daring."[54]

They fell in love.

They moved into a two-bedroom apartment, filled with books and decorated with Shakespeare posters. A bust of Julius Caesar, purchased at a flea market, welcomed guests. Sally made a mean guacamole and strawberry pie. On holidays, they visited each other's families. They were like any young lovers in the first blush of romance, except almost no one, not their families, not most of their closest friends, knew they were a couple—nor did they consider themselves gay. Yes, they loved each other. Yes, they slept together, but, as Molly put it, "In our ignorance we probably would have put [gays] in the category of Those Icky People . . . We didn't want to be associated with our idea of what gay people were like."[55]

Sally was a "super compartmentalizer" who wanted to live under the guise of friendship instead of coming out as a couple. After nearly five years together, the duplicity became too much for Molly. "I was very burdened by the secret," Molly said. "I didn't feel like hiding a big part of myself anymore." Molly told Sally she was leaving her, and headed to New York to start a new, open life.[56] Sally's heart broke.

The government agent, however, had not asked Molly about Sally's romantic history. He wanted to know if Sally was a tidy roommate. "I figured that the dust and dirty dishes wouldn't accumulate in a space capsule the way they had in our apartment," Molly said.[57]

Sally must have been relieved. Yes, the astronaut corps was now open to women and people of color, but it most certainly would not be open to gay people. In 1977, being gay was no longer defined as a mental illness by psychiatrists in the American Psychiatric Association's Diagnostic and Statistical Manual (DSM), as it had been prior to 1973, but was still considered a "sexual orientation disturbance." Most states still criminalized sodomy and sexual orientation was not a protected class in employment.

No, Molly had not told anyone their secret. Whatever questions the agent posed Molly handled with aplomb. Molly's interview, along with Sally's other recommendations, catapulted her into the "yes" pile, a thin stack of applications that ultimately found its way on to the desk of NASA's selection committee for the final cut.

Johnson Space Center, Houston, Texas. August 1977.

After a massive, year-long campaign, the NASA Selection Committee had been inundated with over eight thousand applications from highly qualified, credentialed candidates across the nation for both their mission specialist and pilot astronaut positions. All candidates had to have at least a bachelor's degree in engineering, science, or mathematics; pass a physical; and be between sixty and seventy-six inches tall. Pilot candidates needed at least one thousand hours of flight time. The committee read every single application. Slowly, they began weeding out the least qualified, based on school transcripts, in-field experience, and personal recommendations. After following up on those recommendations, as they had with Molly, they whittled the pile down to hundreds—and then debated.

Ultimately, they invited 208 lucky souls to NASA's Johnson Space Center, where each applicant would undergo a series of tests and interview with the Astronaut Selection Board. Of those, NASA would choose only thirty-five for NASA's newest group of astronauts. In the final week of July 1977, Jay Honeycutt of NASA's selection committee began calling the finalists.[58]

In Malibu, Ron's phone rang. The news from Honeycutt did not take him by surprise. When Ron told his brother that he had been chosen to interview, Carl shook his head and laughed. *Of course.* "Send me a postcard from the moon," Carl said.

In Palo Alto, Sally picked up the phone before class. Honeycutt invited her to the weeklong interview and evaluation at Johnson Space Center in Houston. *Bring jogging clothes for a treadmill test.* Judy, now living oceanside in Redondo Beach, California, got the same call as she was running off to work. Her strategy with Michael Collins had worked.

A few miles away in Los Angeles, Anna was sitting at her kitchen table with her fiancé, Bill, when her phone rang.

"Dr. Tingle, are you still interested in the astronaut program?" Honeycutt asked on the other end of the line.

"Yes, sir!" Anna said, smiling from ear to ear.

Bill looked at his fiancée expectantly—perhaps the call was for them both? No, the call was for Anna alone. Honeycutt wanted her to fly out the following week when she and Bill had planned to get married. Anna asked Bill what she should do. Ask NASA to postpone the evaluation, or fly to Texas? He did not bat an eye.

"Do it," he said. "Go to Houston. We'll figure out the rest."

CHAPTER 2

LIGHT THIS CANDLE

Over Houston, Texas. August 1977.

As Anna's plane dipped below the storm clouds, she got her first look at Space City.

Once sparsely populated coastal plains, prairie, and timber forest, the expanse now known as Space City, or Houston, Texas, had been the fertile home of Native peoples, who hunted game and fished in the plentiful waterways that snaked through the marshlands. In the 1500s, Spanish explorers crept north from Mexico, forever shattering the order of these pastoral communities. American colonists, pushing west in the nineteenth century, all but destroyed their way of life, transforming the area into a trading hub for cotton and agricultural products. So important was the commercial center that Houston was named the capital of the Republic of Texas when the state won its independence from Mexico.

At the turn of the twentieth century, a far more mercurial resource was discovered as developers in Beaumont drilled a water well, but got the world's largest oil gusher instead. Producing seventeen million barrels in its first year, the Gulf Coast of Texas was saturated in black gold. Fortune seekers arrived by the tens of thousands, turning the marshlands into oil fields and the port city of Houston into a boomtown. Another half-century passed, the oil ran dry, and Houston began looking for its next big adventure. In the 1960s, Texas congressman Albert Thomas and Texas-born Vice President Lyndon B. Johnson provided an answer, championing Houston over twenty-two other cities to win the massive government contract to build NASA's new center for manned space flight. Now, as Anna looked down upon the sprawling metropolis, the city bloomed inland from the Gulf of Mexico for forty miles and boasted over two million residents.

As Anna deplaned, she pushed through the hot and humid winds of August, then boarded a small Twin Otter jet that took her the last forty miles to her destination: the Lyndon B. Johnson Space Center.[1] In truth, the center was closer to Galveston Bay than Houston. Humble Oil and Refining Company (now Exxon Mobil Corporation) had donated sixteen hundred acres south of Houston to Rice University, which then conveyed the land to NASA. The flat stretch of sun-beaten cow pasture and malodorous swampland was now called "Clear Lake" for its proximity to a body of water of the same name. Clear Lake was more of a harbor than a basin, a brackish interlude between Clear Creek and Galveston. Just over six thousand people lived in Clear Lake City before NASA's arrival in 1961. By the late 1960s, however, forty-five thousand people called Clear Lake home. Restaurants, shops, motels, and other businesses sprouted like weeds to accommodate the space boom.

Stepping out of the small puddle-hopper airplane and on to the black tarmac, Anna steeled herself for the week ahead. Honeycutt had told her the selection process would include a physical fitness exam, so she had spent the weeks prior exercising to prepare. She had always been at the top of her class academically but had never worked out in her life. Bill blasted the *Rocky* theme song in their apartment to get her blood pumping. Anna put on sneakers and a T-shirt and jogged along busy Crenshaw Boulevard, huffing and puffing on the hills as Bill cheered her on.

A few days before arriving in Houston, she and Bill had canceled their large Florida wedding celebration in favor of a small ceremony before close friends and family at Wayfarers Chapel in Rancho Palos Verdes, California, a stunning wood-beam-and-glass church designed by the son of Frank Lloyd Wright. On a cliff overlooking the Pacific Ocean, Anna Tingle became Anna Fisher.

At Johnson Space Center, Anna would compete against the other 207 highly qualified, highly ambitious candidates selected to participate in NASA's weeklong evaluation. The selection board split the applicants into ten interview groups to be assessed over a course of ten weeks, from August to November 1977.[2,3] The civilian finalists ran the gamut of professions—MDs like Anna, engineers like Judy, and physics PhDs like Ron and Sally. Still, NASA planned on recruiting at least half its new class from its tried-and-true astronaut pool: military test pilots.

Some of those candidates represented a departure from the white male prototype of the past. They included Fred Gregory and Guy Bluford, two Black Air Force pilots with impressive combat records in Vietnam. Another finalist was thirty-one-year-old Japanese American Air Force pilot Ellison "El" Onizuka, who had emerged from the coffee fields of Kona, Hawaii, to become one of the top instructors at the US Air Force's elite test pilot school. If chosen, he would be the first Asian American astronaut.

Inevitably, some of the pilot candidates resented the incursion by civilians into what had always been the turf of the military. Air Force captain Mike Mullane had flown 134 missions in Vietnam as a weapon systems operator in the back seat of a fighter jet.[4] He had seen the blood and guts of enemies and friends hit the windshield of his gunship in the jungle. From age sixteen, he had dedicated his life to flying, with pie-in-the-sky hopes of one day becoming an astronaut. "As soon as there were astronauts," Mike proclaimed, "I wanted to be one."[5] Now he finally had a shot, and he had to compete against *these guys*? "Perpetual students [whose] greatest fear had been getting an A- on a research paper," he said. Eggheads who "a few months earlier [were] be-bopping through the Student Union in . . . save-the-whales T-shirt[s]"? Who had "*accidentally* [seen] the NASA astronaut selection announcement on the bulletin board"?[6]

It was a joke.

Now that he, Anna, Judy, Sally, Ron, and the rest of the finalists were heading to Houston, the time for dreaming and hoping and speculating was over. They would all have an equal shot to prove they had what it took to be America's next astronauts.

Anna tried to steady her breathing as NASA technicians fixed electrodes to her chest to monitor her heart rate. She strapped on an oxygen mask and slowly crawled into the airtight "Personal Rescue Enclosure," also known as the rescue ball. The rescue ball was a white sphere of spacesuit material, less than a yard in diameter, not much larger than a beach ball, and with only a small window to avoid complete sensory deprivation. NASA had designed the rescue ball to transport astronauts through space from one spacecraft to another in case of emergency.[7]

Anna tucked into the fetal position as they zipped her in. If this were a real mission, there would be nothing separating her from the cold vacuum of space but fabric and a tow line. Anyone with an ounce of claustrophobia would not last a second, but Anna lasted longer than a second. *Had it been ten minutes? Twenty?* The technicians did not tell her how long the test would last. After a while, she started relaxing. *It's like being back in the womb.* The next thing she knew, a technician was gently shaking her awake. Apparently, her long hours in the emergency room, whirlwind marriage weekend, and preparation for Houston finally caught up to her. She conked out. No need to worry about an elevated heart rate in Anna's case. She aced the test.

The night before, Anna and the other candidates had checked in to their NASA-appointed accommodations at the Sheraton Kings Inn, an English Tudor-style motel with a medieval theme. Its kitschy sign loomed high above NASA Road 1, topped with a plumed knight's helmet and supported by a fifty-foot lance. As Anna and the other finalists milled about in one of the hotel's briefing rooms, George Abbey, a tall, broad-shouldered man with drooping eyes, a military crew cut, and a permanent five o'clock shadow, called them to their seats. He spoke so softly and indistinctly that Anna could barely make out what he said; she sized him up as a midlevel bureaucrat.

George led with a description of the vehicle the candidates would be flying. Many of them, including Anna, knew little if anything about the new space shuttle. Most of them did not care what they flew so long as they went to space. Anna examined a sketch of the wide, blunt-nosed spacecraft. Shaded black-and-white with short, stubby wings at the back, the shuttle looked more like a plane than the teardrop-shaped Apollo and Gemini capsules of the decades prior. Its systems were much more complex. The shuttle would launch like a rocket, ferry payloads like a transport vehicle, and return to Earth like an airplane. The shuttle would be the most advanced spacecraft ever created.

Getting this behemoth to space required two sentry-pillar rockets, which would be bolted to the orbiter, and a wide fuel tank that fed the onboard engines. Astronaut Story Musgrave compared the improbable configuration, known as the shuttle stack, to "bolting a butterfly on to a bullet."[8] A fragile engineering feat, the orbiter housed even more delicate humans, tethered to over six million pounds of explosive thrust. Despite the clear dangers, none of the finalists batted an eye.

George explained what the finalists should expect for the week ahead. They would be evaluated based on rigorous medical and psychological tests, as well as a final interview with top NASA brass. They should report to Johnson at 8:00 AM—and know there might be reporters.

George's comment proved an understatement. Reporters, especially eager to glimpse the women, swarmed the candidates as they stepped off the bus at Johnson the next morning.[9]

"If women join the astronaut corps, will there be romance in space?" one journalist queried.

"Will women be too emotional to fly missions in space?" shouted another.

"Isn't being an astronaut too dangerous of a job for a mother?"

"Most astronauts have children. They're dads," finalist Shannon Lucid retorted.[10]

Shannon, a thirty-four-year-old chemist and flying enthusiast from

Oklahoma, had three children herself; that was not going to stop her from her heart's desire: "becoming an explorer of the universe."[11] The women then had to listen to their male counterparts wax poetic on their presence in the selection. One offered that having women in the space program could add "spice to life." A few others agreed that all the women in the group were real "lookers."[12]

After jousting with the press, the finalists hurried to their physicals. The "body invasions," as one candidate called them, included hearing and vision exams; heart rate monitoring and treadmill tests; ear, nose, and throat checks; and numerous blood draws.[13,14] A twenty-four-hour urine test required finalists to fill a large jar as they went about their day. The proctological exam with preparatory enemas brought out at least one candidate's competitive nature.[15]

Mike Mullane wanted to prove his superiority over the weakling civilians. Before his proctology exam, Mullane overheard a civilian finalist complaining that he had failed because his colon was not clean enough for an accurate reading. *Pathetic.* No way was that going to happen to Air Force captain Mike Mullane. Mullane crouched on the toilet reading the pre-exam instructions. Hold in the colonic for a measly five minutes? *Screw that.* Mullane did fifteen. Take no more than two enemas? *Hysterical.* Mullane used four. He sat in that bathroom gritting his teeth for an hour. When the proctologist finally performed the exam, he told Mullane he had never seen a colon so "well prepared."[16]

That was how it was done.

NASA required the candidates to undergo physical tests designed to measure "exercise capacity and muscular strength."[17] Ron, a black belt in karate, demonstrated his power and agility. Sally, who ran twenty to thirty miles a week, tired out her observers on the treadmill. "One of the best performances by a female candidate seen at this lab," an impressed doctor noted.[18]

Many of the women applicants, though, had not made exercise a priority. "I was in the middle of my medical training. Working out was not something you ever had time for," Anna said.[19] Some of the female candidates gave the "pull-ups, push-ups, and other physical exercises" the old college try, but few could do a single pull-up.[20] "Of all the women . . . they collectively did three chin-ups. Collectively. I did thirty-five," grumbled finalist Jim Bagian, an exceedingly fit medical doctor cum mechanical engineer.[21] Ultimately, NASA believed that "weightlessness [could be] the great equalizer" and eliminated the test as a criterion.[22]

Not everyone was happy about the decision. "You got chicken shitted out of this," someone on the selection board later told Bagian, explaining that he had been cut from the class after higher-ups prioritized including more women in the group and encouraging him to try again next time.[23] Some NASA managers would deny ever altering the original selection to include more women.[24]

If the physical requirements had been a bone of contention, the psychological tests were equal opportunity confusion. Through one door, Dr. Terry McGuire welcomed the candidates and tossed out softball questions like, *If you were reincarnated as an animal, what would it be?* Anna said she would be a lion for its power and independence. Sally went with dolphin; they are social and smart. The military candidates saw themselves as stallions or American bald eagles—*fast, free, and powerful.*

Through the other door, finalists encountered Dr. Edward Harris. Playing bad cop to McGuire's good cop, Harris aimed to disorient, provoke, and otherwise destabilize the interviewees. He invited candidates to sit in a misshapen office chair: its arms were too far apart, its seat too high off the ground. He alternated the temperature in the room between frigid and hot. Harris asked blunt-edged questions: *Have you ever been sexually assaulted? Have you ever contemplated suicide? Do you fear death?*[25]

His methods knocked some finalists off their game. Astronomer Steve Hawley froze for several seconds when asked *Do you want a Coke?* He gave the question some serious thought, turning over all the potential implications in his head: "Do I want a Coke? If I say yes and he brings me a Coke, should I offer to pay for it?"

"No," he finally said, deciding the question was too loaded.[26]

The most consequential of the tests, though the candidates did not know it at the time, was their interview with the fourteen-member Astronaut Selection Board, composed of famous astronauts, NASA brass, and top scientists. For Anna, the day got off to a bad start.

Anna stumbled across campus in a jewel-green velour jumpsuit. It was not her stylish wedge shoes that made her wobble, but her dilated pupils. She had just finished an eye exam with the ophthalmologist and was rushing to her final interview. Thankfully, a NASA employee took pity on her and guided her to the hallway outside the interview room. After a few minutes, she heard her name called and found herself in front of the interview board. The dozen or so individuals sitting before Anna, the very people who would decide her fate, looked like indistinct blurs.

Anna relied on each board member's verbal introduction. There was senior astronaut John Young, who led the office and had walked on the moon.[27] Dr. Carolyn Huntoon, the only woman on the panel, headed up the biomedical labs at NASA.[28] As the top female manager at NASA, she had traveled with Dr. Joseph Atkinson Jr., director of Johnson's Equal Employment Opportunities programs office, to recruit women and minority candidates from universities across the country. Atkinson was the only person of color on the board.[29]

The man that seemed to be running the show, however, the gentleman to whom the other members of the panel were exceedingly deferential, was the mumbling crew-cut bureaucrat Anna had pegged as a midlevel administrator on the first night at the Kings Inn—George Abbey. She and the others would come to know him by his other epithets: the Godfather, the Dark Lord, and UNO, for "Unidentified NASA Official." George was director of flight crew operations, the all-powerful force behind the Astronaut Office, responsible for operational planning and flight crew activities for NASA's human spaceflight missions and the man who would run the finalists' lives if they were lucky enough to be accepted. He was singularly understated and mystifyingly anonymous for a man in charge of one of the most high-profile teams in America.

Anna's mind raced. Had she treated him with the proper reverence? Had she missed opportunities to impress him? Was she the only one who had not realized who he was?

During the ninety-minute interview, board members quizzed her on technical matters, her personal life, and what she did last summer. The last question always belonged to George, and it always came out of left field to throw candidates off their game. *What do you think about the Panama Canal?* was his go-to for a while. Then word got around that everyone should bone up on their Panamanian history, so George changed his question. *What do you think about the Suez Canal?*[30]

After George finished, he would stand up abruptly and mutter a barely perceptible thanks—and like that, the interview was over. None of the finalists knew where they stood, even if, like Anna, they happened to know that the US had recently voted to return the Panama Canal to Panama.

Although the official testing had concluded, Anna's toughest challenge lay ahead, and her old insecurities were about to resurface at the most inopportune time. The final event of the week was an evening cocktail party, a chance for candidates to mingle with NASA brass. The most notable guest was Dr. Christopher Kraft, director of Johnson Space Center and a legend of the American space program. Kraft had seen the center through the historic Apollo era. He was the father of Mission Control, the system that had guided Neil Armstrong and Buzz Aldrin during mankind's first lunar touchdown and saved the Apollo 13 crew from catastrophe. A good word from Dr. Kraft might change her fate. The other finalists all seemed to sense that as well. One after another, they walked right up to Dr. Kraft to introduce themselves and snag a couple minutes of invaluable face time with the center director.

As Anna stared at this NASA icon, with his serious blue eyes and his dusty brown hair, shyness overcame her. Suddenly she felt like a little girl again, shrinking in a room full of strangers. After all, wasn't that what she was, compared to these great men of history? Just a little girl standing in their shadows. Who was she to think that she could follow in their footsteps, perhaps even earn a few pages in the history books herself?

Suddenly, Anna sensed someone at her side.

"You should come talk to Dr. Kraft."

It was George Abbey. The Godfather himself. He spoke softly as always but had a kind look. Anna smiled and nodded, and George led her over to Dr. Kraft.

"This is Anna Fisher," he said. "She's one of the doctors."

After he made the introduction, George tactfully faded into the background. Anna and Dr. Kraft went on to speak for some time. They got along splendidly.[31]

Midway through selection week, Anna called Bill and insisted he fly out to meet her, so she could share everything she learned about NASA and show him Houston. Bill had not heard about whether he would be selected as a finalist, but he was dying to come tour NASA.[32] Each night, he waited for Anna to return to the Kings Inn and regale him with stories of her day. Anna discreetly mentioned to George that her husband, Bill, applied to be an astronaut, too; maybe a little gentle lobbying would help him get an interview.

Before the Fishers returned to California, they rented a car and toured the neighborhoods of Clear Lake. They frowned at the more traditional homes and the gaudy Texas oil mansions, dreaming of tearing one down and creating their own California open-floor-plan paradise here in Houston. Together, they would *make new magic in a dusty world.*[33] As they headed to the airport, they took one last swing by Johnson Space Center, the row of unassuming, low-slung buildings, where a bold new chapter in space exploration was beginning.

Anna thought about how good it would feel to be part of this place, to pass through its halls each day, soaking in its knowledge, to be among the titans of space, to sit in the giant aluminum frame of the shuttle, atop thousands of pounds of liquid fuel on fire, and be thrust at an unimaginable speed to the heavens above. To skim the surface of the ether, to see what was beyond, and then to float quietly among the stars.

Back in California, Bill got good news: Honeycutt invited him to a later round of interviews in Houston. Maybe Anna's lobbying paid off? Any other couple might have felt the sting of competition, but after Bill returned home from his weeklong interview, he swapped intel with Anna. They rooted for one another. Christmas rolled around, then the new year, and still no word from NASA. Had they been passed over? How was Anna going to support herself now that she had blown up her surgery career?

On Sunday, January 15, Anna got a call—not from NASA, but from Roy Neal of NBC News. A little bird had told him Anna would be hearing from Houston early tomorrow morning. Could the film crew come over to the Fishers' apartment to document the occasion?

That night, she and Bill had dinner with Judy Resnik, who lived only a half hour away in Redondo Beach. They spent the evening speculating over a bottle of wine on Judy's apartment balcony. Reporters had also called Judy and were arriving first thing in the morning.

Bill, on the other hand, had heard nothing at all.

At 6:00 AM Monday, the NBC camera crew arrived on the doorstep of Anna and Bill's apartment. The couple waited nervously as the crew set up in their kitchen. An hour later, the phone rang. The cameras started to roll as Bill picked up the receiver.

"Honey, it's for you," he said, handing Anna the phone.[34] George Abbey was on the line.

"Remember that job you applied for at NASA?" George said. Anna grinned widely. Of course Anna remembered "that job." It had only occupied her every waking thought for the last six months.

"Yes," she answered.

"Are you still interested in being an astronaut?" he asked.

"Oh, you know I am," she said.

Anna thought of being eleven years old again, unsure and alone in a strange land. She thought of the crackling of the transistor radio her teacher put out on the field, of Alan Shepard, of the anticipation he must have felt as he sat, cold in his spaceship, yearning to break free. She was going to be an astronaut, just like him.

Yeah, Anna thought, *light this candle.*

CHAPTER 3

TEN INTERESTING PEOPLE

Teague Auditorium, Johnson Space Center. January 31, 1978.

"Judy Resnik," Dr. Christopher Kraft said into the microphone.

As Judy walked across the carpeted stage, she blinked back the bright lights from the reporters' flashing cameras. *This isn't a dream,* she reminded herself, heading to one of the plastic scoop-back chairs set out on the stage. Earlier in the day, Judy dressed for the event, straightening her naturally curly hair and putting on a white button-down shirt, a below-the-knee skirt, and conservative heels. Now, seated next to Sally Ride, Judy crossed her ankles and folded her hands gently in her lap, as prim and proper as a charm school student. As the rest of the thirty-five astronauts filled the stage in order of last name, the scene looked more like a high school graduation photo than the announcement of NASA's next astronaut class.

Two weeks prior, on January 16, NASA had released the names of its newest astronauts.[1] All three major networks—ABC, CBS, and NBC—headlined the story.[2] "NASA chose the thirty-five persons who will ride the space shuttle to orbit and back," reported CBS News's Walter Cronkite. "Among them are three Blacks, one Oriental, and six women: Ronald McNair, Fred Gregory, Guion 'Guy' Bluford, Ellison Onizuka, Anna Fisher, Shannon Lucid, Judy Resnik, Sally Ride, Rhea Seddon, and Kathy Sullivan. Godspeed to the newest generation of astronauts."[3] For weeks, print, radio, and television reporters had hounded the new recruits, calling them at their homes and showing up unannounced at their workplaces and universities, but today was their official introduction to the world.

Born during and after World War II, these men and women had come of age in the revolutionary 1960s. Their teenage and college years began with the optimism of John F. Kennedy's presidency, but faded into disillusionment during Vietnam, Watergate, and the assassinations of their young leaders—Martin Luther King Jr., John and Robert Kennedy, and Malcolm X. They lived through and participated in political protests and upheaval on their university campuses and in their cities. Many were upending the social rules established by former generations and challenging revered institutions. Compared to their predecessors, they were more open-minded, free-thinking, and resistant to patrician rules.

Astronaut Class 8 looked like none before it. Gone were the rows of buzz cuts and dark suits that typified every prior astronaut group. Yes, there were the usual military pilots of old: Twenty-one were military officers, nineteen of whom had served in Vietnam. But now, there were civilians, too: doctors, engineers, chemists, physicists, earth scientists, and astronomers. They came from twenty-six states and twenty-seven different academic institutions. They were atheists, Jews, Buddhists, Protestants, and Catholics. They ranged in age from twenty-six to thirty-eight. At least one was gay, although no one knew it at the time. They wore flared collars, polyester suits, and turtlenecks. Some of the women donned skirts, heels, and jewelry. Others wore smart slacks and colorful scarves. Some of the men sported full beards.[4] When Judy's class walked on stage for the first time as a group, everyone could feel the new energy, the uptick in tempo. NASA was about to be transformed.

The moment was decades in the making. Since its founding twenty years prior, NASA had excluded women and minorities from its astronaut corps for reasons as complex as the era itself.[5] The way America's space agency came into existence lay at the heart of the issue.

NASA was born out of conflict. Less than a decade after the USSR developed an atomic bomb, its successful launch of *Sputnik* all but shattered America's faith in its own technical superiority.[6] In 1958, President Eisenhower responded with an ambitious plan to match and eclipse the USSR's technological prowess, creating the National Aeronautics and Space Agency: NASA.

NASA's goals loomed large. Funded with over seven billion dollars for its first five years, NASA would coordinate all American space activity. Eisenhower designated NASA as a civilian organization, privately hoping to avoid infighting between the military agencies over space and curb the military's ballooning budget.[7] Publicly, he promoted NASA as an agency *for all mankind*.[8] In practice, the notion that space would be accessible to all types of people would be a long time coming.

After much debate, NASA administrators decided that its astronauts should be culled exclusively from the armed forces, whose members had experience handling dangerous conditions and classified information. More important to Eisenhower, NASA trainers could test military candidates quickly and secretly, shielding the agency from scrutiny or embarrassment in front of the Soviets.[9]

Administrators further narrowed their qualifications for the astronaut position in a brief announcement: "men less than 40 years of age, under 5 feet 11 inches, a graduate of a test pilot school, holder of a bachelor's degree or equivalent, having a total of 1,500 hours of flight time, and qualifying in high-performance jet aircraft."[10] These test pilots, who assessed the military's experimental aircraft, were the subject of Tom Wolfe's *The Right Stuff*—adrenaline junkies who thrived on putting their "hide on the line" every single day.[11] Within four months of the announcement, NASA selected the Mercury 7—a homogenous group of seven white military men.

The choice to limit astronaut selection to the military defined a generation of astronauts and precluded women from applying altogether. Following World War II, women could join the military, but they were prohibited from aircraft combat and could only comprise 2 percent of the enlisted force and 10 percent of the officer class.[12] In 1967, Congress repealed the later restrictions, and by the mid-1970s female pilots were allowed in the Navy and the Air Force, but still restricted from the combat positions.[13] People of color fared little better; the armed forces remained segregated as late as 1948. In the 1970s, there were relatively few Black officers and even fewer Black test pilots.[14]

The white military pilots that filled the astronaut corps quickly entrenched themselves, demanding that the Mercury space capsule prioritize pilot decisions over automated systems. They cultivated an image that their test pilot skills embodied the beating heart of space travel.[15]

As the program matured beyond the Mercury 7, NASA stretched astronaut qualifications to include scientists and pared back its flight experience requirements. The agency recruited Neil Armstrong even though he was no longer active military and accepted John Glenn without a college degree.[16] Still, NASA managers refused to open the door to women and people of color. "I do not think that we will be anxious to put a woman or any other person of a particular race or creed into orbit just for the purpose of putting them there," said NASA administrator James Webb in 1962.[17]

On April 12, 1961, the Soviets wowed the world and embarrassed America with yet another space triumph. Yuri Gagarin achieved global renown as the first human to orbit Earth. Coupled with the calamitous Bay of Pigs invasion, John F. Kennedy's presidency was gasping for air. In urgent need of political capital, Kennedy looked to the stars. On a warm, sunny day in September 1962, Kennedy spoke to a crowd of more than forty thousand people at Rice University and challenged the country to put a man on the moon before the end of the decade.[18]

"We set sail on this new sea because there is new knowledge to be gained, and new rights to be won, and they must be won and used for the progress of all people," Kennedy told the crowd and the nation. "We choose to go to the moon in this decade and do the other things, not because they are easy, but because they are hard; because that goal will serve to organize and measure the best of our energies and skills."[19]

The challenge captured the imagination of a generation. Congress followed suit with financial support, sending NASA's budget soaring to $5.25 billion in 1965.[20] The new Apollo effort meant new jobs as NASA built out its facilities: the Manned Spacecraft Center in Houston, Texas (later the Lyndon B. Johnson Space Center); the Marshall Space Flight Center in Huntsville, Alabama; and the Launch Operations Center at Cape Canaveral, Florida (later renamed after President John F. Kennedy).[21] The Apollo program employed four hundred thousand Americans and required the support of over twenty thousand industrial firms and universities.[22]

As America embarked on the Apollo program, President Kennedy, who campaigned on the promise of equality, worried that racial inequities in the space program were becoming too glaring to brush aside. The executive director of the National Urban League, Whitney Young, urged Kennedy to consider a Black astronaut to get African American youth interested in science and technology. Meanwhile, Edward R. Murrow, a broadcast journalist who parlayed his experience to lead the United States Information Agency, further nudged Kennedy. With racial protests roiling at home, Murrow thought the launch of the first Black man might paint a better image of America abroad. "Why don't we put the first non-white man in space?" Murrow wrote to NASA administrator James Webb. "We could retell our whole space effort to the non-white world, which is most of it."[23]

For President Kennedy, whose steps toward civil rights progress were cautious but resolute, the inclusion of African Americans in the astronaut selection process was a sine qua non. "For symbolic purposes in crossing the frontiers of space," JFK aide Fred Dutton wrote to Adam Yarmolinsky, special assistant to the secretary of defense, "this country would have qualified members from minority backgrounds who would find great response throughout the world."[24] After receiving an unsatisfactory reply,

Dutton gave Yarmolinsky a deadline of November 1, 1961, to recruit a minority astronaut candidate.[25]

That was how a young, charismatic second lieutenant in the US Air Force named Ed Dwight came to receive a curious letter on Kennedy White House letterhead early that year. "I'm inviting you to become America's first Negro astronaut," it began. "If this project succeeds, you will end up being the greatest Negro that ever lived."[26] Dwight was only twenty-six years old but had already logged more than two thousand hours in high-performance jet aircraft. He had three consecutive years of outstanding pilot ratings and a cum laude engineering degree.[27] Thinking the letter was a practical joke, Dwight ignored it. When none of his colleagues stepped up to claim credit, he finally took the letter to his superiors.

"Don't do this, Ed," they warned. "It's a whole different ball game over there. They're gonna eat you up and spit you out." Within days, however, Dwight privately completed his astronaut application. "If I was successful, I would be madly successful," he said. "But if I failed, there was no coming back."[28] The news of NASA's first Black trainee at the Air Force test pilot school made a big splash in the press. Dwight appeared on magazine covers and in newspapers across the nation. He received fifteen hundred fan letters a day.[29]

Not everyone was thrilled.

Chuck Yeager, the Air Force officer and test pilot who first broke the sound barrier in 1947, ran the school at Edwards Air Force base where Dwight trained. Yeager was a national hero who reached mythic stature thanks to Tom Wolfe's depiction of him in *The Right Stuff.* Dwight, however, saw a very different side of Yeager.

"Washington is trying to cram a [racial slur] down our throat," Yeager complained.[30] He hazed Dwight relentlessly. "Don't talk to him. Don't invite him to parties. Don't invite him to drinks," he instructed his other test pilots. "He'll be gone in six months."[31] Yeager called Dwight into his office weekly to ask if he was ready to quit. "Who got you into this school? Are you some kind of Black Muslim out here to make trouble?" Yeager ranted. "Why in the hell would a colored guy want to go into space anyway? As far as I'm concerned, there'll never be one to do it," he threatened.

"You've done nothing more," Dwight told Yeager evenly, "than make me more determined to prove a Negro can do anything a white man can."[32]

Yeager later denied he mistreated Dwight and claimed instead that he had a problem with Dwight's qualifications.[33] By other accounts, Yeager took it as a personal affront that there were even Black men in the Air Force.[34] In the end, when the Group 3 astronauts were announced, Ed Dwight was not among them. He had not even been among the final thirty candidates considered.[35] His exclusion created a stir in the national press. The Soviet Union's news agency eagerly sent a report to outlets around the world that Dwight "was rejected for astronaut duty because he is a Negro."[36]

In 1966, three years after his rejection from the astronaut corps, Dwight resigned from the Air Force.[37] America got its next Black astronaut candidate, Major Robert "Bobby" Lawrence Jr., in June 1967. Tragically, six months later, Lawrence died in a plane crash when his Air Force trainee crashed their F-104 Starfighter into the desert outside Edwards Air Force Base.[38] Some agitators sent his devastated widow letters of unconscionable cruelty. "No coons on the moon," read one. Another said, simply, "I'm glad Bobby Lawrence is dead."[39]

Like their Black male peers, women pilots like Geraldyn "Jerrie" Cobb were banging on the door, only to be shut out. A feisty and press-savvy blond with a movie star smile, Cobb boasted a resume that rivaled the Mercury 7. By twenty-eight, she had amassed seven thousand hours of flight time, held three world aviation records, and served as an executive for Aero Design and Engineering Company, a California-based aerospace corporation.[40]

In 1960, Dr. William Randolph Lovelace, who assessed NASA's male astronauts, wanted to know how women might perform on the same exams. He invited Cobb to be his trial case. Cobb underwent the same grueling physicals, psychological screenings, and flight simulations the Mercury 7 had. She passed all three phases, finishing in the top two percent of all candidates (male or female) and even surpassing some of the Mercury 7.[41]

Cobb's outstanding performance inspired Lovelace to recruit other women for testing.[42] Knowing that NASA and the nation might not be ready for women astronauts, he ran the program in secret, without the agency's knowledge. He tested twenty-five women; twelve passed the physical examinations and went on to further testing.[43] The women candidates, later nicknamed the "Mercury 13," were forestalled when the military refused to allow Lovelace to use their equipment for flight simulation testing. After all, the military argued, NASA had no intention of sending these women to space.

Undeterred, Cobb and fellow Mercury 13 pilot Jane "Janey" Briggs Hart, wife of Michigan senator Philip Hart, went on a letter-writing tear. They shot off pleas to President Lyndon Johnson, the Senate and House committees on aeronautics and science, NASA administrator James Webb, and Manned Spacecraft Center director Robert Gilruth. "Manned spaceflight is a serious scientific endeavor," Gilruth chided. "We are in competition with the Soviets, not for the accomplishment of propaganda stunts."[44]

Nevertheless, Cobb and Hart made their case before Congress in 1962 as part of a larger investigation on NASA's discriminatory hiring practices. The pair pointed out that since women were not allowed to be test pilots, NASA's test pilot requirement made it impossible for them to meet the qualifications. Cobb hesitated to label the practice discriminatory despite the textbook definition. Instead, she argued that the agency had waived other requirements, like its college degree stipulation, for John Glenn.[45] Couldn't one requirement be waived for her, too? Besides, Cobb had flown high-performance jets, and Glenn's five thousand hours of flight time paled in comparison to her seven thousand.[46] Lastly, Cobb argued that women, in many ways, were better candidates than men for long-duration space travel: "[Women] weigh less and consume less food and oxygen than men," Cobb said. "They are more radiation resistant and less susceptible to monotony, loneliness, heat, pain, and noise."[47]

Cobb's argument swayed some in Congress. Then John Glenn testified.

Five months prior, aboard *Friendship 7*, Glenn had become the first American to orbit Earth. "The men go off and fight the wars and fly the airplanes and come back and help design and build and test them," Glenn explained to the senators, the saintly glow of his recent triumph still about him. "The fact that women are not in this field is a fact of our social order."[48] Congress bowed to this founding father of American spaceflight and told Cobb to patiently wait her turn.[49] Cobb never got her chance. In 1963, a year after Congress denied Cobb's request, the Soviet Union won another victory in the space race, sending Valentina Tereshkova to space aboard *Vostok 6*, making her the first woman in space.[50,51]

Despite the blow to the Mercury 13 and the rejection of Black astronaut candidates, the tide of social progress in the women's and civil rights movements could not be stopped by Glenn, Yeager, or anybody else. A year later, President Lyndon B. Johnson signed the Civil Rights Act of 1964 into law. A watershed in the long struggle for equality, the law ended segregation in public places and banned employment discrimination based on race, religion, national origin, or sex. A year later, a series of peaceful protests in Selma, Alabama, turned bloody, convincing Washington that more legislation was needed. The Voting Rights Act of 1965 outlawed the discriminatory voting practices adopted in many Southern states after the Civil War and helped ensure fairness for Black voters at the polls.

On September 24, 1965, President Johnson issued an executive order prohibiting federal contractors from discriminating on the basis of race, color, religion, or nationality for their employment decisions.[52] By the end of 1967, the law was amended to include "on the basis of sex," thanks to lobbying by feminists like the National Organization for Women's Betty Friedan and Pauli Murray.[53,54] The new laws helped advance women and minorities in business, education, and politics. The earnings gap between Blacks and whites narrowed from fifty cents on the dollar in the early 1950s to nearly seventy cents by the late 1970s.[55] By 1970, ten African Americans were serving in the US Congress; one of them was Shirley Chisholm, the first Black woman elected to the House of Representatives.[56,57] In 1967, the Supreme Court protected interracial marriage at the federal level with the landmark *Loving v. Virginia* ruling, and that same year, Thurgood Marshall ascended to the US Supreme Court as its first Black justice.

Depictions of women and minorities in the media were also transforming. In *The Jeffersons*, an affluent Black couple moved on up to a tony Manhattan high-rise. Mary Tyler Moore modeled a successful and single newspaper woman who did not need a man, and bigot Archie Bunker was continually taken to task by his hippie son-in-law, Meathead, on *All in the Family*. *Roots*—which followed one family's journey from eighteenth-century enslavement through the Civil War—became the most popular miniseries on TV in 1977. Tennis legend Billie Jean King proved that women could best men, winning the Battle of the Sexes against Bobby Riggs in 1973. By 1975, Arthur Ashe became the first Black tennis player to win three Grand Slam titles. The US was far from being a paradise of equality, but the country was making strides. Now NASA was desperately trying to catch up.

In 1972 NASA's minority employees made up only 5.2 percent of its full-time workforce, while women made up 16.2 percent, making NASA the worst government agency in terms of employment equality.[58] The next worst agency, the Atomic Energy Commission, had 12.6 percent minority employment—over twice that of NASA.[59]

One notable exception to NASA's lackluster record on equity was the agency's hiring of women mathematicians, as early as 1939, to perform the computations from which NASA launched rockets and satellites. Black women mathematicians, known as "human computers," like Katherine Johnson, Mary Jackson, and Dorothy Vaughan, did the calculations that would send the first Americans to space and land the first man on the moon.[60] The public at the time did not know about these "hidden figures," and NASA had yet to hire a woman or person of color in its most visible role—as an astronaut.[61]

"Whenever astronauts appeared on television or as speakers before selected audiences, someone raised the question, 'When will NASA send a woman astronaut, or a black astronaut, into space?'" said Dr. Joseph Atkinson, then director of the Equal Employment Opportunity program. Equity was a "major issue" facing the space program, he told Congress.[62]

Given its exclusionary practices, NASA was becoming unpopular with a citizenry that thought their tax money was better spent on Earth. A 1973 General Social Survey poll of Americans found that 61.4 percent of respondents felt too much was spent on the space program. In July

1969, the Poor People's Campaign stormed Cape Canaveral in advance of the Apollo 11 launch, protesting the US government's use of tax dollars.[63] "If we can spend $100 a mile to send three men to the moon, can't we, for God's sake, feed our hungry?" they puzzled.[64] A popular spoken-word poem by soul and jazz artist Gil Scott-Heron encapsulated the anger of the era, especially among people of color:

> A rat done bit my sister Nell; with Whitey on the Moon . . .
> Her face and arms began to swell; and Whitey's on the Moon
> I can't pay no doctor bills, but Whitey's on the Moon . . . [65]

As a result of waning public sentiment, Congress's financial support wavered. The agency needed a fresh image. Administrator James Fletcher established the Equal Employment Opportunity Office in 1971, tapping prominent Black activist and human relations administrator Ruth Bates Harris to lead it. She immediately became the highest-ranking Black woman in the agency and the only one in a senior leadership role.

Harris was sidelined from the beginning. Even before she started the job, she was demoted to deputy director, instead of director. The director job promised to Harris was given to a white man.[66] Harris's budget was trimmed, and her effort to stop using a tracking station in apartheid-practicing South Africa was stymied.[67] Deputy administrator George Low privately called her 1973 affirmative action operation a "dumping ground for poor people."[68]

On their own time, Harris and two aides prepared a comprehensive investigation of NASA's employment record.[69] Their findings, released in September 1973, were damning. Harris found that NASA only had a quarter of the minority representation of the rest of the federal government. Even its contractors hired three times more minority employees than NASA itself. Regarding women astronauts, Harris observed sardonically, "There have been three females sent into space by NASA. Two are Arabella and Anita—both spiders. The other is Miss Baker—a monkey."[70] She sent the forty-page report directly to her boss, Administrator Fletcher. A month later, he asked for her resignation. When she refused,

Fletcher accused her of "divisiveness" and fired her under the pretense that she was an ineffective administrator.

The press decried Harris's termination, and dozens of civil rights groups rushed to protest.[71] Congress called on NASA officials to testify and found Harris's dismissal "showed a contempt for the law." Congress stepped in to oversee a thorough "house cleaning in the area of employment."[72]

In turn, Administrator Fletcher promised that America's brand-new spacecraft, the space shuttle, would finally democratize space with its "safer design" and larger crew cabin.[73] "Space flight will no longer be limited to intensively trained, physically perfect astronauts, but will now accommodate experienced scientists and technicians," NASA's 1972 brochure for the shuttle enthusiastically proclaimed.[74] The unfortunate subtext was that women and minorities could have a turn now that NASA was lowering its standards. "The resulting changes in modes of flight and reentry will make the ride safer, and less demanding for the passengers, so that men and women with work to do in space can 'commute' aloft, without having to spend years in training for the skills and rigors of old-style space flight," Richard Nixon boasted to press.[75]

Nixon was right insofar as the orbiter guaranteed far more space for scientists and experiments than the cramped Apollo missions, but in insinuating that anyone off the street could go to space, he was far off the mark. The astronauts that flew, whether pilots or scientists, would have to endure as many years of grueling training and personal sacrifices as the Mercury, Gemini, and Apollo astronauts had. In fact, the shuttle would prove to be the most complicated and dangerous space transport ever created. "Unlike the hardened Apollo capsule heat shield, the shuttle crew compartment used fragile tiles; unlike the Apollo crew module, the shuttle crew compartment was next to rather than above the dangerous rockets; and unlike Apollo, the shuttle had no launch abort system," NASA Ames systems engineer Harry W. Jones pointed out.[76]

Even though the shuttle itself was years away from completion, NASA needed to make good on broadening its astronaut pool. Dr. Christopher Kraft, then Johnson's center director, fast-tracked the recruitment plan for shuttle astronauts, which he promised would "enable a representative number of minority and female candidates to be selected."[77] By December 1975, the Astronaut Selection Board worked out criteria for the new mission specialist role: doctoral degrees in engineering, life sciences, physical sciences, or math; an age maximum of thirty-five years; an ability to pass a Class II flight physical; and the recommendation of an employer. The plan also included a potential waiver on the physical requirements for individuals with "outstanding scientific credentials."[78]

The selection board issued an open call for applications, reaching out to the National Organization for Women, the National Association for the Advancement of Colored People, and the League of United Latin American Citizens. They sent announcements to colleges and universities, women's conferences, and even personal letters to prospective candidates listed in the book *Who's Who Among Black Americans*. Johnson managers and current astronauts stumped before special interest audiences and on television, but after a year of these efforts, the flow of applications remained anemic.[79] NASA's history of mishandling race and sex had ostracized the very candidates it hoped to attract.[80]

To repair the agency's standing with potential applicants, NASA brought on a celebrity spokesperson—none other than Nichelle Nichols, the former Lieutenant Uhura on *Star Trek*. Relentless in her recruitment efforts, Nichols made her case across fifteen colleges and universities, thirty-four professional organizations for women and minorities, and in nearly forty radio and television appearances, like the one Ron McNair had seen.[81,82] By the end of Nichols's campaign, NASA was buried under eight thousand applications.

George Abbey, Jay Honeycutt, and the rest of the selection committee winnowed these down to the 208 finalists invited to interview at Johnson Space Center, from which the selection board chose twenty pilot astronaut candidates and twenty mission specialist candidates and forwarded their recommendations to Dr. Kraft—who in turn gave them to NASA administrator Dr. Robert Frosch at headquarters for final approval.[83] In winter 1978, the *Washington Post* reported that Frosch was scrutinizing the list submitted by the Astronaut Selection Committee in part because it listed only "two black pilots and three women scientists."

Frosch wanted "to be darned sure there aren't more qualified blacks and women who can be added."[84]

This account resonated with NASA's less than stellar history of recruiting women and minority employees (let alone astronauts) and reflected the agency's desire to do better. The gossip also had the unfortunate effect of amplifying a negative narrative: that NASA's final selection underwent last-minute, politically motivated changes. Mercury astronaut Deke Slayton reported that the original selection only included one woman and that qualified male pilot astronauts were dropped to make room for more female mission specialists, calling the rumored changes "some last-minute political bullshit."[85] Slayton would not necessarily have reason to know, since he walked out of the selection process in protest of NASA's decision to recruit women.

As for George Abbey and Jay Honeycutt, they maintained that the original selection always included the six women and four men of color. Frosch had only questioned the number of pilots (twenty) because he felt that the Astronaut Office already had too many (seventeen), most of whom had never flown in space. He communicated this to Dr. Kraft and the selection board, cutting five pilot candidates, all white men.[86]

Regardless of how they had been selected, the chosen ones now sat on the auditorium stage. After Kraft's introduction at the press conference, the new class was released to reporters for interviews and photographs. If there had been any doubts as to who the press would most want to talk to, they vanished at that moment. "I could have mooned the press corps and I would not have been noticed," said Mike Mullane.[87] While the white male astronauts finished answering questions within a few minutes, the press hounded the women and minority men for hours.[88]

"We eventually came to refer to our class as 'ten interesting people and twenty-five standard white guys,'" said Kathy Sullivan, one of the six women chosen.[89] At twenty-six years old, Kathy was just finishing her PhD in geology.[90] Sporting a dark pageboy haircut, Kathy adjusted the jaunty red-and-blue scarf around her neck and set her strong Irish jaw for the many photographers who wanted pictures of her.

Nearby, the Black men answered questions about how they felt regarding the historical significance of their selection and diplomatically fended off suggestions that they were less qualified.[91] "We're going to go up and do a job as professionals, not as Blacks," Ron told a reporter.[92] He did not want the focus on his race to diminish his achievement.

"I would hate to think that I was chosen—or any of the women or minorities were chosen—because of tokenism," Fred Gregory echoed. "I think my qualifications were adequate—super."[93]

Japanese American Ellison "El" Onizuka received short shrift from the mainstream press. Reporters barely mentioned El in their coverage of NASA's new diverse astronauts. When they did acknowledge him, they referred to him as "Oriental."[94]

The women also received their share of insulting queries.[95] Older and wiser Carolyn Huntoon—the only woman in NASA's top management—watched as the six women fielded the press's flatfooted questions. At a break, Carolyn pulled them into the women's room, safely away from the male-heavy reporter pool.

"You're not obliged to answer every question. This is a negotiation, not an exam," Carolyn advised. "So, you're allowed to answer the question they *should* have asked."[96]

Carolyn was so respected at NASA that Kraft had asked her if she wanted to be an astronaut in this new class. She loved her research position too much, she told him, but she did have one request. She wanted a seat on the selection committee. Now that the women candidates had been chosen, she would be their advocate. A formidable presence, Carolyn had a slow Southern way of speaking and a level gaze that held people's attention, especially if she did not like what they were saying. Carolyn had a strong protective instinct and a way of smoothing ripples in the road for those in her charge.

Who are you dating? What kind of perfume do you wear? Would you give up this job if you got pregnant? "They'll ask it all," she said. "How you answer will set the tone for other women and how all of you are treated."[97]

Judy looked around the women's room at the faces of her new

colleagues—Kathy, Sally, Anna, Shannon, and Rhea—as they draped themselves on sinks or leaned against the stalls. In the years to come, the women's room would become their bailout spot, a place to regroup and strategize away from the guys. Today, Carolyn taught them that their fates depended on each other. More importantly, the future success of women in space depended on them as a group.

Indeed, the entire astronaut class—women and men, people of color and whites, pilots and PhDs—would rise and fall based on each other's performances. They would need to bridge their differences and work together to build the future of space travel and win admittance to the most exclusive club on Earth: those who had left it.

CHAPTER 4

BAPTISM BY FIRE, WATER, AND AIR

Johnson Space Center. July 1978.

"Hey! We've got a fire in the cockpit!" a man screamed, then his voice cut out. Within seconds, another desperate voice cut through the static.

"We've got a bad fire . . . !" the second man shouted in pain.

"We're burning up . . . !!!" a third howled.

Then the transmission faded into nothing but static.

In one of the many tiered seats in Mission Control, Ron McNair and his new classmates listened to a recording of the Apollo 1 fire. During a preflight test on January 27, 1967, astronauts Gus Grissom, Ed White, and Roger Chaffee had burned alive. Even though over a decade had passed since the accident, the pain and fear of the astronauts who perished was palpable to the room of new recruits.

The instructor surveyed the faces of the astronaut candidates. *Are you sure you're ready for this?* The audio was a wake-up call, especially for those like Ron who had not served in the military and had never had a job with life-and-death consequences. *If this reality was too much for any of them to accept,* the instructor suggested, *now was the time to go.* No one budged.[1]

A few weeks earlier, as Ron moved his family across the country from left-leaning Malibu, California, to the Lone Star State, the summer sizzled. Disco hits from the Bee Gees, "Night Fever" and "Stayin' Alive," blared from the radio. Billboards advertised the new Hollywood blockbuster *Grease*, starring John Travolta and Olivia Newton-John. In the nation's capital, almost a hundred thousand demonstrators marched in support of the Equal Rights Amendment—at the time, the largest march for women's rights in US history. Muhammad Ali was on the verge of making history at the Louisiana Superdome, becoming the first man to win the World Heavyweight title three times in a row.[2]

When Ron and his wife, Cheryl, arrived in Houston, they found a little starter apartment before moving to a Clear Lake suburb along with the Onizukas and the Gregorys. Everyone that had kids—or planned to—wanted a lawn for football and a cul-de-sac for bike riding. That and the neighborhood's proximity to the middle and high schools made it the obvious choice for families. Single astronauts like Sally Ride, Kathy Sullivan, and Steve Hawley settled into apartments right outside Johnson's back gate with a short commute, volleyball court, and communal barbecue pit.

On the Monday after the July 4th holiday, Ron drove through the gates of Johnson Space Center for his first day of work. Looking up from his baffling acronym-filled schedule, Ron spotted a few of his classmates and followed them to Building 4, the home of Johnson's Flight Crew Operations. Everyone was rushing to the Monday morning all-hands meeting, a staple of the Astronaut Office since the Mercury days.

Standing watch from their office doors, Sylvia Salinas, Mary Lopez, and Estella Hernandez Gillette, all in their twenties, took in the excitement as the new astronauts stormed the hallways. The Hispanic American administrative staff—working in and around the Astronaut Office—came to be known as the Mexican Mafia. As the liaisons for George Abbey and

John Young, Sylvia and Mary, and later Estella, ran the show behind the scenes, making sure things went smoothly in the Astronaut Office. Up until then, the astronauts they worked for were military men, older in age and more conventional in style; they did not fraternize with support staff. Now, "kids like them" were rolling in.[3] The arrival of Astronaut Class 8 was like a breath of fresh air.

A large conference table surrounded by two rings of chairs dominated Room 3025, the locus of the Monday meeting. Assuming the first ring was reserved for administrators and senior astronauts, Ron took a seat in the back row, as did the rest of his class. Everyone, that is, except the blond, mustachioed Rick Hauck, a US Navy commander who by military standards was the most senior-ranking pilot of their class.[4] Hauck took a seat at the table. Some in the room gasped. Others eyed him with suspicion. *Wow, he must either be a fool or the most confident bastard among us. Maybe both.* Either way he made an impression.

Like Hauck, the fifteen fighter pilots in Ron's class had plenty of swagger and bravado, and mixed easily with the veteran astronauts. The old guys, twenty-eight in all, included moonwalkers John Young and Alan Bean, whom Ron met during interview week. They filled the inner circle. Among them were men still itching for their first trip to space, like Bob "Crip" Crippen, the baby of the group at forty years old, and Richard "Dick" Truly, both career military pilots who had flown for both the Navy and Air Force.[5] These yet-to-fly guys were caught between programs, too late for Apollo and—so far—too early for the shuttle. Crip and Truly were part of Astronaut Group 7, who had been transferred to NASA after the cancellation of the Manned Orbiting Laboratory (MOL), a classified Cold War military project developed to acquire surveillance images from space.[6] After a decade at the agency, the former MOL astronauts had only ever flown a desk.[7]

Everyone here wanted a ticket to space, but the ten interesting people would be setting historical precedent, breaking barriers that in the past restricted people like them from space travel. Of the six women in the room, one would be the first American woman in space. While the Soviets had flown the first female astronaut, Valentina Tereshkova, being the first *American* woman in space would earn a prominent place in the annals of history. In 1978, no Black person had flown to space. Ron, along with Guy Bluford, and Fred Gregory would compete to be the first, while Ellison Onizuka would almost certainly be the first Asian American to fly. Guy and Fred, both Vietnam vets, and El, an Air Force test pilot, all spoke the military language of the old guys. Ron was an outsider even among outsiders.

John Young, chief of the Astronaut Office, began the meeting, mumbling "a few forgettable words of welcome" while staring at his shoes.[8] Though he had braved the depths of space four times, on both Apollo and Gemini, Young had not conquered public speaking. Compact, with a jockey's build, Young was a handsome Navy devil with big ears and an aw-shucks demeanor that belied how truly meticulous he was. He preferred solving thorny engineering problems to dealing with management issues, and yet here he was as head of the Astronaut Office. He explained to the new class that they were not yet astronauts; they were still astronaut candidates, or "AsCans" for short. Only after two years of training would they earn the title astronaut and a silver pin to mark the achievement.[9]

Inspired by Navy and Air Force aviator badges, the pin depicted a trio of rays merged atop a shining star and encircled by a halo denoting orbital flight.[10] The silver pin meant you were flight-ready, but the gold pin meant you had flown to space. *That's when you make it.* Young then left the group with a bit of sage advice: "Don't talk about nothing you know nothing about."[11] *Got it. So basically, keep our mouths shut.*

As the old guys left the room, they once-overed the new guys. Quite simply, the old guys were a different generation. They were veterans, test pilots, and men who had never worked with women or civilian graduate students. Underneath their pique was also perhaps a tinge of fear. The line to ride the bird just got a whole lot longer; maybe they would miss their chance altogether.

Who are these new guys anyway? Hell, half of them are civilians, wet behind the ears, fresh off their mother's teat. They traded in high grades and accolades, not in life-or-death. The old guys shook their heads. *Those Fucking New Guys.* "The Fucking New Guy," a military term for the newest grunt in the unit, seemed to suit Astronaut Class 8 perfectly. So was born the official class nickname: TFNG. In polite company, the TFNGs referred to themselves as "Thirty-Five New Guys," but everyone knew what the term really meant.[12]

After the meeting, secretary Sylvia Salinas handed the "New Guys" their official NASA portraits and asked them to create signatures for the auto-pen machine.[13] The agency would print thousands of autographed photos. *Do thousands of people want our autograph?* Ron wondered.

It's astronaut insurance, a veteran astronaut quipped. *If you die, your family will have something to sell.* The joke did not get any laughs.[14]

Biscayne Bay, Florida. August 1978.[15]

"Ready?" a voice called from the deck of the landing ship.

Ron looked out at the speedboat revving its engine in the choppy south Florida water. A long rope connected the harness around his chest to the rearing beast. He glanced down at his black sneakers. He had been told not to wear white shoes as they would attract sharks—his first clue the day would be a tough one.[16]

Ron had been ferried, along with other New Guys, to Homestead Air Force Base outside of Miami, Florida, for a water survival training course. Before they learned how to fly airplanes or space shuttles, they would have to learn how to ditch-out of a plane during an emergency. Deep into the Florida summer, the temperature hovered at nearly 100°F. Sweat collected under Ron's helmet. His flight suit stuck to his tacky skin.

As soon as Ron gave the go-ahead, he would have to run off the back of the ship as a parasail swooped him four hundred feet above Biscayne Bay. Then the man in the speedboat would wave a red flag and Ron would release his harness, parachuting toward the water as though he had ejected from a distressed plane.[17]

Physically fit at five feet, eight inches and 160 pounds, Ron could accomplish nearly any feat NASA could ask of him, but he did have one limitation.[18] Ron did not know how to swim. Ron had not divulged that detail in his astronaut interview, not that anyone asked. Having grown up in the segregated South where public pools were for whites only, Ron had never learned. He had largely been able to avoid the issue . . . until now.

"Ready!" he shouted over the wind and the boat's motor.

Ron jogged off the back of the ship. His parasail lifted him into the sky, his feet pedaling as if he were still running. As he drifted high above the cerulean waters churning below, he hoped this would not be the last of him.

Lake City, South Carolina. 1959.

"You need to leave," a librarian told a nine-year-old Ron McNair. Even at this early age, Ron was not one to back down from a challenge. He needed these calculus books to feed his math obsession, so he stood his ground. *Impudent little boy,* thought the librarian, as she picked up the phone. She made two calls: one to the police department, and another to Ron's mother. Like the swimming pools in Lake City, South Carolina, the library was for whites only.

"Mrs. McNair, I think you'd better get over here. As a courtesy," she feigned. "I'm calling to let you know the police are on their way." The threat to Pearl McNair was implied, but the librarian was more direct with Ron. "You won't be able to sit for a week. Or you'll be in jail," she taunted.[19] Ron patiently waited for the police and his mother to arrive.

When the police saw young, precocious Ron in the library, they shrugged. "Maybe you can let him check out the books?" they asked. A few library patrons who had pretended not to notice the standoff nodded in agreement. Then Pearl stormed in like a lioness to protect her cub.

"Ron," she whispered to him. "What are you doing?" She sighed when he gestured to his math textbooks.

"Why can't I have them?" Ron pleaded. "I'll take care of them."

Pearl taught school and had long impressed on her sons the importance of education. Prevented from progressing past the eleventh grade because of the South's Jim Crow laws, Pearl talked her way into a local college and finished her undergraduate degree.[20,21] While raising Ron, she worked toward a master's degree at South Carolina State College. Three days a week, Ron watched his mother return home from a full day of teaching, climb into a car with a few fellow teachers, and make the hundred-mile journey to night classes in Orangeburg, South Carolina. Pearl maintained this grueling pace for six years to earn her master's in teaching.[22] She understood where Ron got his persistent streak.

Pearl turned to the librarian and offered to cosign for the books. Backed into a corner with unnerved patrons and no support from the police, the librarian stamped the books and handed them to Ron.

From then on, Ron was allowed to borrow books from the library whenever he wished. His curiosity and courage made the Lake City Public Library that much more public.

Biscayne Bay, Florida. August 1978.

As Ron splashed down into briny Biscayne Bay, he dug deep for a dose of the same courage that he had displayed at the library almost twenty years earlier. His parachute swirled around him while he worked to free himself, careful not to kick his legs and entangle himself in the fabric and suspension lines.[23] He had a knife tucked in his vest in case he did. Instructors had cautioned them aplenty with stories of parajumpers who got trapped in their shroud lines and drowned as the 'chute sank.[24] Freeing himself, Ron grabbed at his thigh for the handle to release an orange lifeboat, pulled the ripcord, and inflated it. He threw a leg over, but his waterlogged suit made it difficult to climb aboard. With a herculean effort, he muscled his way into the slippery raft and tumbled into its center.

Ron fired his flare gun, and the rescue helicopter headed his way. Floating below the chopper's frantic wake was like riding a boogie board through a hurricane. The helicopter dropped a rope, which swung wildly in the gale-force winds. Ron clipped his harness to the line and the aircraft yanked him up. Back on the landing ship, Ron collapsed. *Did anyone notice how shaken he was?* As he caught his breath, he noticed his classmate Rhea Seddon looking just as rattled.

Although they both hailed from the South, Rhea—a petite woman with bright blonde hair—could not have been more different than Ron. "I was raised to be a fine Southern lady," Rhea said. Her mother made sure she "had ballet lessons from Mitwiddies' Dance Studio and learned piano from Sister Alvera. I knew which fork to use at a formal dinner and had had a grand tour of Europe. My scariest moment should be whether the soufflé at the dinner party would puff or not."[25] Rhea's Harvard-educated, Yankee-lawyer father bucked the tradition of grooming women only for marriage and encouraged his daughter's interest in science. Rhea became a doctor. Her friends thought she should just get married; her sister said she was "crazy." Her mother shook her head with a bemused smile at her daughter's grandiose plans.[26] Rhea, like Ron, would not be defined by what others expected.

Ron and Rhea had something else in common. Neither could swim. When Rhea was twelve, a teenage boy "jokingly" held her underwater at a friend's pool. She thrashed and choked, but the boy held his hand firm on her head. When she became a doctor, she learned the process of drowning begins after only sixty seconds underwater. Could she have died that day?

Despite her fear, Rhea played it cool in front of the reporters who watched from a flotilla nearby. The press had caught wind of the AsCans' visit and showed up to sneak a peek of the first women astronauts in particular.[27] America's image of the astronaut had long been associated with a macho affinity for risk. Even though the Mercury 13 had proven women to be equally suited for space travel, many in the American public still doubted that women could hack it. This "trial by water" was the first real physical test for the six women. Rhea did not want to let herself or the other women astronauts down.

"You know how to keep your head above water, if the 'chute starts draggin' you?" asked the young sailor who secured her harness to the parasail. Rhea nodded, but fear swelled up in her. *Sixty seconds is all it takes to drown.*

"Any *last* words for your folks back home?" The young man's tongue-in-cheek question threw her off guard as the speedboat took off.

In the blink of an eye, Rhea found herself dangling from her parasail as it lifted her much higher into the air than anyone expected. The Air Force trainers had not taken into account the women's lighter weights and smaller frames when doling out equipment. Rhea, barely breaking a hundred pounds, in an oversized flight suit and helmet wobbling sideways, drifted ever higher into the ether. Onlookers' mouths hung open. Finally, a friendly wind directed her back toward the sea. She braced for her turn at rescue.

Minutes later, Rhea—wet, exhausted, and shaking—plopped back down next to her classmates. Her performance had been anything but graceful, but she had survived. A photographer in the distance angled for the perfect snapshot of her.

"Can you put on a happy look for the camera?" the reporter yelled.

"No," yelled back Sally Ride, who sat next to Rhea.[28]

"Hold it, miss!" another reporter called to Rhea as she prepped for the next exercise. She had faced her deepest fear. A snarky reporter was not going to push her around.

"It's *doctor,*" Rhea shot back before smiling for the camera.[29]

"Female Astronaut Makes a Big Splash!" ran the United Press International headline the next day. A picture of Rhea in her parachute harness, getting dragged through the water, appeared in *Time* magazine.[30] "Lady Astronaut Margaret Seddon . . . sports a big grin after her success!" another reporter wrote.[31] In addition to giving a gee-whiz novelty to the proceedings, *Newsweek* noted the exercises showed women could "take most of the same physical pressure" as male astronauts—not a ringing endorsement for the women, but not an embarrassment either.[32]

Johnson Space Center. September 1978.

Weeks later, Rhea sat in a stuffy library room looking up at pictures of all the astronauts who had died in service. Aside from the casualties of the Apollo 1 fire, most had perished during training, a flight instructor told the New Guys. Four had died while flying in T-38 Talon jets, the world's first two-seat, supersonic twinjet trainer. *Any mistake in a machine like that and you could wind up a smudge on the runway.* It was clear why they taught flight prep in what was colloquially known as the "dead astronaut room." *Pay attention or this could happen to you.*[33]

Rhea, Ron, and the other mission specialists who had limited aviation backgrounds would be "backseaters," training under the tutelage of the experienced military pilots in the group and using a duplicate set of controls located in the jet's back seat. *Flight training,* thought George Abbey, *could be a team building exercise for pilots and scientists,* unlikely bedfellows who would need to trust each other on shuttle missions.[34]

Ron met his fifteen-hour-a-month flight requirement with pilot and fellow TFNG Guy Bluford.[35] A native of Philadelphia, Guy was the eldest of three boys. When Guy's father developed epilepsy and lost his job, his mother became the family's sole breadwinner. A determined student, Guy graduated from Penn State and joined the Air Force. On a scale of one to ten, with one being the most introverted, Guy rated himself a two. Behind his reserved exterior lurked a sharp mind and fast reflexes. Guy had flown combat missions in Vietnam, earned a doctorate in aerospace engineering, and rated as a top Air Force flight instructor.[36]

Not only did Ron hope Guy could show him the ropes, but he also saw a compatriot in a mostly white world. Ron and Guy liked to burn off their flight hours doing midnight runs. They would "snatch a plane" and jet to Brooklyn, Albuquerque, El Paso, or Vegas—wherever the mood took them. Once, in Las Vegas, they taxied alongside the commercial jets at the international airport. "I always took pride in the fact that me and Ron would climb out of the machine and people would look at us," Guy said. "They'd be amazed at two Black guys, flying a hot machine like a T-38." Guy loved the independence of flying. "If I am in the plane, I'm in charge. I'm the boss."

On the way home, Guy would turn over the controls to Ron as they soared through the night. Mach 0.7, Mach 0.8, Mach 0.9. The plane would shake as they made their way through the transonic buffet, pushing into a bow-wave of noise. Below the skies, earthbound mortals on the Vegas strip were treated to their sonic boom.[37]

Not everyone was thrilled with George's training arrangement. Some thought that T-38s should remain the domain of men. When Sally Ride proudly hung a new poster on her office wall that said A WOMAN'S PLACE IS IN THE COCKPIT, some of the male astronauts taunted her. "No, a woman's place IS a cock-pit," snickered one TFNG as his buddy Mike Mullane chuckled. Mullane, who self-admittedly traded in sexist barbs, and Sally barely spoke for the next decade.[38] Underneath the office tension ran a long-held belief that women did not have the right stuff to become high-flyers.

The legendary Chuck Yeager himself balked at the notion that women should be breaking the sound barrier. "She's flying chase," Kathy Sullivan's T-38 trainer hollered out to Yeager, trying to introduce the young woman to the flying legend. Kathy beamed with pride at her newly earned skills. Yeager paused midstride, looked back over his shoulder.

"Riding, maybe. Ain't flying," he sneered before stalking off. With his comment, he took the wind right out of Kathy's sails.

The wives of the male pilots disliked the training approach for a totally different reason. They did not want their husbands spending hours of one-on-one, focused attention with the new women astronauts. For their part, the women tried to fly under the radar; Sally and Anna went shopping at the mall to buy khaki pants and button-down shirts so they would look like the guys.[39] *No matter. There was something about those sexy little T-38s.* The freedom, the danger, and the rush of being the fastest thing in the air made a passionate tonic. More than one supersonic love affair bloomed between the pilots and their trainees.

Somewhere over Houston. Fall 1978.

Sitting in the cockpit of a T-38, Robert "Hoot" Gibson, a hot-shot Navy test pilot and Vietnam vet, looked like a dashing Hollywood action hero. Blond with a touch of ginger in his hair, he had sharp features, windblown red cheeks, twinkling blue eyes, and an ample mustache. A popular flight instructor, Hoot was clear in his direction, kind to the rookies, and, above all, unflappably calm. As Rhea climbed into his back seat, Hoot powered up the jet. They raced along the runway then took off, soaring into the sky. Rhea felt the Gs pile on, pushing blood from her head to her feet. To avoid blacking out, Rhea knew to perform the "anti-G straining maneuver" that she had been taught. She tensed her legs and abdomen to constrict the veins in her lower extremities and force blood flow back to her brain.[40]

As they reached forty thousand feet, they spotted their target, the Shuttle Training Aircraft (STA), a modified Grumman Gulfstream II airplane that pilots use to practice flying and landing a shuttle. In preparation for the first shuttle flights, the New Guys were learning how to escort the shuttle back on its cross-country descent from space. In their T-38s, the astronauts would record important flight images and data of the shuttle for study later.[41] For the real thing, astronauts flying chase would coordinate with Air Force controllers to spot the shuttle as it blazed into the atmosphere at one hundred thousand feet. Pilots would fly in a "lazy racetrack pattern" until the aircraft met up and flew to the descent. Flying chase was a carefully orchestrated dance with life-and-death stakes.

With the STA above them on their port side, Hoot pulled the T-38 to forty thousand feet and to the left to keep the plane in sight. Most commercial airlines do not go above thirty-five thousand feet; since the air above that has so little oxygen, engines have trouble creating the combustion they need to run. Above forty thousand feet, the air is even thinner.[42]

Sure enough their plane went dead silent.

"Both engines are out," Hoot reported. Rhea could barely breathe. At that altitude, the engines generated cabin pressure. Oxygen seeped from the cabin and the plane sank through the sky like a stone falling through water.

As a former emergency room surgeon, Rhea prided herself on keeping her cool in life-and-death situations, but this time her own life was at stake. *Not today. Not like this,* she thought. She was not ready to die.

"Why don't you pull out the emergency checklist and talk me through the procedure?"[43] Hoot said. Rhea knew the checklist by heart, but she was grateful for the distraction. She calmed down, got focused. As the plane slipped into free fall, Hoot and Rhea discussed how they might get one of the engines running again.

The plane dropped to thirty-five thousand feet.

If Hoot and Rhea could not restart the engines, Hoot could attempt a dead-stick landing by gliding unpowered to an impromptu landfall, a nearly impossible maneuver in a T-38. By all predictions, they would crash. There was just one other option.

"You remember how to bail out, don't you?" Hoot said. She did. Rhea had passed ditch-out training after all.

They descended just below thirty thousand feet, the height of Mt. Everest.

"Sit up straight, knees and elbows close to the body, head back, pull up the handles, and squeeze the triggers," Rhea recited.[44] *Hope to God the hatch opened. Hope to God you cleared the plane. Hope to God the parachute released.* Yep, she knew the drill.

Bail-out survival rates were miserable. That unhelpful fact popped into Rhea's mind, along with the image of her own official NASA portrait being hung in the "dead astronaut" room.

Jesus Lord. They were at twenty-five thousand feet.

Hoot gave the engines another go. Would they start?

A whirring sound buzzed in Rhea's ear. The engine blades began spinning fast enough to restart. Hoot brought one of the engines back to life. *God bless him.*

Rhea exhaled with relief. The second engine started up soon after.

"Well," Hoot said on landing. "We cheated death one more time."[45] Rhea gained a new appreciation for the old fighter-pilot maxim.

That night at happy hour, the beers tasted especially good and conversation flowed. Having gone to the edge of life and back, the two skipped the small talk and began to open up to one another.[46] Hoot, married at the time, confessed that he had recently moved out of his home. His pilot, and now astronaut, lifestyle had strained his marriage to the breaking point.[47] Rhea listened empathetically.

On a class trip later that year, the New Guys visited Patrick Air Force Base, south of the Cape. While lunching at the beachside officer's club, Rhea noticed Hoot, dressed in his coat and tie, wandering aimlessly along the beach.[48] Rhea heard from another classmate that Hoot's wife had left town, taking their two-year-old daughter and everything they owned with her back to San Diego.[49] "I had never seen anyone so down," Rhea said.[50] "Watching him cope and still handle his job responsibilities made me admire him even more."[51]

Rhea and Hoot found themselves spending more and more time together at work events and then dawdling afterward, at classmates' homes, at bars, in parking lots. One late night, she drove Hoot home to his apartment in her brand-new silver Corvette. She idled her car, he lingered, and then finally Hoot reached over and kissed her.[52]

Ellington Air Base, Houston. Winter 1978.

Dressed in a flight suit, Ron lay glued to the floor of a Boeing KC-135, an airliner-sized military plane, as it climbed into the sky. Padded on all sides, the jet's hollowed-out interior resembled a rubber room in a psychiatric ward. Ron was feeling a little unhinged himself at 1.8-G, a fraction of what an actual shuttle ascent would feel like. *But it gave you an idea.* It was hard to move his legs. Lifting his head was impossible.

After taking off from Ellington Field, the plane peaked at an altitude of thirty-three thousand feet. Coming over the top of a curve, the aircraft leveled off before nose-diving toward the ground, matching the speed of gravity, to cancel out the pull of gravity. The descent provided Ron and his cohorts a brief period of zero-G. To an outside observer, a military plane plummeting to Earth at a thirty-degree angle looked like pure madness, but parabolic flights were the best way NASA trainers could simulate weightlessness without leaving Earth's atmosphere.[53]

Ron's feet and hands floated up, followed by his whole body. He watched his classmates do flips. Rhea playfully held Hoot over her head, looking like Mighty Mouse. Anna did a tumble. Flipping, gliding, and drifting, Ron laughed and did a karate chop in midair. *This wasn't so bad.* Thirty magical seconds passed.

Then the pilot pulled out of his dive at a full 2-G, pinning the inhabitants to the bottom of the plane. *How many parabolas are we doing?* Someone grunted. *Thirty,* an instructor answered. The first couple of goes were a blast, like a roller coaster without the track, but soon the sound of retching filled the air, along with an odor. The smell of a regurgitated Air Force base lunch permeated the cabin, setting off a chain reaction that resulted in a group barf-off.[54]

Despite the nausea, Ron would have flown the aptly named "Vomit Comet" all day long if it meant he did not have to be in the water again. In parabolic flights, zero-G could only be maintained for thirty seconds at a time. To be able to train for longer periods, NASA managers would lower the astronauts into an enormous pool: Johnson's Weightless Environment Training Facility (WETF). Even though the underwater environment did not recreate space, the pool could provide neutral buoyancy, creating a close analog. Here the New Guys would don nearly three-hundred-pound spacesuits to practice spacewalks, or EVAs (Extra-Vehicular Activity), during which they worked on submerged shuttle hardware.[55]

Before they could perform these technical underwater ballets, they had to become scuba certified and pass a series of tests, including treading water for twenty minutes, swimming underwater, and fetching bricks from the bottom of the pool.[56] Ron fretted. If he could not overcome his aversion to water, his astronaut career would be over before it began. *I've already come so far*, he thought.

If Ron's life had taught him anything, it had shown him how to overcome seemingly insurmountable obstacles. He thought back to his freshman year at North Carolina A&T University, where Ron struggled to keep up with his studies. He was not as well prepared as the "big-city" students to face college-level courses. Paralyzed by self-doubt, he visited his guidance counselor, Ruth Gore, determined to switch majors from physics to music. "I'm in over my head," he told her. "I'm so far behind, I don't know if I'll ever catch up."[57]

Gore listened to his confession, quiet as a priest. "Before you make up your mind," she suggested, "take a few aptitude tests."[58] Ron agreed. As Gore scrutinized his results, Ron waited anxiously in his seat. "Ronald, I think you should try physics, because I believe you're good enough."[59] *You are good enough*. Ron had needed to hear those words. Gore's belief in him helped him recommit to physics.[60]

At MIT, Ron not only struggled academically, but also socially. On campus, his peers insisted that MIT had relaxed its standards for him. Off campus, locals stopped their cars to hurl racial slurs at him. Once, a gang of white men jumped him in Harvard Square. Another time, a friend's neighbor sicced his dog on him. *So much for the liberal Northeast.*[61]

Ron likened MIT's PhD program to a "five-round prize fight," with each round lasting one year.[62] The mostly white study groups that would have helped him catch up with his classmates excluded him. Feeling isolated and outmatched and nearing the end of his rope, he repeated Gore's advice like a mantra: *You are good enough*. Instead of giving in to anger, Ron decided to reach out for help and create his own community, forming a Black student study group with others who also felt alone and exasperated with their coursework.

Confronting NASA's swim test, Ron considered a similar solution. *I need an ally.*

Rhea, still terrified of drowning, could only tread water for three of the twenty minutes required by the test and only swim about a third the length of the pool before "getting panicky."[63] Hoot, a lifelong surfer, tried to help her, suggesting that she "relax and float."[64] *Easy for him to say,* Rhea groused. Even when she followed his instructions, she flailed, then sank like a stone. At home, Rhea wept from exhaustion, frustration, and embarrassment.[65]

Ron and Rhea each watched the other struggle through the exercises and commiserated during their breaks. Ron confided in Rhea that he had been shut out of the "whites only" pools of his childhood; as a fellow Southerner, Rhea lamented the hateful history of discrimination.

Rhea learned that Ron had been forced to struggle against the policies of segregation, which prevented him from getting the childhood education or the opportunities he had so richly deserved. He had worked in the fields for extra money and had put himself through MIT with scholarships. He planned to return to South Carolina someday to teach and mentor young Black men like himself.

Ron discovered that Rhea had earned her chops at "the John," the nickname for the John Gaston Hospital in downtown Memphis, Tennessee, where "the poor, downtrodden, and just plain bad" were routed for their traumatic injuries and medical care.[66] Rhea's patients, most of whom were low-income and underprivileged, would explain away their injuries with far-fetched tales. A patient with a gunshot wound would say he had been walking home from "choir practice" when "some dude jumped out and shot him."[67] After surgery, Rhea would realize the bullet had come from a police revolver and that law enforcement had shot the patient after a burglary. At the John, "one learned to be a skeptic," Rhea quipped.[68] "Churchgoing teetotalers did not often visit our world."[69] Women were not allowed inside the surgery doctors' lounge, so Rhea resorted to napping in folding chairs outside the nurses' reception area. Both doctors and patients assumed she was a nurse.

"We were both determined to overcome the past," Rhea said.[70] Their shared phobia was not a physical limitation, but a mental block. They made a pact: While their classmates practiced swimming twice a week at the scheduled training sessions, Rhea and Ron returned to the Clear Lake

Recreation Center community pool every day after work, away from the watchful eyes of their colleagues and NASA instructors.[71]

Ron held on to the edge of the pool, staring at the mothers and their dog-paddling children, and all the athletes who were there to improve their swim times. He heaved his breath in and out and then plunged into his lane, swimming like a madman, flailing and splashing as far as he could go underwater, then stood up and did it again. Rhea did, too. The onlookers must have thought they were "nutjobs," but each day they got farther.[72] The WHITES ONLY signs on the Southern swimming pools they had grown up around were specifically meant to keep these two apart. Even if they were no longer living in the Jim Crow South, they turned some Texan heads at the community pool.

After mastering one pool length, Ron and Rhea focused on the last element of the swim test: retrieving a brick from the bottom of the pool. They started with a rock and worked their way up. Soon it was game day.[73] Standing on the edge of the pool, Ron visualized the outcome, telling himself, *I am good enough.* He plunged in, gliding smoothly below the surface of the water for the entire length of the pool. When he confidently tapped the ledge, a broad smile spread across his face.[74]

This is not going to be my undoing, Rhea told herself as she used a combination of breaststroke and whip kick to power to the bottom of the pool and retrieve the brick. Moments later, her head popped up above the water line, as she thrust the brick in her hand high above her head with all the gusto of an Olympic athlete.[75] The challenges of NASA's more daunting WETF and EVA training were ahead. At this moment, Ron and Rhea could not have cared less. By joining forces, they had prevailed.

Not all the New Guys, however, would find strength in teamwork. After all, a competition was afoot to earn a seat on the bird. To emerge victorious, they would have to navigate the sometimes complicated politics of the Astronaut Office and win the favor of the one man who seemed to wield all the power over it: George Abbey.

CHAPTER 5

I'LL BE YOU

Cocoa Beach, Florida. Summer 1978.

Kathy Sullivan peered out from the back of a long receiving line, waiting to meet *Florida Today* newspaper founder Al Neuharth and his wife, Florida state senator Lori Wilson Neuharth, at their grand oceanside estate in Cocoa Beach, Florida. Neuharth, whose newspaper provided heavy coverage of NASA launches during the Apollo era, was well on his way to making Gannett Publishing the largest newspaper company in America. His home, eccentrically named the "Pumpkin Center," after a remote South Dakota crossroads where hunters stopped to rest, was nestled on over an acre of Floridian jungle and was a hotspot for the Space Coast haut monde.[1,2]

Standing on their front doorstep, Al and Lori warmly welcomed the new astronauts, knowing NASA's return to space would bring excitement back to their corner of the world. For nearly ten years, with the end of the Apollo era, Cocoa Beach—the once bustling spaceport—had been a near ghost town. Now laughter pealed out from behind the green-and-orange pumpkin-themed stained-glass windows at their backs, where a powerful collection of local business leaders, state politicians, and the press mingled, making idle conversation as they waited for the evening's real guests of honor: Kathy's class of "soon-to-be-famous rocket-riders."[3]

George Abbey, who was rumored to have back channels in Washington, had a keen understanding of politics and helped arrange social events like this.[4] He wanted his new astronauts to form personal connections with those who signed off on NASA's budget and penned headlines about the program. The young AsCans had little to no training for their new role as NASA's emissaries. They may have had an understanding that their jobs would put them in the public eye, but they had no idea the degree to which political forces would dramatically shape their lives. *Who are the power brokers? What do they control? What should I say to win them over?* "All you could do was be nice," Rhea said, "or as we say in the South, be sweet."[5] Kathy, for one, had little interest in gladhanding politicians or hobnobbing with the press. The schmoozing that came so naturally to others never felt right to her.

As the line inched forward, Kathy spotted Sally slip out of the queue and duck back toward the exit. *Is she trying to sneak out?* Kathy wondered, aghast at Sally's quiet rebellion. When Sally walked past, Kathy unceremoniously grabbed her by the arm and stuffed her back into line.

"If I have to do this, you have to do this," Kathy said pointedly. Sally shot Kathy an annoyed look; she did not much shine to these kinds of events either. Sally shuffled in beside Kathy—a glint of mischief twinkling in her eyes. *Fine, but how about we mess with them a little bit?* she suggested.[6] *They only see us as the "women astronauts," not as individuals. Let's put them to the test.* Sally peeled off her nametag and stuck it on Kathy. *You be me and I'll be you.*

Kathy thought for a moment. Sally's suggestion challenged her every rule-following impulse. *Why not? At this point, we know each other's spiels as well as our own.*

Kathy peeled off her nametag and gave it to Sally.

When the two made it to the front of the line, Al pulled Kathy in for a greeting, bellowing, "Oh, Sally Ride! I'd love to get you out on those tennis courts." Lori enthusiastically welcomed Sally: "Kathy Sullivan, so nice to meet you. I recognize you from your picture."[7] Kathy and Sally both graciously accepted the warm welcomes, stifling laughter at their perfectly executed ruse.

Stepping past the Neuharths, Kathy entered the home's labyrinthine interior. With its Florida pine paneling, stone accent walls, and bamboo furniture, the place resembled a very large tiki lounge or, with its maze-like geometry, an acid-dipped Hawaiian resort. Fanning out from the living room, Kathy stepped onto the patio dominated by a lagoon-style pool, surrounded by a Jacuzzi, a koi aquarium, and an illuminated tree house that overlooked the Atlantic. Guests spilled outside, entwined in boisterous conversation, smoking and sipping cocktails.

Kathy worked the veranda, smoothly introducing herself as Sally and even signing "Sally Ride" autographs. As Kathy mingled as Sally, she heard one of her male classmates, who was not privy to the joke, exclaim to another guest, "Well, there's Kathy Sullivan over there! Go talk to her." To her horror, his extended index finger pointed squarely at her. The guest regarded the nametag on Kathy's chest with a mixture of confusion and ire as he came to grips with the unhappy fact that he, along with every other important partygoer present, had been duped. Knowing the jig was up, Kathy quickly found Sally and switched nametags. The stickers wilted gracelessly, curling at the corners as the two resigned themselves to a party that now felt far less fun.

The New Guys had come to Cocoa Beach as part of their visit to the Kennedy Space Center. To get a sense of the scope of the institution they had joined, they were visiting NASA's ten space centers and many of its major contractors. Over forty thousand people worked for the agency and its partners.

At Kennedy Space Center, they learned why NASA had chosen central Florida for its launches. Flung out into the Atlantic, Canaveral was the southernmost cape in America, closest to the equator, sparsely populated, and facing east—all excellent qualities for getting to orbit. Rockets could launch eastward from the Cape, using the direction of Earth's rotation to slingshot to orbit. With only the vast Atlantic Ocean below as rockets ascended to space, there was minimal risk of crashing into dense population centers if a launch failed.

Anna got chills looking at Launch Complex 5, where Alan Shepard rocketed off the planet in *Freedom 7.* Ron stood speechless before Launch Pad 39A, where Neil Armstrong, Buzz Aldrin, and Michael Collins departed on their historic Apollo 11 flight. He and his brother Carl had watched the television, transfixed, as Armstrong took his first steps on the moon, telling the world, "That's one small step for man, one giant leap for mankind."[8] Now, a decade later, Ron and his classmates would mark their own history on these hallowed fields.[9]

At Marshall Space Flight Center in Huntsville, Alabama, the class met with director William Lucas, who oversaw NASA's propulsion and rocketry program. Lucas viewed himself as "the heir to Wernher von Braun's visionary excellence." Von Braun was a Nazi scientist who had emigrated to the United States after World War II to evade Russian capture and become part of Operation Paperclip, the secret program that brought Germany's scientists to America to join the country's brain trust. He had led the center during the Apollo era and built the legendary rockets that took America to the moon. Marshall emerged as a hard-charging engineering shop with a collective chip on its shoulder: The center had the task of building the engines and rockets that would ferry the shuttle to space, but not the glory of flying the spacecraft. That honor belonged to Johnson Space Center, which housed Mission Control and the Astronaut Office.[10, 11]

Traveling as a group all over the country, on planes, trains, and buses, the New Guys could not help feeling—and acting—like they were in summer camp. During a visit to the Wright-Patterson Air Force Base in Ohio, the New Guys started a food fight in their hotel and nearly got kicked out. At Cape Canaveral, Hoot lured his classmates into doing "flaming hooker" shots—an ounce of whiskey set on fire in a brandy snifter glass. Hoot set his cigar down, slicked his mustache back, and, undeterred by the blue flame hovering over the syrupy liquid, threw the whole thing back like a fire-eater. *No big deal.* The innocent postdocs gathered their courage. The bartenders lined up the shots, torching the glasses. Alas, as each poor grunt attempted to replicate the feat, the flames singed their faces.[12]

After a trip to the National Space Technology Laboratories (later the Stennis Space Center) on the Pearl River in Mississippi, where the shuttle's main engines were tested, they swept into New Orleans, alternating rounds of Alabama slammers with rum-fueled hurricanes, and stumbled down Bourbon Street. The revelers griped that Air Force pilot Dick Scobee brought his wife. Scobee, considered a class leader, hosted many a BBQ at his home with his wife, June. Despite her popularity among the astronauts, they insisted on her discretion. *June,* they hounded, *what happens on Bourbon Street stays on Bourbon Street.*[13]

Like Kathy, many of the New Guys were young and still single. Others were not, but acted like they were. Male AsCans basked in the attention of adoring young women on their travels—a hallmark of astronaut life. "We found ourselves surrounded by quivering cupcakes," Mike Mullane said. "Some were blatantly on the make, wearing spray-on clothes revealing high-beam nipples and smiles that screamed 'Take me!'" The space groupies were undeterred by wedding rings.[14] AsCans absent for the bus ride home became fodder for gossip. Overhearing the buzzy chitchat, naval captain Rick Hauck—the group's de facto chief, who had dared to sit in the front row at their first Monday morning meeting—made an announcement to his classmates. "If you see something that offends you, keep it to yourself. It's none of your business. You could damage someone's marriage." By the 1980s, the fear that adultery or divorce might end one's career had dissipated, but NASA still valued discretion. "Zip your mouths, not your pants," interpreted Mike Mullane.[15]

Though they were the New Guys, many of Kathy's classmates were acclimating to the hard-partying culture of the old guys. The iconic astronauts from the Mercury days (save for John Glenn) were known as macho adrenaline-junkies who liked to chase women and drink beer, and many of the New Guys planned to follow their lead accordingly. Kathy, however, quickly grew tired of these late-night bar crawls and whiskey-drenched evenings.

As the child of an alcoholic, Kathy had a powerful negative association with drinking and did not partake. When Kathy was twelve, the maternal grandmother who had always showered her and her older brother Grant with love died from a short but painful battle with cancer. Her sudden death saddened Kathy, but it sent her mother, Barbara, into a deep grief that Kathy could not begin to understand. "The strong and nurturing mother that I had known since infancy disappeared," Kathy said.[16] She and Grant would often return home from school to find their mother drunk, in a near-catatonic state on the living room couch.

More troubling than her mother's condition was the fear that her father, Donald, might choose to commit her mother and subject her to shock treatments, commonly used to treat depression in the 1960s.[17] To avoid that fate, the children fed their mother, put her to bed, and hid her car keys so she would not drive drunk. With Donald working a demanding job as an aerospace engineer, the children kept the house running smoothly. Barbara stayed at home, but the mother Kathy knew as a child now only existed in her memories. That mother was brilliant but frustrated. Kathy's grandfather had discouraged Barbara from pursuing higher education and instead urged her to marry early and have children. *She settled for so much less than she wanted,* Kathy thought whenever she tucked her mother into bed, a painful reversal of roles.[18]

Whole-brained, brilliant, and venturesome, Kathy had a boundless curiosity about the world. By the time she was seven, Kathy could read coordinates and draw her own maps.[19] She loved reading *National Geographic,* squirreling away dozens of the pull-out maps that came with each issue under her bed. Kathy's flair for foreign languages—French, Russian, German, and Norwegian—paired well with her longing for travel and adventure. She set her sights on becoming a Foreign Service Officer, a job that would steal her away to exotic lands.

Then a general education requirement at the University of California, Santa Cruz forced Kathy into a science class. She fell in love with geology and oceanography. *The open sea called to her.* Her coursework took her to Bergen, Norway, and then to Nova Scotia for graduate work at Dalhousie University and Canada's premier oceanographic center, the Bedford Institute of Oceanography. Kathy likened her research expeditions to great symphonies, with planning the voyage being the writing of the musical score and the journey at sea the concert. Just as the conductor begins the performance with the lift of the baton, the ship lifts its anchor and embarks on the adventure. Each person plays a crucial role to the whole, executing their tasks and improvising when plans went awry, all to make the work a success.[20]

Kathy dreamed of landing a position on the *Alvin* submersible, a vessel that could descend more than six thousand feet and allow Kathy to view the volcanic landscape of the seafloor. Her brother Grant—a corporate jet pilot and flying fanatic—had a different idea for her. As they lounged about at their family's California home over Christmas break, Grant told her that NASA was recruiting women scientists to their astronaut corps. *She'd be perfect for the job.*

"How many twenty-six-year-old female PhDs can there be?" Grant said. Kathy quickly dismissed her brother's idea as silly. Back in Nova Scotia, she resumed her PhD work, analyzing data and building maps for an upcoming voyage that would help her decode the geological history of the Newfoundland Basin for her dissertation. As she flipped the page of a scientific research journal, she saw a NASA advertisement for the astronaut program staring back at her. *Was it fate?* The prospect of understanding Earth from a new perspective grew on her. *Isn't NASA just building a research ship in space?*[21]

Kathy jokingly told her mother after applying to NASA, "When I finish my degree, I'm either going two hundred miles up or six thousand feet down."[22]

"Isn't there anything on the surface that excites you?" her mom shot back wryly. For Kathy, a born adventurer determined to explore the least-known recesses of our world, the answer was no. Grant had been right: Kathy was a natural fit for NASA. Athletic—a formidable racquetball opponent by all reports—with an imposing intellect, Kathy spoke in a way most people wished they could write. She was unafraid of the "dog work" space missions might entail, having turned a wrench or two on her deep-sea missions.[23] As Kathy embarked on her journey to NASA, she thought of something her mother told her before she had disappeared into depression: "Don't you stop short of what your full potential is." Kathy would take that advice to heart.[24]

Johnson Space Center. Fall 1978.

It's stable in two attitudes! Max Faget excitedly told the AsCans as he stood atop a desk, holding a model plane made of balsa wood, paper, and tape. *Ten degrees angle of attack!* he shouted, launching the plane into the air. It gently glided across the classroom like a folded paper plane, before skidding to a stop.

Okay, so the guy can make a model airplane, Kathy and her classmates thought.

Of course, Max Faget was not a hobbyist. Despite his five-foot-six frame, Faget loomed large at NASA. A native-born Belizean with large elfin ears, a bow tie, and a Cajun accent he acquired at Louisiana State University, Faget was an icon, having contributed to the design of the Mercury, Gemini, and Apollo spacecraft.

And . . . stable at forty degrees. He tilted the plane's nose up, revealing its underside, before launching it once again across the room. This time, the plane fell through the air, belly first. His point was taken. The spacecraft had various angles of attack. At say, forty degrees, the shuttle could press through the Earth's atmosphere. Like the tear-shaped capsule of Apollo, the shuttle's underside would be covered with heat-resistant materials to protect it from the high temperatures of reentry. Once the orbiter entered Earth's atmosphere, the shuttle could shift its angle of flight, gliding like a plane onto a runway. The shuttle was large enough that it would emit two sonic booms—one for the nose and one for the tail.[25,26]

George Abbey, John Young, and Alan Bean had enlisted luminaries like Faget to instruct the new astronauts. Since there was no precedent for the brand-new shuttle, the men made up the curriculum as they went along.[27,28] The New Guys noticed that their bosses were at times winging it. "I was expecting to come in and have it be like a military flight school," said Anna. "But really they were inventing how they were going to train us on the fly." For Guy Bluford, the training seemed haphazard at times, even though "a lot of effort was provided to give us the technical knowledge that we needed."[29] The New Guys were on the ground floor of the development of the new aircraft, cocreating its protocols.

Instruction covered a wide gamut, including physics, engineering, aerodynamics, avionics, and electronics.[30] Lecturers taught in NASA-speak, which meant a lot of acronyms and very little explanation. A CDR was a commander. A PLT was a pilot. But what on Earth was a "puck-a-moo"? As it turned out, that ridiculous piece of jargon referred to a Pulse Code Modulation Master Unit, part of the shuttle's avionics suite. *Obviously.*

Lessons often spilled out of the classroom. Geology experts took them to the desert plains of Arizona to study Earth's origins. Leading astronomers provided the latest thinking on the genesis of the universe underneath the projected night sky at Houston's planetarium. "I taught them four years of undergraduate and two years of graduate astronomy in twelve hours," said their astronomy professor.[31] To Kathy, the experience was like drinking from a fire hose.[32]

George wanted to level the knowledge field, to bring everyone up to speed in all disciplines. "We [the scientists] took aerodynamic classes, which [the pilots] slept through," Sally said, "and they took science classes, which we slept through."[33] *Or skipped out on.*

On the first day of a weeklong class on the aerodynamics of Advanced Stability and Control, Sally stifled a yawn from the back of the classroom. Sally already understood what was being taught—*the physics of a flying object.* She was after all a PhD in physics from Stanford. She hated to have her time wasted, especially when they were working twelve-hour days. Shortly before lunch, Sally stood up, grabbed her books, and left, never to return.

"Nobody was exempt from any of the courses, no matter how much prior expertise they had," Kathy said. Kathy followed the rules and never skipped a class. Even though she held a PhD in the subject, she was forced to take introductory geology. She loved it, "because of the men who taught it," she said. "Both had trained Apollo geologists, and one had active research projects about Venus." *Venus!!!* "Plus, I was curious what aspects of geology they'd pulled together for us—not merely Geology 101, but with a slant toward the observations we'd eventually get to make out the window." To Kathy, every moment of instruction was a gift, and she was not going to miss anything—*not one minute of it.*[34]

The Outpost Tavern, Clear Lake, Texas. Any Friday Night.

A storm silenced the cicadas. Pickup trucks lined the muddy parking lot of the Outpost Tavern. In its past incarnation, the joint had been a World War II pilot barracks, relocated from Ellington Field to a vacant lot on Egret Bay Boulevard and NASA Road 1. Located a mile from the gates of Johnson Space Center, the new bar, painted red and white to look like a barn, was emerging as a hotspot for NASA folks who were short on venues around Clear Lake, still largely a bedroom community surrounded by cow pastures.

Dodging the rain, Kathy slogged through the muddy parking lot, landing her boot in one of the many potholes. She pushed her way through the saloon-style doors, fashioned in the shape of red-leather-bikini-clad women, to dim yellow lights and the familiar buzz of chitchat and chortling from coworkers already a drink or two into their night. A stench wafted toward her—stale beer soaked into cheap carpet, cigarette smoke, and hamburger grease. For this young astronaut, it was just another Friday night in the low country.

Before the Outpost became the de facto bar of choice, the Mexican Mafia had chosen meeting spots for the Astronaut Office's Friday-night happy hours, which George considered a hallowed tradition. These dives included the Turtle Club, a floating bar off a Clear Lake dock with a cramped karaoke stage, and Maribelle's, a Day-Glo pink saloon whose clientele included local shrimpers. Maribelle's was famous for its Naughty Nighties parties, for which chesty waitresses would dress up in Frederick's of Hollywood lingerie, as well as mud-wrestling competitions (the women astronauts opted out) and everybody's favorite, the crustacean races (*you had to be there*).

By comparison, the Outpost was tame. The old barracks had previously housed Fort Terry's Universal Joint, which served barbecue and proudly hosted the Gemini and Apollo splashdown parties. When Fort Terry's closed, former garbage truck driver Gene Ross bought the place, revamped it, and reopened it as the Outpost Tavern. Between wives at the time, Ross threw all his energy and his entire savings into making the Outpost a success. When the bar first opened, the flight operations director himself stopped in: Moseying up to the bar and sitting next to a wooden Indian sentinel, George nodded to Gene, who slid him a beer. Gene then asked the director if next time he would bring a box of astronaut pictures for the wall. George obliged, and soon the astronauts themselves followed.[35]

The pub became so popular that NASA folks nicknamed it Building 99—a place where work could continue with a beer in hand. The place transformed into a veritable space museum. A banner over the bar announced FLIGHT'S OVER—IT'S MILLER TIME AT THE OUTPOST! A spacesuit floated from the ceiling with a can of beer in its hand.[36] A neon LONE STAR sign dappled red and blue light across the framed astronaut photos that lined the walls.

Kathy took a sip of Tab and looked out at her TFNG colleagues as they gabbed and caroused in the dim, smoke-filled interior. Gene flipped patties and served ale. On a good day, he would sell forty cases of beer and a hundred hamburgers. Across the room, astronauts would shoot pool while "guzzling Cuervo Especial tequila out of a pint bottle and cutting the burn with a Budweiser," according to journalist Donald Myers.[37] Apollo astronaut Joe Engle sometimes played the electric organ and sang "Jesus Wants Me for a Sunbeam." "I will ask Jesus to help me / To keep my heart from sin," Engle crooned. The song might have sounded moving on a Sunday at church, but in this setting it was strikingly incongruous, and utterly hilarious.[38]

Every Friday night George held court at his favorite booth. Tonight, George's basset hound eyes looked a little more droopy than usual—a clear sign someone would have to drive him home. The astronauts would draw straws for "Abbey duty" and see who of those sober enough to drive would have the privilege of taking George home. *Not it,* Kathy thought.

There's only so much diet soda you can drink, watching other people get sloshed. Kathy paid her bill at the bar and headed home.

Partying with the boss, some surmised, came with perks. George's favorites became known as the bubbas (or if a woman, a bubbette). Officially, a bubba was any astronaut George selected to serve as his personal deputy. This meant moving into the office adjoining George's and handling work the boss did not want to deal with personally, such as delivering bad news or processing tedious paperwork.[39] George rotated astronauts through his office every few months, or simply when the mood struck him.

George saw himself as a father figure to his new astronauts, although some more darkly referred to him as the Godfather. A physically large man with a dark crew cut, jet-black eyes, and olive skin, he looked not unlike Marlon Brando as Francis Ford Coppola's Don Corleone. He mumbled like him, too. One had to lean in to hear what George had to say, giving him complete attention. "Quiet, taciturn, undemonstrative . . . in an alarming way," noted NASA artist Robert McCall, adding, "[a]larming . . . in that he's not animated."[40] His deadpan, nonreactive manner made him both peculiar and mysterious to the young recruits.

Though he said little, George possessed uncanny powers of observation and a steel-trap memory. "You could think he was dozing off at the bar," Kathy said. "Two days later he would repeat verbatim something that had been said at 12:30 in the morning when you were pretty sure he was completely asleep."[41]

Once, his astronauts surreptitiously snuck off to El Paso, Texas, to fly stunts. They gunned up four little Pitts Specials, aerobatic airplanes, flying in a diamond shape while a T-38 hovered dangerously beneath them. Another T-38 rolled on top of the quintet inverted, allowing its back-seat pilot to snap a picture of the planes beneath it, giving the illusion that they were flying in an unlikely if not impossible formation. The moment was one of pure exhilaration and extreme danger. *Wow, we got away with it,* they thought. Days later, George called an impromptu meeting with TFNG Jon McBride, who had participated in the stunt. George had a picture of their little antic framed on the wall behind his desk. *How the hell did he get that?*[42] In an age before cell phone cameras, texting, or even digital pictures, George had tentacles in every NASA outpost.

As flight operations director, George had the singular power to determine if and when astronauts flew. The power of that office had been established in the Gemini days. There were no committees or official reviews. Flight assignments were his decision, and his alone. Even George's physical office, which overlooked the astronauts' suites, seemed a symbolic representation of his control over them.[43]

To win the boss's favor, the New Guys were encouraged to participate in George's favorite extracurricular activities: the weekly intraoffice softball games and the annual chili cook-off. These informal, nonmandatory but highly suggested activities could get competitive. Astronauts were there to win on behalf of the Astronaut Office. George once pulled New Guy George "Pinky" Nelson, a former college baseball player, out of his selection-week psychiatric exam to play as a ringer on the softball team. *I'm pretty sure I'm an astronaut because I was a good softball player,* Pinky later joked.[44]

The chili cook-offs and forced team sports "were just another form of competing for George's approval," Kathy said. "I'm not going to turn myself into somebody I'm not on the off chance that that might affect my chances."[45] Instead, Kathy thrived in one-on-one sports, trouncing Fred, Guy, and several other colleagues at racquetball.[46] Sailing a little cutter through Clear Lake on the weekends, she would get lost in her thoughts, enjoying the calm solitude of the water. Kathy preferred to make friends outside the Astronaut Office or outside the agency entirely. She opted out of the constant networking, deciding that her work would have to speak for her. "I know who I am, and I know what I'm doing," she said. "I know what I'm capable of."[47]

Seattle, Washington. Fall 1978.

One thing was clear: Kathy was not a bubbette. Would that affect her getting her silver wings? It certainly meant missing out on the glorious boondoggles George arranged for his favorites. On one such occasion, George sent Judy, Sally, and Anna—plus pilots Dick Scobee, Rick Hauck, and Dan Brandenstein—to Seattle, Washington, to visit the Boeing Company, which made the 747 that transported the shuttle around the country. Far away from their spouses and partners in Houston, the sextet got drunk on flaming Moose River hummers (the West Coast cousin to the flaming hooker) and pranked a Volkswagen driver, picking the Beetle up off the road and "reparking" it on the sidewalk.[48]

Invariably on these trips, someone issued a dare to call UNO.[49] One might expect that a man nicknamed the Dark Lord would be enraged, or at the very least annoyed, by a late-night phone call from his inebriated subordinates. In truth, George loved to hear his AsCans bonding and was flattered to be included in the merriment.[50]

At Boeing the next morning, the six New Guys, a little worse for wear, hopped aboard a 747 to practice touch-and-go's, or looped take-offs and landings. Anna's turn came and she slid into the pilot's seat. Fighting off an aching head and a queasy stomach, she gritted her teeth and brought the massive cargo jet down to the runway. The Boeing pilot gave her the thumbs-up. "Oh, thank you. That was my first landing ever," Anna said. The pilot, who had wrongly assumed Anna was an experienced pilot, turned white as a ghost.[51]

Before the group left Seattle, they procured a classified payload—fresh salmon for Washington State native George. Flying their T-38s, they buzzed over the Pacific low enough to see the jet wash, then dove through the Marslike canyons of Utah, barely skirting the jagged edges. Anna snapped a picture of Scobee and Judy flying dangerously close to the desert floor, framed it, and gave it to George along with the salmon.

Kathy was not the only New Guy standing apart from the pack. Ron and Guy skipped the happy hours, cook-offs, and softball games altogether. Fred indulged the happy hours, although never played softball.[52] These men opted instead to create their own community with a few of the other TFNG families and their neighbors. Ron recruited Charlie Bolden, an African American Navy test pilot who would go on to become a NASA administrator, into the astronaut corps, and they became close friends. Fred and Charlie were both inducted to Ron's Black fraternity, Omega Psi Phi. Though Guy, Fred, and Ron did not complain of any racism at NASA, they did note that there was a divide in some of their social activities.[53,54]

Perhaps the best gauge of how the astronauts felt about race and gender was their class comedy skit. The annual talent show, known for bawdy, off-color skits and blue humor that would never fly at the Monday all-hands meetings, became a rite of passage for every astronaut class. The New Guys parodied their finalist interview week, highlighting NASA's call for diverse astronauts.

One of the women portrayed Carolyn Huntoon as "Dr. No-Balls-At-All." Ron's Dr. Atkinson, the only African American on the selection panel, talked "jive," as Fred Gregory translated his words into English.[55] Drs. Atkinson and No-Balls-At-All interviewed an astronaut candidate composed, totem pole trench–style, of Judy's head, Ron's arm, and a white male TFNG's arm. Judy answered the panel's questions while Ron and his colleague waved their arms in wild, uncoordinated gesticulation. The audience was in stitches. At the final question—*What makes you a good fit for this job?*—Ron and the other New Guy flailed their arms as Judy smiled slyly and answered, "I have some rather *unique* qualifications."[56]

In more modern times, the skit might be considered racist and misogynistic, though the New Guys did not see it that way. "We were doing takeoffs. We were basically teasing with caricatures," explained Kathy, who helped plan the portrayal of Dr. Huntoon.[57] Although nearly all the TFNGs reported fair treatment at NASA, the skit shows that issues of race and gender were not far from their minds.

As diverse as the New Guys were, the group began to assimilate to the iconic image of the American astronaut. They could not help trying on the familiar tropes—leather bomber jackets, aviator sunglasses, and cocksure attitudes. They no longer walked. They strutted.[58] Ron grew beatnik sideburns and donned dark sunglasses, gracing Vegas with his sonic booms. Rhea, with her Marilyn Monroe bob, zipped around town in her silver Corvette, her handsome astronaut beau beside her. Sally, dressed in her crisp flight suit, zoomed off with her new friends in their T-38s to observe a total solar eclipse over Montana. As the moon moved across the sun's surface, they raced at nine-tenths the speed of sound in the opposite direction, giving them an unparalleled view of the totality that lasted four minutes and ten seconds, nearly two minutes longer than anyone on the ground. On the way home, they detoured through South Dakota, buzzing so close to Mount Rushmore they could almost touch Washington's prominent nose.[59]

Their bravado did not sit well with everyone in their lives. Sally came to NASA with her boyfriend, Bill Colson. Her new macho attitude did not square with the woman he had fallen in love with at Stanford. Their old conversations about academia paled in comparison to Sally's boasts of adrenaline-fueled training exercises. "Sally talked about other astronauts as if they were gods," Colson said.[60] Then there was the fact that Molly, Sally's ex-girlfriend from Stanford, was always visiting.[61] Colson was one of the few people who knew what Molly had meant to Sally and increasingly did not understand where he fit into her life.[62] "I had nothing to offer," he said. "I felt out of place more and more."[63] In January 1979, Sally told Bill she wanted to live alone.

Astronomer Steve Hawley noticed Colson moving out of their apartment complex. From their first days at NASA, Steve thought Sally was cute and *just his type of girl.* They literally had the same comment about joining the astronaut corps: "It's your basic once-in-a-lifetime opportunity."[64] Steve initiated a friendship. They worked out together, strategized together, watched sports together. They did everything together except talk about their feelings. "She was a lot like I am in the sense that, when she was worried or upset, she would just be quiet. And that's what I do. So there were lots of times when we wouldn't talk about stuff," Steve said.[65]

Steve was well liked for his quick sense of humor. Even the military guys accepted him into their fold.[66] Once, when Senator Adlai Stevenson met with the astronauts, Hoot introduced himself to the senator as a "Navy fighter pilot." Dan Brandenstein introduced himself next as a "Navy attack pilot." Then Steve Hawley, PhD, followed up with "attack astronomer."[67] The nickname stuck, then morphed into "A^2," or "A-squared," which is how Sally began referring to him.[68] Sally found Steve witty and charming and soon the two fell into a romance.

As the year progressed, the class became *a crew*: Anna and Bill became a center point of social life. On the weekends, they hosted parties in their home, which boasted an open floor plan and vaulted ceilings. Because of their busy schedules (and because Bill would never make a decision on decor), the couple had no furniture, save the lawn chairs on their deck. Propping himself on the fireplace and grabbing a beer, Steve Hawley told Anna she should put a basketball hoop in the living room to take advantage of all that space. At the Onizukas', the group would dine on Hawaiian-style pig roasts and guzzle down barrels of Coors beer. *Who was the old sweaty guy dragging the kegs in and out?* Kathy wondered. *That's Alan Shepard!* Now in his retirement, Shepard had bought a Coors distributorship.[69] Sally learned that Joseph Coors had opposed the Equal Rights Amendment and stopped drinking the beer, but still went to the parties.

After work they would meet at a little Italian restaurant called Frenchie's—a misnomer, but the two Italian brothers, the Cameras, who bought it, never changed the name. Nestled in a nondescript strip mall, Frenchie's, with its white tablecloths and candlelit tables, offered rich Italian fare, from lasagna and veal marsala to homemade gnocchi. On a Friday night, the line for dinner would run out the door and along the side of the building. Cocktail hour unfolded impromptu on the sidewalk as owner Frankie and his wife, Anna, chatted with the astronauts as they came in.

On their one-year anniversary, a blazing-hot July 4th weekend, the group organized a camping trip to Canyon Lake in nearby Texas Hill Country. The astronauts showed up looking like a biker gang in their aviator sunglasses and leather jackets, and promptly piled into tubes to float down the Guadalupe.[70] All the stress of the year melted away with the heat, the water, and the beer.

Ellison Onizuka and Mike Mullane drunkenly lit off fireworks.

"Dad, don't you think this is kind of dangerous?" Mullane's son asked him. He was right, but Mullane and the rest of the New Guys did not care.

"We had become the kids," Mullane said. "We were bulletproof. We were immortal. We were astronauts."[71] While the pilots flipped burgers outside, their wives, preparing a salad in the kitchen, knocked on the windowpane and flashed their boobs. The women New Guys collectively rolled their eyes, while the young pilots screeched with laughter. As a team-building exercise, the TFNGs formed a conga line and danced over those foolish enough to have gone to bed early.[72] Anna and Bill woke up the next morning to find their canoe up in a tree.[73]

The first year of astronaut training had been euphoric for the thrill-seeking, adventure-loving young scientists and explorers. "We didn't walk. We floated along the hallways in weightless glory," said Mullane. "You couldn't have beaten the smile from our faces with a stick."[74] They had raced off into the sunset strapped into the fastest things in the air, their T-38 jets. They had vanquished their worst fears in the rushing waters of the Atlantic Ocean. They had made fast friends and new loves.

A few months after their one-year anniversary, the New Guys spotted handwritten signs scrawled in red marker around Building 4: ALL ASTROS & ASCANS MEET W/DR. KRAFT AND G. ABBEY FRI 31ST 1500 TO DISCUSS NEW ASTRO SELECTION; ISSUES & ANSWERS, ETC. A group meeting with the big bosses George Abbey and Dr. Kraft? That was highly unusual. And what was this about "new astro selection"? They themselves had not flown yet. Heck, the shuttle had not even flown. Why did the program need new candidates? Were they doomed, like the old guys, to be passed over for new hires?

The following week, on August 31, 1979, the New Guys shuffled into the meeting room and settled into their seats. Kraft stood at the front of the room and announced that the AsCans could drop the word "candidate" from their job titles. Though they had only completed half of their scheduled training, Kraft was impressed with all they had accomplished as students, pilot trainees, and as representatives of NASA.

"You all are astronauts," Kraft said.[75]

What did he say? We're astronauts now?

In a press release issued after the meeting, George stated that NASA was "pleased with the performance of their first class" and that the agency would begin recruiting a new class of astronauts to start in 1980. In a small ceremony, Kraft presented each of the new astronauts with their very own silver pin.

No one had washed out, not even the "girls," Rhea thought.[76]

Ron was thrilled. Now he was not only a solid swimmer, but he had his silver wings.

Anna was elated, but not just for herself. The newly announced astronaut class of 1980 might be Bill's ticket to space.

Kathy had made it through the program without being part of the bubba clique. She was carving her own path, one that did not appropriate the persona of the old guys or appease the egos of the power players.

As an outsider, Kathy viewed their sudden promotion with a healthy dose of skepticism. She could not help but wonder why this was happening so soon. The terrific news seemed to come out of nowhere. Had they really done well enough to skip a whole year of training? Something felt off.

Underlying Kraft's praise and the expedited commencement lurked a more immediate reality: The New Guys were desperately needed in the field to contribute to the shuttle's progress. If the first year had been a youthful euphoria, the days to come would sober them up. The New Guys were about to discover the truth about the troubled machine that would carry them to the stars.

CHAPTER 6

GET THE SON OF A BITCH IN SPACE

Rockwell Facility, Palmdale, California. September 17, 1976.

Tucked behind sloping purple mountains whose ridges shield it from the smoggy metropolis of Los Angeles lies the sleepy high-desert town of Palmdale, California. Back when land sold for fifty cents an acre, Swiss-German settlers mistook the Joshua trees here for palms and named it Palmenthal, or Palm Valley. The misnomer stuck, but settlers, unable to grow grains and fruit in the brutally dry climate, moved on. A near century later, Palmdale remained little more than a few scattered stores, single-story homes, and an airstrip, unremarkable in all manners except one. Since the onset of the Second World War, Palmdale had quietly played host to three of America's premier aerospace and defense manufacturers: Lockheed, Northrop, and Rockwell International.

On September 17, 1976, Palmdale was anything but quiet.

Thirty-five thousand people swarmed the small town, thronging its limited-capacity motels and fanning out to neighboring townships. Rousing at dawn, spectators gathered outside Rockwell International's facility, desperate to glimpse what newspapers were calling "mankind's greatest technological achievement."[1] Celebrities and dignitaries, from California governor Jerry Brown to the cast of *Star Trek*, sat expectantly outside a hangar, sipping champagne and smiling for the paparazzi.[2]

As the Air Force band sounded the brassy opening theme of *Star Trek*, reporters and photographers stood on their tiptoes. Onlookers stomped their feet, cheered, and waved tiny American flags.[3] Spectators craned their necks to see a first glimpse of America's new space plane poke around the corner of the hangar. Waved on by technicians in white coveralls and towed by a tractor emblazoned with patriotic Stars and Stripes, the orbiter slowly rounded the corner.

One-hundred-twenty-two feet in length, with a tail that loomed five stories above the gathered crowd, *Enterprise* gleamed white in the bright Mojave sun. Blocky and blunted, with a black snub nose, the orbiter lacked the damn-your-eyes elegance of the needle-nosed Concorde supersonic jet, but its unorthodox proportions meant crews could safely reenter Earth's atmosphere and land gracefully—rather than plop into the ocean like an Apollo capsule. Short, wide delta wings flared out at the back of its long fuselage, resembling the swept back appendages of a moth more than the outstretched wings of a bird. Still, there was something majestic about it, something otherworldly.

Rockwell had contracted to build *Enterprise* and its sister orbiter *Columbia* for $2.7 billion. Though *Enterprise* had been assembled in southern California, its birth was a nationwide endeavor. The wings came from New York via the Panama Canal, the payload bay doors from Oklahoma, and the maneuvering thrusters from Missouri.[4] Nearly fifty thousand people employed by NASA and 240 subcontractors across forty-seven states contributed.[5] The shuttle was an American original.[6]

The rising sun brought the desert floor to a blazing hundred degrees as Mr. Spock himself, Leonard Nimoy, took the stage. "This is a reunion for me," Nimoy shared, referencing the starship *Enterprise* he first flew on air in 1966. The first shuttle had been renamed *Enterprise* (instead of the more patriotic *Constitution*) after fans of the show launched a massive letter-writing campaign. It was a reunion for the show's cast, too: Nimoy gestured to creator Gene Roddenberry as well as former cast-mates George Takei, who played Sulu, and Nichelle Nichols, the former Lieutenant Uhura.

Political allies Governor Brown and Senator Barry Goldwater hob-nobbed with NASA brass like Chris Kraft, Max Faget, and George Abbey. George, who had just been named NASA's new head of flight operations, looked upon the vehicle with deep affection. Despite his bloodless, inscrutable demeanor, the Dark Lord had a heart, and that heart belonged almost entirely to NASA.

George had dedicated the last ten years of his life to the space agency. He witnessed the first moon landing from Mission Control, absorbed the aftermath of the Apollo 1 fire, and helped to coordinate the recovery of the Apollo 13 crew. For years, he had been a talented administrator and Chris Kraft's right-hand man. "George," Kraft would say, "knew where all the bodies were buried."[7]

Flying was encoded in George's DNA. Growing up as a young boy in Seattle, he loved to watch planes take off from the nearby Boeing facility and consumed *Buck Rogers* comics and *Flash Gordon* movie serials. His Welsh mother, Brenta, and his Canadian father, Sam, a WWI veteran, modeled clean and orderly lives—no smoking, drinking, or swearing.[8] As an adult, George threw out the prohibition against drinking and even occasionally indulged in a cigar.[9] Still, he remained faithful to the last rule, mumbling instead of cursing whenever he grew frustrated.

After graduating from the Naval Academy in 1954, George joined the Air Force as a test pilot. He was a natural: "The best pilot I've ever seen," one coworker said.[10] George had loftier dreams. The 1957 *Sputnik* launch had ignited a competitive fire in him that could not be doused. He wanted to go to space. In 1958, he met Mercury 7 astronauts Gus Grissom, John Glenn, and Al Shepard as they passed through Wright Air Force Base in Dayton, Ohio, where George was stationed. Though they were longer in the tooth, George felt his credentials were on par.

He decided to toss his hat into the ring and applied to become an astronaut.[11]

One man stood in George's way: Chuck Yeager.

Yeager, whose formal education ended in high school, resented bookish types like George, a quiet college graduate with an advanced degree. When George applied to be an astronaut in 1966, Yeager was the commandant of the Aerospace Research Pilot School, where most future astronauts trained, and a member of the Air Force's astronaut selection committee. George was a talented Air Force pilot but had never been to Yeager's elite school, nor did he fit the swashbuckling test pilot profile that Yeager cultivated. As soon as George met the requirements for the astronaut program, Yeager and the Air Force decided they would put another requirement over and above the NASA requirement: All finalists had to be graduates of Yeager's Aerospace Research Pilot School, thus keeping George out of the program.[12]

George never forgot how Yeager froze him out. Nor would he forget how he had spurned Ed Dwight for being Black.[13] *Space is for everyone,* George believed. When Kraft offered George a chance to define a new vision for the American astronaut, he leapt at the opportunity to become the new "astronaut maker."[14] George vowed that his applicants—men and women, doctors, engineers, and scientists, of all backgrounds—would get a fair shot at the dream that had eluded him.

On that late summer day in 1976, he stood before the vehicle this new class of heroes would fly. As enthralled as he was with *Enterprise*'s unveiling, George knew better than anyone that the shuttle was still an experimental vehicle. No engines or rockets had yet been approved for launch. The tiles of its heat shield had not been attached, and its flight control software needed more testing.

George didn't even know if the damn thing would ever fly.

The shuttle's origin story was a peculiar tale indeed, with many twists and turns. Like many great American endeavors, its trajectory was shaped by an alloy of politics, money, science, and ego.

In the twilight of the moon landings, NASA's budget was in freefall. The war in Vietnam was growing increasingly expensive, and for reasons as complicated as the era itself—Kennedy's assassination, strained finances, declining faith in American institutions, and growing concerns about economic inequality—NASA's public support was cratering.

Still, the agency had the ambition for another giant leap. In 1969, NASA's blueprint was fit for a science fiction novel: America would build a fleet of shuttles to assemble three Earth-orbiting space stations, a permanent lunar base, and a lunar-orbiting space station. A nuclear-powered shuttle would make trips to the moon commonplace. Wernher von Braun, the mastermind behind the Apollo rockets and first director of NASA's Marshall Space Center, set his watch for a Mars expedition in November 1981 and then charted a course for Venus.[15]

NASA's dreams were starkly at odds with the political will of the moment. President Richard Nixon, ever the tactician, read the polling. The White House issued a statement in March 1970, stressing that the unbounded promise of space must not distract the nation from the "many critical problems here on this planet, [which] make high-priority demands on our attention and our resources."[16] Nixon's budget office slashed the space stations, the lunar landings, and the trip to Mars. *What would be left?*[17]

Luckily for NASA, the specter of the Soviets regaining dominance in space on Nixon's watch and the value of political support from the aerospace industry was enough to keep Nixon at the table. NASA administrator James Fletcher suggested a compromise—let the space shuttle be the first step to the longer-term goal of deep space exploration.[18] For now, the shuttle would have to be enough.

Nixon's budget office balked at the shuttle's $10 billion price tag.[19] Fletcher insisted that the shuttle would be worth it—a multifunctioning, multitasking spacecraft.[20] Not only would the shuttle be the gateway to crucial earth science, astronomy, and physiology—estimating crop yields, tracking air pollution, measuring ocean temperatures, launching telescopes, and studying long-duration space stays—but it could also be a delivery truck to the stars. Commercial enterprises, foreign governments, and the United States Department of Defense (DoD) would all pay to launch satellites from the shuttle's massive payload bay by skilled astronauts.[21]

Nixon bought the idea. "The general reliability and versatility [of] the shuttle," he said from his San Clemente estate in 1972, "seems likely to establish it as the workhorse of our whole space effort, taking the place of all present launch vehicles."[22] From the outset, the shuttle was to be everything to everyone—at bargain-bin prices.

Perhaps more perniciously, NASA's Nixon-era deals signaled the erosion of the agency's independence. NASA would have to accommodate design requirements from the DoD, its biggest customer, for national security payloads.[23] Pentagon officials wanted to double the shuttle's payload capacity to carry large, classified spy satellites. They demanded the orbiter travel farther, faster: The shuttle should be able to take off from Vandenberg Air Force Base in California, fly over Russia, then return to its launch site after a single orbit.[24] None of these requests were simple. NASA engineers had to expand the orbiter, modify its wings and gliding capabilities, and build more powerful (and therefore more dangerous) rockets to get the heavier shuttle into orbit.[25]

Even with these new design requirements, Nixon shrank NASA's budget from $5.5 billion during the height of the Apollo era in 1965 to $4 billion in 1979, a 70 percent decrease when adjusted for inflation.[26,27] The budget cuts caused an "aerospace depression" across the country—two hundred thousand engineers and technicians lost their jobs.[28]

Despite the leaner budget, the shuttle remained a grand vision. Mechanical engineers were designing a system that took off like a rocket and landed like a jet to maximize reusability and lower costs. Engineers at Marshall Space Center were pushing rocketry to new limits to launch the massive space plane. Materials scientists were inventing lightweight ceramics to maintain the space plane's aerodynamics and prevent it from burning up on reentry. Computer engineers were developing futuristic software to perform the shuttle's many sophisticated tasks in a world that did not yet have personal computers or cell phones.

NASA officials publicly touted each milestone, promising that the shuttle would be operational by 1980 and perform 725 missions over the ensuing decade-plus.[29] NASA had only completed thirty-one manned space flights to date, so the claim was a staggering number by any measure.[30] The great fiction—maintained by both politicians and NASA's administrators—was that the Nixon budget was enough to meet all the shuttle's design specifications. With great optimism for such an audacious endeavor, NASA engineers and designers worked around budget limitations with ingenuity and plenty of elbow grease. Still, year after year, until the late 1970s expensive design problems ultimately were deferred to meet the demands of budget officials and Congress. As the first shuttle launch approached NASA's engineering debt came due.

Rogers Dry Lake, Edwards Air Force Base, California. October 1977.

Formed two and a half million years ago, Rogers Dry Lake in California's Mojave Desert was once a plentiful endorheic lake, full of oysters, mollusks, and prehistoric fish. Birds dove from a clear blue sky to feast upon the bountiful aquatic life. Bison and mammoths lumbered along the lakeshore. At twilight, bats flitted through the piñon and juniper trees as shrews and mountain gophers crawled into sandy burrows.

Time and heat robbed Rogers of its life and waters, leaving behind a dense salt-mineral layer, making it one of the hardest and largest dry lake beds in the world. Capable of withstanding pressure as high as 250 pounds per square inch without cracking and exhibiting a remarkably even grade, the lake bed was a perfect landing area for experimental aircraft like the shuttle, which were liable to overshoot shorter and narrower tarmac landing strips during test flights.[31]

One thousand reporters and over seventy thousand spectators had gathered at Rogers for the first of the shuttle's approach and landing tests on August 12, 1977. Enthusiasts staked out choice viewing spots overnight to catch sight of the space plane. Camper vans stretched to the edges of the horizon. Now, at the final test, George stood beside Sally Ride under the VIP tent of Runway 22. He had invited her, as he had Anna, Ron, and a few other finalists to witness *Enterprise*'s approach and landing tests. At the time, none of the candidates knew if they had been chosen by NASA yet. George was over three months away from making those calls, though the invitation seemed to indicate they were still in the running. Sally came that day with her old friend Molly Tyson.[32]

The tests were the first major hurdle for the new orbiter. If *Enterprise* passed, the shuttle would head to Marshall Space Center, where it would be mated with an external tank and inert solid rocket boosters and be subjected to punishing vibrations to ensure the entire shuttle stack could withstand the forces of launch. Then *Enterprise* would be ferried back to Palmdale and refitted with the major systems for orbital flight. That day, *Enterprise* was still very much a prototype, lacking a propulsion system, a heat shield, a life-support system, and most of its flight deck instrumentation.

As the bright Mojave sun warmed the crisp morning air, George and Sally watched all seventy-five tons of *Enterprise*, bolted atop its reengineered Boeing 747 transport vehicle, taxi onto the runway.[33] The improbable configuration looked like an orca riding a humpback whale. Since *Enterprise* was engineless at this point, the 747 would lug it to the heavens until, at peak height, seven pyrotechnic bolts exploded, ejecting the orbiter for the glide down to Rogers Dry Lake below. Without power, the pilots had no second chance at a landing.[34]

The flight would also test the shuttle's new software. In standard planes of the era, a pilot's instruments were wired directly to the mechanisms they controlled. Pilots could get a "feel" for handling the aircraft; the machine responded intuitively to human touch. Not so with the shuttle. Nothing but code linked *Enterprise*'s stick and rudder pedals to the flight controls. For pilots used to gears and switches, the infinitesimal delay between computer input and response was noticeable and made flying the bird all the more difficult.

The main computer had already crashed once during *Enterprise*'s four prior test landings. A backup computer had taken over and prevented catastrophe.[35] Computers were still a fledgling technology. The original shuttle could only handle 1.4 million instructions per second, compared to the roughly sixteen trillion operations a second that a modern-day iPhone processor can handle.[36] In today's computing terms, shuttle pilots were relying on a souped-up pocket calculator to guide their massive spaceship as it hurtled toward the ground at speeds of over three hundred miles per hour.[37]

George, Sally, and the rest of the spectators trained their eyes as the conjoined ships climbed through the cloudless blue sky, barreling toward the sun. Astronauts Fred Haise and Gordon Fullerton sat behind the controls of *Enterprise* as the 747 heralded the orbiter to a peak height of twenty-three thousand feet.

"Go for Sep," the ground radioed. Haise pushed a square white button on the shuttle's control panel, which detonated the explosive bolts that held *Enterprise* to the 747 and released the shuttle from the plane's clutches. The aircraft veered left as the spaceship banked right, pitching forward at a vertiginous twenty-two-degree slope. Commercial jets typically land at about three degrees.[38]

Enterprise initiated its glide toward the lake bed, slicing through the air at 330 miles per hour and traversing a distance of twelve miles in under two minutes. As the space plane approached the runway, its speed melted away, 280 miles per hour, 260, down to 240. Still, as it passed over the threshold of the tarmac, it flew forty miles per hour *too fast*.[39]

Haise quickly deployed the air brake and elevons, long flaps positioned along the back edge of the wings, to slow the bird and bring it down. The elevons only increased the shuttle's lift, and its altitude plateaued as the plane coasted a few hundred feet above the runway.

"Come on, Bess!" Haise pushed forward on the control stick to force *Enterprise* down.

The wheels smacked hard against the concrete and the craft skipped back into the air, wobbling left, then right. Haise gripped tightly on the control stick, fighting to stabilize the orbiter. Like the driver of a hydroplaning automobile wrestling the steering wheel back and forth, Haise only exacerbated the problem with every motion. As he pushed on the control stick, the command fed into the computer control system and, with the momentary lag, furthered the oscillation.

"Let the stick go!" Fullerton barked at Haise. The shuttle would stabilize and bring itself down. The crowd watched with bated breath, an anxious six seconds before the shuttle's three wheels finally hit the tarmac again and rolled to a stop.[40]

What the hell was that? George thought.

"You did as badly as I did!" guest observer Prince Charles of England teased Haise and Fullerton upon their exit. The previous day, the prince, a member of the Royal Air Force, had himself botched two landings in Johnson's shuttle simulator.[41]

If not clear before, Sally now understood that, should she be lucky enough to be accepted to the astronaut program, she would be entrusting her life to technology that was not merely ambitious, but also highly experimental. "Up until this morning, I wasn't afraid at all. Maybe I should have been," Sally confided in Molly. "[S]eeing that landing this morning, knowing that they couldn't pull up for another try—that scared me."[42] Despite the technical hurdles, headlines spoke of a monumental success, saying that "the shuttle had worked flawlessly" and suggesting a launch before the end of the decade.[43,44]

Behind the scenes, shuttle program managers were worried. Insights gained from the approach and landing tests led engineers to alter the design of the shuttle, replacing portions of the vehicle's aluminum frame with stronger, relatively lightweight titanium castings. *Enterprise* would be costly to modify at this stage in the vehicle's development. By the end of 1977, managers decided it would be more efficient to refit one of Rockwell's test models instead of retrofitting *Enterprise.*[45] *Enterprise*, the inaugural orbiter that had crowds spellbound a year prior, would be a test vehicle only and would never fly in space.[46] Rockwell's Structural Test Article 099 would instead become the program's second orbiter and would, to the world, come to be known as *Challenger.*

Ellington Field, Houston. March 10, 1978.

Bob "Crip" Crippen drove George Abbey up to the tarmac in his beat-up red pickup truck. They passed stragglers remaining from the larger contingent of press and space fans who had come to see *Enterprise* that morning. The orbiter was making a pit stop on its cross-country journey from Rockwell in California to NASA's Marshall Space Center in Huntsville, Alabama for vibration testing.[47] Earlier that day, George had called Crip out of the blue and suggested the two go out to see the test vehicle together. The spontaneous invitation from the ever-cryptic George piqued Crip's curiosity.

Born in Beaumont, a tar-rich bayou northeast of Houston, young Robert Crippen fell asleep each night to the sound of crickets and the rhythmic hum of the oil derricks. He was the son of a petrol worker who became a chicken rancher after losing a couple of fingers in an oil rig accident. Crip's mother added a gas station and a beer joint to the family's chicken ranch in Porter, Texas, and then things really got going at Crippen's Drive-In, where a neon sign hung out front welcoming wildcatters and ranchers in for greasy burgers and happy hour drinks. That life was all well and good, but for as long as he could remember, Crip dreamed of soaring high above the Texas tar fields and sought out the nearest point of departure—Houston's William H. Hobby Airport. His father would drive him forty miles south, right up to the chain-link perimeter fence around the tarmac. The cool metal of the railing pressed against his face, young Crip would watch airplanes take off and land. One day, a flight attendant invited the pair to tour a Douglas DC-3 airplane on the tarmac. Crip climbed into the cockpit and sat behind the controls, imagining what it would be like to fly this behemoth through the skies. He was hooked.[48]

Despite being the picture of an astronaut—a deep tan, Texas charm, and all-American good looks—Crip was not in fact an astronaut, at least not a gold pin–wearing one. In 1966, the Air Force recruited Crip to their human spaceflight program, an audacious project that aimed to send military astronauts and reconnaissance payloads to an orbiting station called the Manned Orbiting Laboratory. The Air Force canceled the program in 1969 before Crip could fly, and he transferred to NASA. Ever since, the agency had hung Crip out to dry, grounding him until they could get their next spacecraft on the launching pad. Eager as ever to get his golden pin, Crip was also aware that he was pushing forty and his chances to fly were quickly slipping away.

Crip idled his truck up to the glorious space plane, hoisted high atop its Boeing 747 mule, and parked in the shadow of its wings.

"How'd you like to fly the first one?" George finally broke the silence.

Crip barely understood the words coming out of George's mouth. *Are you serious, George?*

George was. No one knew the shuttle computers better than Crip. When he started at NASA, Crip volunteered to help develop the shuttle's software. At the University of Texas, he had been one of the first students to enroll in the school's fledgling computer science program, back when computers used punch cards. Crip also had a near-perfect memory, allowing him to easily recall and quickly locate the over two thousand switches that commanded the shuttle's more than 2.5 million moving parts. To George's mind, Crip was the perfect astronaut to pilot the first mission.

Hell yeah! Crip whooped. He wanted to turn cartwheels right there on the tarmac. Not only would Crip finally get his shot at going to space, his first flight would be the historic maiden voyage of the shuttle. It was a test pilot's dream. Crip would pilot Space Transportation System-1 (STS-1) alongside commander John Young, the most experienced astronaut in the office. Rockwell was beginning the final assembly of their bird *Columbia* in Palmdale, to be ready by November 1979.

Or at least that was the plan.

National Space Laboratories, Hancock County, Mississippi. December 1978.

Across the bayou from New Orleans, past muddy sloughs clogged with mosquitoes, sat 13,500 verdant acres carved out of what was once pine forest.[49] A pair of twenty-story test stands towered two hundred feet above the alligator-infested Pearl River.[50] Here, at Marshall's test facility in south Mississippi, the shuttle's Rocketdyne-designed main engines would be bolted down and fired up to validate their performance—thrust, structural integrity, and functionality.[51] When all went to plan, the test fires spewed so much steam and exhaust that they generated their own rain clouds.[52]

Huddled on the roof of the test control center, a few hundred yards from the test stand, George Abbey, Bob Crippen, and John Young waited on tenterhooks before the most ambitious test firing yet.[53] While most of the shuttle's main engine tests fired just one of its three engines, today's test would activate all three, to gauge the performance of the complete propulsion system.

The most complex, power-dense engines ever built, these motors were capable of generating thirty-seven million horses' worth of thrust and were part of a system designed to perform complex throttling, one capable of nearly unlimited starts and stops over an astonishing fifty-five-mission, ten-year lifetime.[54] On launch, each of the shuttle's three main engines would devour a swimming pool's worth of fuel in twenty-five seconds, heating liquid hydrogen and oxygen from almost absolute zero to 6,000°F—hot enough to boil iron.[55] The external tank, which fed the engines, was shaped like a cannon and coated with foam insulation. At over a hundred fifty feet long and twenty-seven feet wide, the tank was the largest component of the shuttle stack.[56] Once the tank emptied, it would separate from the orbiter and burn up on reentry, littering its remains over the Indian Ocean.[57]

Today, the three main engines, representing years of design work, thousands of labor hours, and tens of millions of dollars of equipment, would be put to the ultimate test.[58]

The countdown to engine start wound down to the last ten seconds. Over the intercom, George heard the numbers that never failed to make his palms sweat. *10 . . . 9 . . . 8 . . .*[59]

Songbirds dropped cool notes into the meadow, unaware of what was coming. *3 . . . 2 . . . 1.* The roar of an R-25 engine drowned out the ever-present hum of crickets, as white clouds billowed forth from the hangar.[60]

Inside the propulsion system, the engines' trash can–sized turbopumps forced high-pressure liquid oxygen into its lightning-fast turbines. The liquid oxygen was kept separate from the liquid hydrogen until it entered the combustion chamber, where the mixture of the fuel (hydrogen) and oxidizer (oxygen) ignited in a powerful, explosive chemical reaction. To absorb the awesome upward thrust produced by this chemical reaction, the test stand was embedded nearly sixty feet into the earth and stood 150 feet above ground.[61] The engines fired downward into a curved deflector that directed the flames out one side of the hangar. Sprinklers blitzed the deflector with cold water to prevent the metal shield from melting entirely. This mixture of H_2O, engine exhaust, and extreme heat generated the dense, puffy rain clouds the Mississippi test facility was known for.

For over three minutes, the engines seemed to be performing flawlessly, then the unthinkable happened. Orange and yellow flames flashed, then a colossal fireball erupted, enveloping the engines and the test stand. George knew it was bad, but he could not fathom how bad. After the fire was extinguished, the spectators ventured outside to survey the smoldering site with the Rocketdyne engineers, but there was barely anything left to analyze. The event would be the worst fire of the program to date, destroying an entire engine, millions of dollars of technology, and delaying further testing.[62] George suspected the engines' turbopumps were at the root of the failure.[63,64] Tasked with spinning up to thirty-eight thousand revolutions per minute, the turbine blades in these sophisticated pumps could exceed temperatures of 1,900°F. If a blade broke under the extreme pressure, it could tear through the pump, fuel lines, and tanks, triggering catastrophic self-destruction of the entire engine.[65]

As powerful as the main engines were, they only produced 20 percent of the thrust necessary to catapult the shuttle into orbit. The remaining 80 percent would come from a pair of solid rockets larger and more powerful than any humans had ever created, rockets being developed by a midsize chemical company in the northwest desert of Utah.[66]

Brigham City, Utah. Spring 1979.

High in the rolling hills of Utah's Wasatch Mountain Range a few miles northwest of the Great Salt Lake and Promontory, Utah, where the nation's eastern and western railroad lines joined to form the transcontinental railroad over a century earlier, a train one dozen railcars long sat empty, waiting for its space-age cargo.

Four solid rocket booster segments, each filled with one hundred fifty tons of solid fuel, a volatile mixture of aluminum powder and ammonium perchlorate with a rubbery, eraser-like consistency, waited beside the train. The shuttle's two solid motors each consisted of four segments, and each was over twenty-five feet long and twelve feet in diameter, large enough for a person to walk inside. Because the segments traveled by railway, they could be no wider than a railcar. When the four cylinders were stacked on top of each other for launch, they would weigh 1.3 million pounds and rise nearly fourteen stories high.[67]

Technicians carefully hoisted each one by crane and lowered them slowly onto the waiting flatbed railcars. Then they encased each in a clamshell-like hinged metal tube, to protect the delicate metal hardware from damage during the overland trip. Between each segment car linked an empty boxcar, spacing the rocket sections to help prevent unwanted pyrotechnics should the journey to Cape Canaveral get bumpy. Laden with its six hundred tons of cargo, the train now slowly pulled away from the depot, winding its way east.[68]

Seven days and twenty-five hundred miles later, after rattling along America's aging railway system, the cargo neared its destination—Cape Canaveral, Florida. There was one final hurdle—the drawbridge over the Indian River to get to the Space Coast. The train conductor carefully engaged the throttle and the engine slowly eased forward, never exceeding twenty-five miles per hour. The track was designed to handle four times that speed, but with cargo like this? No way. The operator wanted to keep vibrations to a minimum; any ding might affect the motor's integrity and performance.[69]

The rockets' long trip from the mountains of Utah to the lowlands of the Cape left some shaking their heads. Morton Thiokol, the small Utah company that had won the contract, had not been a giant in the aerospace industry. More experienced East Coast companies could have moved the rockets in larger pieces by barge, eliminating the need to send smaller individual rocket segments jostling over the railways. Some suspected Thiokol was a political decision: Senator Frank Moss (D-Utah), the newly appointed chairman of the Senate Aeronautics and Space Science Committee, and Dr. James Fletcher, the new NASA administrator, both had deep ties to the state, and both believed in returning wealth to Utah.[70]

Thiokol developed a joint system to fasten all the parts together. A tang-and-clevis connection, not unlike the tongue-and-groove joints of many hardwood floors, connected the segments' steel casings together.[71] Naturally, gaps remained at each of the three joints, as airtight seals are impossible where metal meets metal, so Thiokol engineers installed two rubber O-rings surrounded by plastic putty at each coupling to create a hermetic seal.[72] O-rings, commonplace in engineering design, can be found nestled inside wristwatches, preventing water from entering and corrupting the fragile gears inside. Hardware stores stock O-rings up to two inches in diameter, but the boosters' quarter-inch-thick O-rings measured over twelve feet in diameter. If the first O-ring failed to seal the joint, the second served as backup, blocking hot gasses from escaping the rocket and damaging the shuttle stack.

Though the solids packed a punch, they came with significant launch risks. Once lit, they could not be throttled down, unlike the main engines. Over two million pounds of propellant burned in an unstoppable chain reaction while connected to the liquid fuel–filled external tank and the crew-inhabited orbiter, making the first two minutes of liftoff while the boosters remained attached the most dangerous. If one booster lit without the other, the shuttle would go catapulting to the side. To remedy the issue, engineers hollowed out the booster centers, like a pencil without its lead. They installed garbage can–sized rockets at the tip of the boosters, which sent flames down the center to evenly ignite the solid fuel from the inside out.[73]

Prior to the shuttle program, solids had only been used to launch ballistic missiles. They had a reputation for self-destructing. Other rocket technologies might have been safer, but they demanded longer lead times and higher upfront costs.[74] In the end, the solids won the day.

Over Edwards Air Force Base, California. March 9, 1979.

A moderate 70°F with clear skies over the Mojave Desert, a beautiful day for flying, thought Mercury 7 astronaut Deke Slayton, as he zipped his T-38 up to *Columbia,* which was hoisted atop a 747.[75] In the back seat, Deke ferried an Associated Press photographer to record the orbiter making a final test flight before beginning its journey east to Cape Canaveral.[76]

As Deke flew closer to the orbiter, a bewildering sight came into focus. His jaw dropped as hundreds of the heat-protecting tiles came loose one after another, scattering through the air. After a few minutes, *Columbia* looked like it had suffered several rounds of artillery fire. If the tiles were coming off now during a routine test flight, what would a trip to space do? Deke could already hear management screaming about costs and delays. Meanwhile, the photographer behind him snapped away, happily having stumbled into a whopper of a story.

Reporters nationwide pounced on the embarrassing snafu, recounting the onset of a zipper effect as tiles tore off in strips along the tail and fuselage as the shuttle reached the desert sky. The full extent of the damage was visible on landing: "Gaping holes along the leading edge of

Columbia's five-story-high tail, more gaps below the cockpit and shreds of insulating tape scattered across the runway."[77]

Four weeks earlier, on February 3, 1979, George Abbey, Bob Crippen, and John Young had flown to Rockwell for the complete combined systems test of *Columbia*.[78] The results were nothing to write home about. Computers, electronics, and hydraulic systems proved finicky. Even with improvements, the complex controls would require the full attention of expert pilots during flight.[79]

Nevertheless, John Yardley, NASA's pipe-smoking associate administrator for Space Transportation Systems and chief program manager for the shuttle, decided to certify *Columbia* as complete.[80] Yardley was under tremendous time pressure from Congress. "Get the son of a bitch in space," he barked to his subordinates, "or we're going to lose the program."[81] Yardley then directed Deke Slayton to transport the orbiter to the Cape, where an army of Rockwell engineers waited to suit it up for a real launch, the date of which—November 1979—fast approached. Now looking at the gaping holes in *Columbia*'s heat shield, he regretted his decision.

Two Weeks Later. Kennedy Space Center, Florida. March 24, 1979.

"A star-spangled mess," Yardley groaned as he paced the ground floor of the cavernous Orbiter Processing Facility, examining the gaping holes in *Columbia*'s heat shield made even worse from its cross-country journey when it encountered storms.[82]

Am I supposed to fly that thing? Crip thought when he caught sight of the pockmarked shuttle.

NASA engineers always understood that the shuttle's heat shield would be one of its major design hurdles. The orbiter's speed, topping out at 17,500 miles per hour, would convert into heat on reentry, generating temperatures north of 3,000°F.[83] Where the tin-can Apollo capsules were outfitted with heavy, costly epoxy shields that burned away on reentry, the eighty-five-ton reusable orbiter required a lightweight option that did not need to be replaced after every flight.[84] Ceramics—strong, light, and good at insulating heat—emerged as the best option. For the leading edge of the wings and nose cap, which absorbed the worst of the reentry plasma and friction, NASA used reinforced carbon-carbon (RCC), a cutting-edge product from a former Air Force program. Reinforced carbon-carbon fortuitously became more durable at higher temperatures.[85,86]

To cover the shuttle's belly and the rest of its surface, technicians produced over thirty thousand six-inch-long square silica tiles. To affix the brittle ceramics to the flexible aluminum skin of the orbiter, engineers glued the tiles to cushiony felt pads.[87] "Aluminum, glue, adhesive, felt, adhesive, tile" went the sandwich, in the words of Thomas Moser, chief of structural design at Johnson Space Center.[88] The delicate interplay between tiles and the surface of the orbiter, thought Moser, who was in charge of the shuttle's tile system, was proving maddening. The felt layer was cross-stitched for structural integrity, but in a twist of pure irony, those same stitches ended up concentrating stress into the tiles and cracking the ceramics.

One workaround involved strengthening the tiles with silica, but there was no workaround for the biggest problem of all: the sheer scale and complexity of relying on humans to precisely lay, cure, and bond thirty thousand unique tiles onto the orbiter. Damage to any one tile could potentially doom the shuttle.[89]

No one knew how many tiles streamed off the orbiter during its cross-country voyage. Some had fallen off where the glue failed, others had cracked horizontally.[90] A Sisyphean drama unfolded at the Cape, as workers scrambled to repair the damage. Over the next few weeks, Yardley frantically hired anyone and everyone off the street to help retile *Columbia*. Technicians pulled the orbiter into the enormous Orbiter Processing Facility (OPF) hangar that adjoined the even more massive Vehicle Assembly Building (VAB), a fifty-story-tall rectangular structure with a volume almost four times that of the Empire State Building—large enough to generate its own internal weather. On humid days, condensation coalesced near the ceiling rafters and rained down on the technicians busy assembling various components of the shuttle stack below.

Outside the VAB and OPF, workers set up twenty trailers to provide food and restrooms to Yardley's massive new workforce of twelve hundred, many of whom were in their late teens and early twenties. Inside the facility, the techs craned their necks up at the shuttle's dark belly. Wearing white gloves, they slotted thick ceramic squares onto the shuttle's aluminum skin with ruddy silicone caulk.[91] Technicians worked around the clock, three shifts a day, seven days a week. The project went badly. Relatively untrained workers installed new tiles incorrectly, while more experienced techs found failing tiles that needed to be removed and replaced. Some tiles had to be switched out more than once. In the end, over thirty thousand tiles were swapped in and out on the orbiter.[92]

The rust-colored adhesive tended to dry too quickly, which further complicated matters. In frustration, one technician spit in the mix. His momentary lapse in judgment led to what seemed like a providential discovery: The saliva kept the adhesive flowing for a few extra seconds, enough time to adhere three or four more tiles. The news spread to the other technicians, and soon enough, everyone was spitting into their silicon-rubber concoctions. Unbeknownst to the technicians, however, the spittle eroded the glue's adhesive ability. It would be another ten years until NASA discovered this practice and rooted it out.[93]

As weeks turned into months, and the technicians sweated it out, *Columbia* remained cratered. The rocket engines, still in Mississippi, had logged only half the total operation time required for their certification. The solid rocket booster segments were still furiously being assembled at the Cape.[94] "Progress has taken considerably more time and resources than had been planned," Yardley leveled with NASA administrator Bob Frosch. NASA needed $185 million more to continue the shuttle's development. Frosch took the news badly, fearing the Senate appropriations committee was going to flay him alive and send the shuttle to the chopping block.[95]

George Abbey was burning the candles at both ends. Having recruited the first class of shuttle astronauts, George stood on the front lines of this battle to get the shuttle in space. He was often the last to leave the office. Most nights, he did not get home until midnight. After a four- or five-hour nap, he returned to work at dawn. The long hours took a heavy toll on George's family life. He barely saw his wife, Joyce, or their five children. His two eldest, George Jr. (twenty-two) and Joyce Brenta-Kathryn (nineteen), had left the nest for college, but his wife struggled raising their youngest three—Suzanne (thirteen), James (twelve), and Andrew (seven)—while George pursued NASA's higher calling.[96]

George graduated his new class of astronauts a year early, not because they had done remarkably well—even though he thought they had—but so they could help save the shuttle program. He assigned New Guys to every aspect of the shuttle's development, from working out the kinks in its software through endless simulations to triaging its hardware through vehicle integration. He sent the New Guys to monitor contractors' progress and impress upon them the seriousness of their job: *We'll die if you don't get this right.*

George even put his team on the tile issue. Yardley expressed confidence that the new tile configuration would stick while separately contracting with the Martin Marietta Corporation, an aerospace and materials manufacturer, to deliver a tile repair kit that astronauts could use in space.[97] George balked. How might astronauts fix tiles on orbit when no one had ever attempted a spacewalk from a shuttle before? George sent his best and brightest, including Anna Fisher, to Martin Marietta's Denver laboratory to test the kit.[98]

Meanwhile, Center directors urged Frosch to approach President Jimmy Carter directly for additional funding.[99] The odds were long: The president was at best apathetic toward the shuttle and at worst hostile. In a diary entry, Carter belittled the shuttle as "a contrivance meant to keep NASA alive."[100] In late 1979, President Carter's budget chief recommended scuppering the space shuttle program. The president planned to notify Frosch of his decision on November 14 at the White House.[101] By all accounts, as Frosch walked into the Oval Office that day, the shuttle should have been dead on arrival.

To Frosch's great surprise, President Carter began the meeting by announcing he had decided to *support* the shuttle. Carter asked for a status report on the shuttle program to help inform the White House's forthcoming official announcement of support. Frosch told Carter the main engines were a "pacing item," but they were making progress. The president, a scientist himself, got into the details. He asked about the tiles. Frosch assured Carter they were coming up with a better way to affix them. Frosch asked for $8.5 billion, about 25 percent above the original estimate, for the next five years, to complete the program and get STS-1 on orbit. NASA had already missed its original launch date in 1979, but Frosch assured Carter NASA would be able to launch the first shuttle by 1980.[102] Carter promised to give NASA his full and vocal support in front of Congress and get him the additional funds he needed.[103]

Carter's sudden about-face was due in part to the behind-the-scenes work of an unlikely savior: Secretary of the Air Force Hans Mark.

Mark, a tall, genteel man with a pepper-gray crew cut, had fled Nazi-occupied Austria as a youth. As a Cold War hawk, he believed the shuttle would give America an advantage against the USSR. Mark, who came to wear shuttle-themed ties, knew that the orbiter would have the ability to launch larger and therefore more powerful surveillance satellites. As far as the Americans knew, the Russians had nothing like the shuttle.[104]

In 1979, Yardley, Faget, and Frosch approached Mark for help. Mark went to work convincing Carter administration skeptics.[105,106] He held weekly "space breakfasts" in his office, inviting high-ranking NASA and DoD officials to discuss the shuttle's importance in national security matters. These meetings, which came to be known as the Corn Flakes Club, included Secretary of Defense Harold Brown, who had the power and influence to sway President Carter.[107]

Carter was primed to hear the message, having recently signed the SALT II agreement resulting from the Strategic Arms Limitation Talks with the Soviets in June 1979. The SALT II Treaty banned new US and Soviet missile programs and limited the number of ballistic and long-range missiles in the arsenals of both superpowers. To enforce the agreement, America needed state-of-the-art spy satellites to monitor the USSR. The shuttle program offered both a vehicle that could carry such satellites to orbit and a team of highly skilled astronauts that could help deploy, triage, and even retrieve them on orbit.

After Carter's promise of support, Secretary Mark and Secretary Brown testified before the Senate in February of 1980 that the shuttle would be instrumental to enforcing SALT II. The DoD *needed* the shuttle in its Cold War battles. Carter's budget passed, and the shuttle was saved, but not without a price. Mark had a list of demands tied to his support of the shuttle. The Pentagon would get priority on shuttle flights when national security was in play.[108] In addition, the military could recruit their own astronauts, separate and apart from the Astronaut Office, in order to maintain secrecy for their classified missions.[109,110]

The Pentagon would also develop and require the shuttle carry an Inertial Upper Stage (IUS) rocket that could push spy satellites into geosynchronous orbit 22,236 miles above Earth's surface. From this distance, satellites could remain over a single location on the planet and monitor targets, such as Soviet military bases, continuously. Since the shuttle itself only reached a maximum altitude of 330 miles, the upper-stage rocket would be strapped to a satellite in the shuttle's payload bay. After deployment, spacecraft controllers at the Air Force's Satellite Control Facility in Sunnyvale, California, would ignite the IUS to boost a satellite approximately 21,900 miles higher.[111,112] Over 60 percent of the national security payloads required this boost. If space travel was not dangerous enough, flying with IUS in the shuttle's payload bay was like driving an automobile with a loaded bomb in the trunk.[113]

In return for these concessions, the DoD agreed to wind down its own program of expendable launch vehicles, single-use rockets like Delta and Atlas, making the shuttle its only way to space. The two agencies would be inextricably linked.[114]

For many in NASA's administration, the partnership was a bitter pill to swallow. The fact that the shuttle was the DoD's only path to orbit pressured an already stressed system. "We sold the shuttle on the basis of the fact it would do all the NASA programs, all the DoD programs, and all the commercial programs," Kraft reflected.[115] Would it deliver all that was promised?[116] George grimaced at the extra risk his astronauts would bear, as well as the culture of secrecy that came with DoD operations. *Science needed transparency to flourish*, George believed. Plus, he was not thrilled about having DoD astronauts—who were not his direct subordinates—aboard the shuttle.[117]

At home, George's all-consuming relationship with NASA finally took its toll. He and his wife, Joyce, separated, then divorced, with Joyce moving out of their shared home. Now George was left to raise his children as a single father while looking after his work family—his astronauts. Before leaving for the office at the crack of dawn, the newly single dad plugged in a Crock-Pot to make dinner while his kids were at school. In the evening, he returned home briefly to eat with them, then headed back to Johnson for the late shift. The Mexican Mafia and even some of his astronaut buddies helped fill in the gaps, picking up George's kids from soccer practice and band rehearsal or, should the slow cooker fail, dropping off a pizza at the Abbey house.[118] George hired a housekeeper, but his daughter Suzanne, oldest of his children still at home, complained that the woman was not much help. Suzanne, now in her first year of high school, told George to fire the housekeeper and pay her instead. She took over many of the housekeeping duties and began looking after her two younger siblings. It was far from perfect, but George made do.[119]

Martin Marietta's Testing Facility, Denver, Colorado. February 1981.

Anna Fisher, blissfully unaware of the politics behind her work, stared up at a giant, 120-foot-long mockup of the orbiter's belly, jet-black with silica tiles. Wearing a hundred-pound spacesuit while strapped to the end of a two-story hydraulic crane, she dangled ten feet in the air. She was attached to the prototype of Martin Marietta's Manned Maneuvering Unit (MMU), a *Buck Rogers*–style jetpack that astronauts would don to fly untethered in the vacuum of space.[120] Anna was testing the feasibility of spacewalking to repair critical tile damage on orbit. Since there were no handles or tethers along much of the shuttle, a spacewalker would have to use the MMU to do tile repair.

Anna eyed the mockup and estimated how far and how fast she would have to travel forward to reach the underbelly. She pushed down on her hand-control toggle, zipping across the room much faster than expected, and crashed into the tiles. *That's no good.* If this was the real deal, instead of a simulation, she would have destroyed part of the shuttle's delicate heat shield and any chance her crew had of safely returning to Earth. Anna toggled back, this time whizzing halfway across the room in the other direction. *Wow! Fun! But boy is this thing sensitive.* Adapting, she tapped the toggle gently and slowly made her way back to the mockup. When she was an arm's length from the surface, she steadied the MMU, and stuck circular, suction cup–like pads to the orbiter's mockup to hold herself in place.[121]

Anna pulled out the kit Marietta had devised. She would use a small gun to shoot out a pink, gooey adhesive that could fill any small gap or bind a replacement tile to the orbiter. In vacuum chamber tests, the goo often bubbled and outgassed as it was applied. Anna imagined it would be awful to have that toxic sludge float toward you in space or eat through your spacesuit.

More challenging still, each one of the shuttle's tiles was unique and carefully assembled like a jigsaw puzzle. If she tried to jam in a tile that did not fit, she threatened to damage all the others around it. *No, it would be better to use more glue.* After filling in the hole, she grabbed a wire brush to scrape off the excess material and make the surface smooth again. But the brush stuck to the surface and when she tried to pull it loose, the tile came off too. Anna was making the hole worse. *If I can't do this in a warehouse in Denver, how is Crip, by himself, going to do this out in the great void of space?*[122]

Anna gave the bad news to Dr. Kraft. Her team found they often caused more damage than they fixed. What was more, if the untested MMU jetpack failed, with nothing to hold on to, an astronaut would simply float away into the hereafter. On STS-1, there would only be two crew members. John Young could not spot Crip on the spacewalk. Anna predicted they would need two years to create the kind of technology to make a tile repair in orbit operable.[123]

Crip, who had also been evaluating the tile repair procedure at Martin Marietta, had come to the same conclusion. *But two years?* With a delay like that, NASA would lose the confidence of Congress and the entire shuttle program would shutter. "We were already too far behind," he said.[124] Crip had waited fifteen years for his chance to earn his golden pin. He understood that the shuttle was an experimental vehicle and with that came peril. *Wasn't this the price explorers might pay? Wasn't this the life he had chosen?* So when George asked him, *Are you okay with the risk?* His answer was simple: *It's time to fly.*

CHAPTER 7

THE DREAM IS ALIVE

Kennedy Space Center, Merritt Island, Florida. April 11, 1981.

A little less than halfway down Florida's eastern coast, a strip of land juts into the warm waters of the Atlantic Ocean like a kneecap on the state's long peninsular leg. Carved off the mainland by the Indian and Banana Rivers, Cape Canaveral and the lesser-known Merritt Island to its west are crisscrossed by tiny creeks and man-made channels. Both are barrier islands like many off Florida's Atlantic coast, formed at the end of the last ice age, when the ocean submerged coastal marshes and created the intercoastal waterway—a system of natural inlets, saltwater rivers, and tidal lagoons that run the length of the eastern seaboard. Tufted by rough scrub grass and saw palmetto shrubs, this land was once passed over by the Seminoles in favor of shadier inland groves, where the soil was richer and the air cooler.

In the absence of serious human settlement, the land teems with wildlife. Alligators patrol the murky depths of the Cape's lagoons, dolphins play beyond the breaking waves, and turtles bask in shallow pools dappled by afternoon sun. Crabs scuttle across sandbars, and wild pigs squeal and rut in the dense, low-lying foliage. More than 360 species of bird use the islands as a migratory pit stop on their way to the warmer turquoise waters of the Caribbean. Despite the prevailing misconception, it is Merritt Island, not Cape Canaveral, that hosts Kennedy Space Center. From this mound of earth, astronauts are flung from the green and blue embrace of this planet into the inky black of space.

On the eve of the first space shuttle launch, nearly a decade after the last Apollo launch, there were signs of neglect. Empty rocket casings and abandoned launch segments lay buried in the tidal mud, overrun with grasses, homes to new creatures. The meadows looked like the ruins of a forgotten future. But with the launch of STS-1, Kennedy Space Center was about to awake from its slumber.[1]

Columbia sat on Launch Pad 39A, its black nose pitched toward the heavens at a vertiginous 180 degrees. Its external tank, bolted to the belly of the space plane, filled with over five hundred thousand gallons of liquid hydrogen and oxygen fuel that would soon rocket the ship to space. Attached to either side of the tank were the two slender solid rocket boosters. Together, the orbiter, external tank, and two boosters (the shuttle stack) rose from the launchpad like a space-age basilica bridging Earth and heaven. Beside it stood the Fixed Service Structure, a towering steel latticework that provided access to the shuttle's flight deck.[2]

Tomorrow, moonwalker John Young and rookie pilot Bob Crippen would climb this tower and take command of the ship. But tonight, one of the New Guys, Fred Gregory, sat watch over *Columbia*'s cockpit. He stifled a yawn, pausing to look outside the orbiter's window at the waxing crescent moon.

As a "Cape Crusader," Fred was part of the shuttle support team—a coveted position, especially for the shuttle's maiden voyage. He had been with the crew every step of the way—shadowing their training, running through simulations, and palling around with them after work at the Mouse Trap—the only restaurant in Cocoa Beach open past 9:00 PM.[3]

Now, with T minus ten hours until launch, Fred guarded the cockpit. As launch control ran final system checks on *Columbia,* Fred made sure every switch on the orbiter's sweeping dashboard was returned to its correct launch configuration.

Fred's ascent to the cockpit of *Columbia* as a Cape Crusader began shortly after his class's first-year graduation. Fred thought the assignment was an excellent indication of how well he fared in the competition among the New Guys to fly first. When Fred earned his silver wings in 1979, no nation had yet flown a person of color, although the Soviets were secretly training Arnaldo Tamayo Méndez, an Afro-Cuban, in Star City, Russia. Though NASA wanted to break the barrier, the Soviets got there first; Méndez launched aboard Soyuz 38 on September 18, 1980, six months before STS-1. Still, Fred, Ron, or Guy could make history as the first African American astronaut. Fred had an excellent shot. After all, he was winning in every way—a confident, tall, handsome man with a calming baritone and a disarming smile. He was also one hell of a pilot.

Fred had grown up in Washington, DC, in the shadow of Howard University. His neighborhood was a haven for Black intellectuals like the Gregorys. Fred's father held a degree in electrical engineering from MIT, his mother an education degree from Miner Teachers College, and his paternal grandfather attended Amherst College and Yale Divinity School in the early 1900s, among the first African Americans to do so.[4] His maternal grandfather was a professor of surgery at Howard University.

Despite their distinguished academic pedigrees, both of Fred's parents faced limited job opportunities because of the color of their skin. Like other Black professionals at the time who were discriminated against, Fred's parents got jobs teaching at the local Black schools. The injustice benefitted Fred and his peers who knew their education was "at least as good" as the education white students received in their separate school system.[5]

In 1954, Fred's neighborhood became one of the first test cases for school integration, and Fred was rerouted to a previously all-white high school. On his first day at Sousa High School, Fred walked to the top of a small hill overlooking the school and saw the entire student body gathered on the school's front lawn. *Had the bell not rung? Am I early?* No, the entire school was on strike. They were protesting him. Fred was the first and only Black boy in the freshman class.

"Goodness gracious, the student to teacher ratio is going to be amazing," Fred quipped to himself.[6] Despite the humor with which he approached the moment, the event became life-defining. Fred looked at all those angry students from his place on the hill, white boys and girls with whom he played neighborhood games of baseball and pickup football, now demanding his exclusion. He wondered what was going to happen next. Fortunately, the principal ordered the kids to get their "butts back in the school" or risk expulsion, and the unrest soon quieted. Fred excelled academically and was soon asked to be captain of the Hall Patrol, though he never felt "accepted socially" by his white high school classmates.[7,8]

During his time at Sousa, Fred began dating Barbara Archer, a student from a rival high school. "I had a brand-new driver's license, a brand-new girlfriend, and we drove to Andrews to watch an air show," Fred said. He admitted, "I don't think she really understood or appreciated at that time the love and passion that I had for both the military and the airplanes," or where that passion would take them.[9]

After high school, Fred matriculated at Amherst College, but after a year in the ivy-covered academic bastion, he decided that he did not want to become a lawyer or banker. Fred realized, "What I really wanted to do was fly."[10] Fred's father, still in Washington, DC, began walking the halls of the Capitol to get his son a congressional nomination to the Air Force Academy. Usually, the nomination would come from a candidate's own senator or congressperson. However, Fred—like all residents of the nation's capital—had neither. Fred's father's persistence soon paid off, and Fred secured a nomination from Harlem Congressman Adam Clayton Powell Jr., New York state's first black congressperson.

When Fred graduated from the Air Force Academy, he married Barbara and became a fire jumper, a helicopter pilot, and a test pilot. If it had wings or a rotor, he wanted to fly it. The year 1966 took him into the heart of the Vietnam War, where he flew a Kaman HH-43 Huskie helicopter as a combat rescue pilot. Fred flew over five hundred missions into some of the war's worst conflicts. Even as enemy bullets ripped into unlucky servicemen around him, Fred remained calm and courageous. In one daring rescue, Fred balanced his Huskie on a cliff, hanging one wheel off the edge as a unit of Marines rushed into the shelter of its cargo bay. Almost a dozen soldiers clambered on as Fred held the chopper steady. Unable to all fit inside, some clung to the outside as Fred flew the overloaded helicopter to safety.

After returning from duty, Fred enrolled at the United States Naval Test Pilot School in Patuxent River, Maryland, then took assignments at Wright-Patterson Air Force Base in Ohio and NASA's Langley Research Center in Virginia. At Langley, he flew over two dozen different aircrafts as a research test pilot for rotary and fixed-wing planes. "I was like a kid in a candy store," Fred said.[11]

Fred was not looking for a new job, but, one night, *Star Trek*'s Lieutenant Uhura "spoke to him" through the television, just as she had to Ron, imploring men like him to join NASA. Fred, a serious Trekkie, took note.[12] Not until General Benjamin Davis Jr. came knocking, though, did Fred request an application. The renowned commander of the World War II Tuskegee Airmen and a family friend who knew Fred's background, General Davis personally asked Fred to apply to NASA as a way of paying it forward. As the first African American military pilots, the Tuskegee Airmen had opened the door for Black men in the field. Now Fred had the opportunity to open the door for African Americans in the astronaut corps.

"Until then I thought I was carrying a torch I'd lit," Fred observed. "I realized I was carrying a torch they'd handed me."[13]

Sitting in the orbiter on the night before *Columbia*'s first launch, it was easy to believe that destiny had brought him here. *The boy on the hill, the Air Force Academy graduate, the Vietnam vet, becomes the first Black man in space.* The headlines wrote themselves.

The shuttle's first flight would come almost three years after Astronaut Class 8's selection. If the new vehicle performed well tomorrow, the New Guys would be one step closer to their own crew assignments.

America's triumphant return to space was exactly what the country needed after a decade of tumult. The Vietnam War had rolled into Watergate. The economy sputtered into stagflation. The Iran hostage crisis closed the decade. So far, the 1980s had not been much better. Two weeks before the shuttle's launch, on March 30, newly inaugurated President Ronald Reagan was shot and nearly killed in an assassination attempt outside the Hilton Hotel in Washington, DC. The United States needed a win.

In the days leading up to launch, spectators flocked to Cape Canaveral. The roads were clogged with traffic—cars, buses, and droves of people on foot. More than forty thousand gathered to watch and thousands more camped out on Florida's beaches.

On April 12, 1981, Bob Crippen climbed into the shuttle beside John Young, the "wise old man" of the Astronaut Office.[14] To say Crip was eager barely captured the half of it. Still, Crip knew this moment was not without danger. One month earlier, during the orbiter's countdown rehearsal, nitrogen gas poisoned five technicians while teams tried to pinpoint the source of a leak near the shuttle's main engines. Breathing pure nitrogen gas is like inhaling carbon monoxide—the body continues to respirate normally despite the lack of oxygen to the brain. Loss of consciousness comes within seconds. Death follows soon after. Two of the technicians died after being airlifted to area hospitals. A third passed away from ongoing health complications years later.[15]

The tragic accident underscored the risks the astronauts faced. Young and Crip were about to embark on NASA's first crewed experimental space flight. Previously, NASA's spacecraft launched in unmanned tests before astronauts ever climbed aboard. Too complicated to operate without a crew, *Columbia* had never left Earth's atmosphere until now.

"I think we might really do it," Crip told Young as the countdown reached T minus one minute.[16]

"You're go for launch," said Hugh Harris, NASA's public affairs announcer, over the PA system to the spectators and television audience. At T minus 3.3 seconds, liquid hydrogen rushed into the main engine turbopumps, which spun up to twenty-eight thousand revolutions per minute. "We have main engine start," Harris continued. The twin booster rockets ignited simultaneously, lifting the shuttle with over five million pounds of thrust.[17] Young and Crip were pinned to their seats. The entire flight deck began to shake.

"Smooth sailing, baby," said Harris.[18]

From their outposts across the country, the New Guys felt the same rush of anticipation that was coursing through Crip's and Young's veins. Guy and Kathy worked media support, providing commentary to ABC News, while Judy overlooked the launchpad from NBC's media booth.[19] Rhea hunkered down at the Cape with Vietnam vet parajumpers, ready to provide medical rescue to the crew in case of an abort. From behind the controls of a helicopter parked on a Kennedy runway, Fred watched *Columbia* rise, ready to move if there was an emergency.[20]

High above Cape Canaveral, Sally circled Launch Pad 39A in the back seat of Scobee's T-38, poised to meet the shuttle in case it had to perform a risky return-to-launch-site maneuver.[21] In awe, Sally watched *Columbia* rise, leaving a 750-foot-long, 250-foot-wide plume in its wake.

Inside *Columbia,* Crip and Young saw the lightning conductor at the top of the launch tower drop out of view as they cleared the gantry. Young called "Tower clear!" and the roll program initiated, rotating the orbiter belly-up toward the sky, hanging upside down below the external tank. The roll maneuver reduced the pressure on *Columbia*'s delta wings as the orbiter rocketed toward the stratosphere.

"I'll be damned!" TFNG Pinky Nelson hollered, from a nearby T-38, snapping photos of the rising space plane with his Hasselblad camera. "It worked!"[22]

With *Columbia* clear of the launch tower, control of the flight switched over from Kennedy's Launch Control Center to the Mission Control Center at Johnson, the home of the country's manned space flight program.

Thirty-five seconds after liftoff, as *Columbia* neared thirty thousand feet, its main engines throttled back to 65 percent. The shaking inside the cockpit intensified. The shuttle was approaching maximum dynamic pressure—max q—the moment right before it broke through the sound barrier when the pressure on the vehicle was at its greatest. Here, the shuttle's upward velocity pushed against the dense lower atmosphere, subjecting the entire frame, that discordant grouping of external tank, solid rocket boosters, and orbiter, to the greatest mechanical stress during ascent. Coupled with the thrust from the boosters, it made for one hell of a bumpy ride.

As *Columbia* punched through max q, TFNG Dan Brandenstein from Mission Control gave the command for the main engines to throttle back up to 100 percent.

At 150,000 feet, the shuttle began its planned dissolution.[23] The boosters, now fully expended, separated from the external tank in a flash of bright light. Like a butterfly shedding the spent carcass of its own chrysalis, *Columbia* rose as its two boosters fell away. Inertia would keep the boosters traveling upward in a parabolic arc until they reached the peak of their flight and dropped to the Atlantic Ocean, each deploying a massive parachute to help guide it to splashdown.

Up until this moment, Crip and Young had been experiencing 3Gs, or three times the force of gravity. When the boosters separated from *Columbia*, the cabin went dead quiet. Crip scanned the instrument panel in front of him to make sure that yes, the main engines were still putting out nearly one and a half million pounds of thrust.

In Mission Control, shuttle designer Max Faget, the man who had designed the stubby glider and had the audacity to will it to space, jumped from his seat at the back of the room and excitedly exclaimed, "They're off!"[24]

Columbia continued to rise, accelerating now from a mile per second, to two miles per second, and still pushing faster as the atmosphere thinned around them. In Mission Control, Brandenstein declared that *Columbia*'s flight thus far was "right on the money."[25] At eight minutes and thirty seconds, the main engines cut off, having done their job. Rocketdyne's motors had come a long way from their disastrous first test firings in Mississippi. Around Crip, mission checklists began hovering in microgravity. Now moving at 17,500 miles per hour, the speed needed to maintain lower Earth orbit, their bird had become the fastest winged vehicle ever flown.[26]

"What a view, what a view!" Crip exclaimed joyously.[27] He had worked for this moment for fifteen years.

Crip removed his helmet and floated across the flight deck to open the payload bay doors, exposing two massive aluminum radiator panels to the cold vacuum of space. The panels would help cool *Columbia* on orbit and prevent her avionics from overheating. Through the tempered glass window, he watched the radiators begin to unfurl. As the aft section of the orbiter became visible, Crip saw something that sent a chill down his spine.

To the left and right of *Columbia*'s tail, on the shuttle's curved surface, he saw patches of black where the white silica tiles should have been.[28] *Jesus Christ. This is exactly what they had feared. How many of these tiles are missing?*

The news ricocheted around Mission Control. A video review of *Columbia*'s launch pinpointed the cause of the tile damage: Powerful acoustic waves generated by the ignition of the booster rockets had rebounded off the launchpad and knocked the tiles off the orbiter.[29] Anna's heart sank.[30] She knew the danger Crip and Young faced, especially if additional tiles had been shorn from critical areas on the shuttle's underside, areas Crip could not see from the middeck. The shuttle would return to Earth at a forty-degree-angle of attack, with its underside meeting the atmosphere first.[31] The black carbon-carbon tiles on the shuttle's belly protected it from the worst of the heat of reentry. If those tiles were missing, it could be a death sentence for Young and Crip.

Hans Mark, the former secretary of the Air Force who had been a strong proponent of the shuttle in its development days, listened from the back of Mission Control. Reagan was about to appoint him as NASA's new deputy administrator, but that day Mark was there as a spectator. Because he had been head of the National Reconnaissance Office (NRO), leading the military's classified surveillance program, Mark knew something most did not: the DoD had top-secret spy satellites on orbit that were capable of high-resolution imaging of an object in flight.[32]

Mark conferred with Dick Truly, Crip's classmate, who was a vice admiral in the Navy and also understood the potential of using DoD satellites to photograph *Columbia*. Then Brigadier General Ted Twinting, DoD manager for Space Shuttle Support Operations, picked up Mission Control's red telephone and called the Pentagon.

Aboard *Columbia,* Crip and Young pushed the thought of catastrophic tile damage from their minds. They had a mission to complete. For the moment, Crip turned his attention toward a pressing problem with the development flight instrumentation (DFI) recorder. The machine would not turn off. Left unfixed, it would churn through the limited supply of magnetic tape long before reentry when it was most needed to record vital thermal data that would help ensure the shuttle could return future crews to Earth safely. Crip had to cut power to the machine every half hour until Mission Control could find a permanent solution.

Nine hours into the mission, Crip picked up a television camera and beamed the shuttle's very first in orbit status report back to Earth. The rookie pointed the camera at Young in the commander's seat, and the veteran astronaut heaped praise on the shuttle: "The vehicle has been performing beautifully. Much better than anyone expected on a first flight."[33] If Young was worried about the tiles, his voice did not betray concern.

While Crip and Young slept, the DoD and Mission Control worked to assess the threat posed by the absent tiles. A team of flight controllers modeled reentry to determine the likelihood of catastrophic heat failure due to the missing tiles near *Columbia*'s tail. At around eleven at night Houston time, the team determined the sixteen missing tiles would not present a problem on reentry. One major question remained: *Were more tiles missing from the belly of the orbiter?*

To answer this question, Mission Control, the DoD, and *Columbia* began a delicate cosmic ballet that required the shuttle to sync up with a DoD satellite—both traveling tens of thousands of miles an hour in different orbits—while there was enough sunlight to illuminate the belly of the shuttle. There would be only three opportunities for such synchronicity over the remaining thirty-five hours of *Columbia*'s flight. On the second day of the two-day mission, Crip and Young awoke to find flight changes from Mission Control. They spent most of the day executing a series of orbital maneuvers that would help position *Columbia*'s belly toward the imaging satellite.

After a friendly call from Vice President George H. W. Bush, they floated to bed.[34] As the two astronauts slept, one of the classified DoD imaging satellites captured high-resolution photos of the shuttle's heat shield, in what seemed like a technological miracle at the time. The images showed that the critical tiles on the bottom of the orbiter were all intact. It was a triumphant moment for the shuttle program and for Hans Mark, who had helped forge the relationship between NASA and the DoD years earlier. *The two institutions were better together.*

The shuttle was not out of the woods yet, however. Reentry was the most uncertain period of *Columbia*'s journey. Using scale models and careful analysis, engineers predicted how the orbiter *might* respond to temperatures exceeding 3,000°F, but the vehicle itself had never been subjected to the extreme thermal conditions of reentry. Fred Gregory waited with Ellison Onizuka and George Abbey in a transport off the tarmac at Edwards. If the orbiter landed, they would be the first to meet and secure it. Fred and everyone else who had a hand in getting the shuttle to space now held their collective breath, hoping it would return safely to Earth.

As *Columbia* plunged through the atmosphere, descending to 330,000 feet, the shuttle ionized the air by shearing electrons from molecules, creating a superheated plasma. A pinkish glow formed outside the window of the spaceship. "A bunch of little angry ions," Crip said, signaled things were heating up outside. It felt like "flying down a neon tube."[35] The plasma engulfed the orbiter for sixteen minutes, cutting off all communication with the ground. They were "in the blind." At the point of greatest danger, Crip and Young were entirely on their own. Would a minor chip in the tiles initiate a zipper effect, peeling off rows of tiles? Would superheated plasma melt through the heat shield? Then Fred heard Crip's voice.

"What a way to come into California!" Crip exclaimed over the radio.

Columbia blazed across the California coastline at Mach 4, and one hundred thousand feet in altitude. TFNG Jon McBride flew the lead chase plane, with Pinky Nelson in the back seat snapping photos once again.

"I have a Judy on the target," McBride reported, letting air controllers know that he was beginning his midair rendezvous with orbiter.[36] The sleek little T-38 aligned itself with *Columbia* and pulled its nose up to glide with Young and Crip home. "Tallyho!" shouted McBride as he buzzed up beside the bird.[37]

Down below, hundreds of thousands of spectators waited at Edwards to welcome the shuttle home and celebrate the country's return to human spaceflight.[38] In a VIP area close to the runway, California Governor Jerry Brown, *Star Trek*'s Leonard Nimoy, film director Steven Spielberg, singing cowboy Roy Rogers, and Mercury astronaut Scott Carpenter, among others, sat with NASA officials. The atmosphere crackled with anticipation.[39]

As Young angled the eight-five-ton shuttle for a landing on Rogers Dry Lake, his heart rate increased to 135 beats per minute. He had one shot to land.[40]

"Nice and easy does it, John," radioed astronaut Joe Allen from Mission Control. "We're all riding with you."[41]

In his T-38, McBride kept parallel with *Columbia* as he narrated her descent: "Twenty feet, fifteen feet, ten feet." The steady narration allowed Crip and Young to know exactly how high above the runway *Columbia* was. The orbiter was flying beautifully. McBride called out "Five, four, three, two, one, touchdown!" as Young brought the shuttle down for a smooth landing.

"Welcome home, *Columbia*. Beautiful! Beautiful!" McBride cheered. It *was* beautiful. Young's landing was smooth for passenger plane standards, let alone for a first-of-its-kind space plane returning after its very first trip to space, the type of landing that would have fighter pilots nodding with admiration: *He really greased it.*

Wheels stop, Columbia.

"You have fifteen seconds for unmitigated jubilation," flight director Don Puddy told his ground team. "And then let's get this flight vehicle safe."[42]

Behind a chain-link fence, thousands cheered *Columbia*'s return. Fueled by adrenaline, they rammed their RVs into the fence, knocked it over, and stormed the runway. Helicopters checked them, like border collies corralling sheep.

"This is the world's greatest all-electric flying machine, I'll tell ya that!" Young hollered.[43] Yanking off his flight helmet, the commander jumped out of his seat on the flight deck and out the hatch, bounding down the steps and around the back of the orbiter to get a look at the tile damage for himself. Crip followed, grinning widely. He heartily shook George's hand, who was the first to greet them.

"Good job," George said, a typical UNO understatement. Still, George, like Young and Crip, beamed. A long road led to this victory, a decades-long voyage filled with bureaucratic stalemates, political backrooming, millions of labor hours, billions of taxpayer dollars, the bravery of astronauts tough enough to fly the darn thing, the deaths of committed technicians, and the near tragedy of having the fruit of all that labor—*Columbia* and its crew—incinerated because of those missing tiles. But *Columbia* made it. The heat shield held. Despite the dangers, "known and unknown," as Kraft would say, their little ship survived. America was back in space.

"The dream is alive," Young happily proclaimed.[44]

Houston. September 13, 1981.

After a scorching summer, the heat still gripped Houston. Wet, warm air refracted in waves off the viscid asphalt as Ron McNair drove his car home on I-45 south of Houston. Ron glanced over at his wife, Cheryl, in the passenger seat. At three months pregnant, she was starting to show. He wondered how long it would be until he could teach his kid karate.

Ron's eyes flicked up to the rearview mirror. Something had caught his attention. Another vehicle on the horizon churned up the road behind him. Within moments a Texas-sized pickup was tailing him. *What's the matter with this guy? Is he a lunatic? Or is he a good ol' boy wanting to send a message?* The grill of the pickup now filled his mirror. Before he could react, the truck roared violently, rear-ending them.

In a time before airbags, Ron and Cheryl slammed into their dashboard headfirst. The impact knocked them both unconscious. Their car careened off the road. Minutes later, as the two lay slumped in their seats, the sound of ambulances swelled in the distance. The paramedics rushed Ron and Cheryl to the emergency room at Southeast Memorial Hospital, seven miles from Clear Lake. *Eight fractured ribs, internal bleeding, a bruised lung.* Ron was fading. Assessing his condition, the on-call doctor whisked him away to the intensive care unit.

Cheryl suffered injuries herself, but her vitals remained strong. When she regained consciousness, she learned that her unborn baby had survived the crash, perfectly healthy. *But what about Ron?*

Ron spent the night in the ICU in critical condition.[45] His torso ached; he struggled to breathe. By morning, though, his condition stabilized, and he was moved into a hospital room for recovery.[46] As he regained awareness, his head throbbed. He started seeing double. His head injuries had almost certainly caused the diplopia, but when and if the condition would resolve itself, no one yet knew. As Ron's pain flared, so did his anger.

The Sunday of the accident, the McNairs, being devout Christians, had attended services at the Wheeler Avenue Baptist Church. The couple had met at a church event in Boston, back in Ron's MIT days. When they moved to Texas, they vowed to find "a church home."[47] Though in downtown Houston, a forty-minute drive from their house, Wheeler Avenue Baptist Church, a parish with a predominantly African American congregation, felt like that home.

The McNairs had lived in Clear Lake going on four years. Despite having a busy work schedule, Ron found time to play saxophone with a group of amateur musicians at Johnson, teach physics courses at Texas Southern University, and start a karate class for the children at Wheeler Baptist. He and Cheryl were growing into their new lives, and ready to start their family. In six months, Cheryl would be having their first child.

The accident not only endangered their lives, but also threatened Ron's chances of getting his first flight assignment. Despite being shy in small gatherings and in the classroom, Ron thrived as a public speaker. The young, beautiful McNairs had an undeniable star quality that suited Ron's new profession perfectly. He knew that an astronaut's public presence mattered to NASA and that he ranked highly on George's list. Had today's misfortune cost him the honor of a flight assignment, just as similar injustices had cost Ed Dwight and Bobby Lawrence their shots before him?

Lake City, South Carolina. May 8, 1966.

Ron and his brother Carl sat on folding chairs in a rain-soaked field a few miles outside of Lake City, among five thousand other well-dressed, mostly Black spectators. Despite the downpour, the crowd had gathered in the early morning hours of Mother's Day to hear Reverend Dr. Martin Luther King Jr. speak on the importance of voting. Days before, Ron and Carl learned that King would be flying to their hometown, landing at a tiny local airfield usually reserved for crop dusters. The event would take place ahead of the local South Carolina primaries for the United States Senate seat.[48]

Ron listened attentively as local civil rights leaders spoke and sang hymns on the makeshift stage, a flatbed tractor trailer. As Dr. King ascended the stage, the crowd erupted in applause. Ron felt a chill run through his spine. Sixteen-year-old Ron had learned of the Montgomery Bus Boycott that took place ten years earlier, when Dr. King first rose to national prominence, and about the Freedom Riders five years after that. He knew about the Bloody Sunday massacre of peaceful protestors in Selma and the Voting Rights Act of 1965, passed in large part due to their efforts. Speaking to the South Carolina crowd now, King delivered an inspiring call to action that gripped Ron for the rest of his life.

"We have power represented here today," Rev. King proclaimed loud and clear in his deep, sonorous voice. "And I come to ask you to go all out, to get every Negro in this county registered to vote . . ." King urged.[49] Sitting in the field that day, Ron felt Dr. King was passing the torch of history to him, Carl, and the rest of the crowd. "Our God is marching on and so I say, walk together children. Don't you get weary."[50] Shortly after the speech, Ron walked into his local record store and purchased Dr. King's speeches on LP. Whenever he needed inspiration, Ron would play the speeches to himself.[51]

After the car accident in 1981, Ron channeled Dr. King's courage. He thought of the sacrifices of those who had come before him. He compared their tenacity to his own, their hardships to the prejudice he himself had faced. He knew he would not let this accident stop him from achieving his dream. Ron was determined to enliven the next generation the way King had inspired him.[52]

NASA issued a brief statement about Ron's condition: Ron had been in an automobile accident; he had been admitted to the ICU; by the following morning he was deemed stable. "No other details of the accident or McNair's injuries are available," the press release concluded.[53] Days and weeks afterward, Ron did not share the news of his misfortune with his classmates. Even though he missed work, few if any of his peers knew why or even that he had been hurt. Knowing his injuries might harm his career, Ron did not want to advertise his physical limitations or the fact that he would have to take time out from training.

While Cheryl prepared for the birth of their first child, Ron began the process of recovery. At the karate dojo, he built back his strength and rehabilitated himself. His fractured ribs slowly healed; but the double vision persisted.[54] By the time his son, Reggie, was born in February 1982, Ron was getting better. Still, he had already been out of the office for six months, a lifetime in the astronaut program. There was no way to sugarcoat the bitter truth: Ron's accident might erase his own place in the history books.

CHAPTER 8

TO HAVE AND HAVE NOT

Building 2, Johnson Space Center. Summer 1979.

Judy stared up at the sixteen-by-seventy-two-foot mural that stretched the length of the curved entrance to the Teague Auditorium visitor center. Wearing blue coveralls splotched with paint, artist Robert McCall stood on scaffolding, working color on to one of the many astronaut figures depicted in full human scale. Dr. Kraft had commissioned the piece, titled *Opening the Space Frontier—The Next Giant Step*, that was to preview America's next chapter of spaceflight.[1,2] McCall had crafted NASA's Apollo murals and created film art for director Stanley Kubrick, but this commission would be his tour de force. After months of work, the piece was finally coming together.

Orbiting space stations, futuristic satellites, and a fleet of pioneering space shuttles circled the globe engaged in every manner of scientific enterprise. George Abbey from the left-hand corner of Mission Control and Dr. Kraft with his familiar headset marveled as rockets launched from around the nation. In the foreground, a handsome John Young—arms outstretched, carrying the American flag—stepped through a glorious field of clouds with his crew trailing him. To his right stood Mercury 7 astronaut Gus Grissom, with his familiar widow's peak and crew cut.[3] To his left, a woman in a lab coat, looking like Carolyn Huntoon, peered through a microscope. Next to her a Black scientist with glasses and a mustache, perhaps Ron McNair, focused on a chemistry experiment.

As Judy scanned the massive work, her eyes stopped on a young female astronaut, peering off into the distance. The likeness was unmistakable. The soft bob that she had come to NASA with framed the astronaut's face. Her warm dark eyes, her profile, the strong chin, even the long fingers curved around the helmet evoked Judy. Judy had had breakfast a few times with McCall while he was working on the mural but had never expected he would use her has a model. Yet there it was—the name on the astronaut's spacesuit was her own.[4]

Wasn't it too early in her career to be canonized in this way? She felt embarrassment creep over her—then curiosity. Was this a sign that George might select her to be America's first woman to space? No matter how hard Judy tried to push the question from her mind, it popped up—in the shower, at the gym, before she went to sleep each night. *Would she be the chosen one?*[5]

Everyone knew that the stakes were high for whoever would be the first American woman to space. Despite the Soviets having already flown a woman cosmonaut, the title would still merit a place in the history books and, whether one wanted it or not, a lifetime of accolades, book deals, and speaking engagements. "We would all get caught up in it now and then . . . and obsess about it," Kathy confessed. "There clearly was some kind of a horse race or beauty contest."[6]

"I can't wait to get up," said Sally Ride when a reporter asked her whether she wanted to be first. "We're all eager for the first flight." But she went on to say, "As far as being the first American woman goes, that doesn't mean all that much to me."

"But you wouldn't mind making history that way, would you?" the reporter pushed.

"I wouldn't mind," Sally admitted.[7]

For some, the competition meant that the women TFNGs were more rivals than teammates. Sure, they were cordial, but Kathy decided that a "bare-your-soul friendship" was off-limits with Sally, Anna, Rhea, Shannon, or Judy. "Everyone wants to fly first, everyone wants to fly a lot, and so sure, we're jovial with each other," Kathy said. "But it felt more to me like a competitive environment."[8]

The best indication of where everyone stood in the horse race was their technical assignments, the rotating jobs given to the New Guys to broaden their understanding of the shuttle and help prepare the bird for her maiden flight. Technical assignments ranged from managing shuttle cargo to working on persistent engineering problems to playing key roles in Mission Control.[9] For some, like Kathy, it was like "starting in the mail room and working your way up the ladder," learning by doing to understand the many aspects that go into making space flight successful.[10]

The method by which George Abbey gave a person a particular assignment was "a complete mystery to us," Kathy admitted.[11] One thing was clear: The Godfather was at the helm, issuing technical assignments as he saw fit and, as far as Kathy could tell, that almost always meant giving the best assignments to his bubbas. The New Guys analyzed their early duties compulsively, studying the cargo lists and the shuttle manifest, trying to parse which jobs were important. They reasoned the closer they were to the core shuttle capabilities, especially those that would be showcased in the first flights, the sooner they would fly. If they were assigned to marginal systems like food creation, personal hygiene kits, or payloads scheduled on later shuttle missions, then they were probably at the back of the line.

One of the most essential shuttle capabilities, satellite deployment and retrieval, required mastery of the remote manipulator system (RMS), also called the robotic arm or Canadarm because it had been developed by Spar Aerospace, a Canadian contractor. The robotic arm was a

900-pound, fifty-foot-long flying crane that cost $100 million to develop. It was designed much like a human arm, with joints that mimicked a shoulder, elbow, and wrist. The six total joints had the ability to pitch and yaw, offering an impressive range of motion.[12] Astronauts would use the robotic arm to pluck and lift multi-million-dollar satellites out of the payload bay and release them into space. Not only would the robotic arm deploy satellites for the US government, but other customers were lining up, including foreign nations and high-paying companies like AT&T.[13] The arm would undergo shakedown tests on the second shuttle flight and start deploying its first small satellites a few flights later. Training on the arm exponentially increased one's odds of being assigned to a shuttle mission.

Of the six TFNG women, Sally was assigned to the arm first.[14] From 1979 to 1981, Sally routinely traveled back and forth from Houston to the Spar Aerospace facility in Toronto.[15] There she would sit intently behind a shiny silver control panel that came up to her chin and work two hand controllers that looked like joysticks. The left-hand controller rotated the robotic arm 360°. The right-hand controller moved the arm up and down, left and right, in and out.[16] Determined to master the arm, Sally practiced tirelessly, sometimes working from 8:30 AM to 9:00 PM. Her eyes, the same shade of blue as her NASA-issued flight suit, stayed laser-focused on the robotic arm as she put it through its paces in simulation after simulation. "Sally loved perfection," said Jim Middleton, the chief engineer of the robotic arm program. "She wanted things to be right."[17] Even if it meant skipping dinner.

Sally's hand-eye coordination, honed by years of competitive tennis, blew everyone away. "It's a skill that you wouldn't find normally," Middleton said. "The ability to do the same thing over and over again and repeat it exactly—like a race car driver doing laps."[18] Carolyn Huntoon agreed: "She took to working that RMS like a duck takes to water . . . It was designed for pilots, but she was doing better than the pilots were doing very early on."[19]

Sally knew that the arm was her golden ticket to space. She was not about to let anyone best her, and she was not above sabotaging the competition.

Like Sally, Kathy understood the importance of the arm. She completed the appropriate background training and convinced Crip and the lead instructor to let her have a go at the RMS simulator in Houston.[20] As Kathy powered up the arm and positioned herself before the control panel, Sally wordlessly brushed by her and climbed down the ladder to the lower deck.[21] Within a few moments, the arm died.

Flustered, Kathy ran through her checklist before descending the ladder herself. She soon discovered that Sally had "nonchalantly unplugged all the circuit breakers that provided electricity to the arm."[22] *What the hell? Was this Sally's idea of a practical joke?*[23] "With Sally you would never know. If you really felt you needed to ask, that showed that she had gotten to you," Kathy said. "She liked that."[24]

Although Judy's first technical assignments involved Spacelab and payload software, George assigned her to the arm next.[25] "[Judy] and I had worked together very closely . . . on the remote manipulator system," Sally said. "There was a period of time when she and I were the two main astronauts working on that system."[26] Judy was as determined as Sally, hell-bent on mastering the arm.[27]

Like Sally, Judy had excellent hand-eye coordination, but from practicing the piano rather than playing tennis. From an early age, Judy's mother, Sarah, made Judy play the piano for hours every day after finishing her homework. The intense training earned Judy's acceptance to Juilliard—but it also took a toll. *My mother stole my childhood from me,* she would later say. Sarah's demanding standards damaged Judy so much that, as an adult, she rarely played the piano publicly, or even for close friends.

Her father, Marvin, an optometrist, appeared to be Sarah's opposite, buying the kids ice cream and spoiling them with affection. Husband and wife often got into explosive fights, with Marvin trying to protect Judy from her mother's flights of fury. Sarah once refused to get fourteen-year-old Judy a pair of ice skates, so Marvin snuck out and bought them anyway. When Sarah found the skates, she burned them. "Can you imagine?" Marvin said. "Next year, I bought her another pair."[28] Sarah's harsh judgments did not stop with the piano. She was tough on Judy in other respects, even down to her disapproval of Judy's natural hair—a mass of raven curls. Sarah demanded that her daughter straighten her hair by rolling it out with empty frozen orange juice cans to achieve the WASPy look popular in the 1950s.

Sarah's critiques haunted Judy. *Do more. Be more. Practice until you ache, and then do it again. Be perfect.* "No matter what Judy did, it was wrong," Marvin remembered.[29] Her mother identified the central exchange of her life: *Love is earned, not given.* Even then, love was elusive. It hid behind phrases like "You can do better" and "That was good, but . . ." The deprivation gnawed at Judy, made her run faster, stretch herself farther—she built walls around the person she really was. Those closest to her, even her own father, said they never really knew Judy.[30]

When Judy was seventeen, her parents divorced. The court granted Sarah custody of the children. Within a year, Judy was back before the judge asking for a "divorce" from her mother. She won and lived with her father until college, staying close to him all her life. When Sarah sent letters, Judy ripped them up without even opening them.[31]

Despite—or maybe because of—her mother's endless criticism, Judy excelled at everything she did. Even in an extended Jewish family of accomplished scientists, doctors, and musicians, Judy stood out for her accolades. She could already read and do simple math by the time she entered Fairlawn Elementary School in Akron, Ohio. She skipped kindergarten. By high school, Judy was a National Honor Society recipient, the only female member of the math club, and a concert-level pianist. She rejected Juilliard and her mother's musical dreams to study electrical engineering at Carnegie Tech.

While at Carnegie Tech, Judy met a Jewish fraternity boy and fellow electrical engineer named Michael Oldak, who became her "only extravagance."[32] Although neither would remember who made the marital overture, in 1970, the same year they graduated, they got hitched.[33,34] The newlyweds moved to New Jersey to start their lives, both getting jobs at the RCA Corporation, a major electronics company. There Judy worked with missile and surface radar equipment. Michael quickly lost interest in engineering and instead applied to law school at Georgetown University. Judy followed Michael to Washington, DC, returning to school herself to pursue her doctorate in electrical engineering at the University of Maryland.[35] When the issue of kids came up, the marriage fractured. Michael wanted to start their family, but Judy did not want to recreate the trauma of her own. In 1975, they divorced.

In 1977, Judy finished her doctoral work on the effects of tiny electrical currents on the dark pigments of the retina, a subtle homage to Marvin, her optometrist father.[36] That spring, while vacationing on Bethany Beach in Delaware with friends, she heard NASA's call for mission specialists. Something clicked: *I wasn't meant to be cloistered in a laboratory.*[37]

Even though Judy had embarked on a brand-new life of her own making, she still worked with the same intensity that Sarah had drilled into her as a child. The piano had been replaced by the robotic arm, which she learned to master with all the fervor and discipline of someone who simply would not fail.

While Judy and Sally became experts on the robotic arm, Anna worked on spacesuits. She hoped the assignment might mean that she would get to perform an Extra Vehicular Activity, or EVA. EVAs, more commonly referred to as spacewalks, were how the astronauts would repair satellites or the shuttle in orbit. Donning Apollo astronaut Pete Conrad's old spacesuit, Anna submerged herself in the WETF pool alongside fellow TFNG James Buchli. There, they developed the first tools used to work on shuttle hardware in space—a gear caddy, a winch to manually close the payload bay doors, and a tethering system for moving across the shuttle.

George assigned Shannon, along with Mike Mullane and Dick Scobee, to Spacelab—a modular science lab that would fly in the cargo bay on designated shuttle flights. The program would grow from small-scale experiments to more complex endeavors in which the laboratory filled the whole cargo bay.[38] The astronauts' job was to manage Spacelab's scientific experiments. Shannon, a chemist, was a logical choice, but Mullane was not. The Air Force officer was experienced in flying combat missions in Vietnam, not handling test tubes. This, of course, was part of George's plan. The technical assignments were doled out "with malice aforethought—to put people where they would be uncomfortable," George explained. "We put scientists in operational jobs, military test pilots in science. They had to be able to cope with assignments that went beyond their experience."[39]

Not everyone was pleased with their assignments. Yes, physician Rhea Seddon was charged with working on the medical kits, but she was also stuck evaluating astronaut food. She rationalized to herself, *It's not sexist; they're leveraging my background in nutrition.* Still, she "feared what it boded for the future."[40]

In preparation for STS-1, Kathy worked for a stint with Crip and Young on the shuttle software system. "I'm getting a chance to be useful for the cool kids," Kathy said. "That's got to be an advantage."[41] However, she was soon whisked away to Alaska to fly two eight-week aerial research missions, mapping Earth to provide NASA with information for future space flights.[42] Did George send her because he needed a geologist like Kathy for the task, or had she unknowingly done something wrong? "It felt like I was banished to Siberia," Kathy confessed.[43]

Kathy was also given the even less glamorous task of assessing the female urine control device for EVA spacesuits. At first, the technicians suggested using a device that attached to the female anatomy, the way condom catheters fit over the penis.[44] *Have you ever seen female anatomy?* Kathy asked. *What exactly are we attaching to here, boys?* She pointed out that the condom device, while impossible for women to use, was not great for men either. *What if your urine bag broke during launch?* The technicians did not listen until a male astronaut sprang a leak on national television. The program ultimately adopted the Disposable Absorption Containment Trunk, or DACT, NASA lingo for an extra-absorbent adult diaper. While the technicians praised Kathy for her prescient advice, she could not help but be chagrined at the work she was doing. *Sally got the hundred-million-dollar arm to deploy million-dollar satellites; I got the diapers.*

Before his accident, Ron was assigned to the Shuttle Avionics Integration Laboratory, or SAIL. SAIL was equipped with the same software the shuttle would use. The mission specialists and pilots were there to work out its kinks.[45] Even though SAIL could be intense—requiring twelve-hour workdays—those who completed the training said they learned more about the shuttle there than on any other assignment.[46] Seven months after his accident, in March 1982, Ron officially returned to NASA. As he continued to recuperate, he resumed "light duties," including flying in the back seat of a T-38 with the chase teams for STS-3 and STS-4.[47]

Fred helped George with vehicle integration, flight data interpretation, and launch and landing support for early missions. In addition to acting as support crew for STS-1, Fred managed the emergency landing contingencies for the early flights, leading a team of more than forty first responders. He was a Cape Crusader for both STS-2 and STS-3. Meanwhile, Guy "bounced around the country," shifting assignments every four months—from SAIL to Spacelab to the robotic arm. He was gaining a little experience with everything.

In private moments at the gym, at home with their spouses, or in a dark corner of the Outpost Tavern, the New Guys tried to decipher who among them were George's favorites. "We had absolutely no idea what his thought process was," Sally admitted. "And the more time you spent trying to figure it out, the less you really understood it."[48] For George, the right stuff never boiled down to a single skill or quality. He believed well-rounded individuals—who could just as easily mingle at a cocktail party as triage in space—made great astronauts.[49] At softball games, George sat on the sidelines and watched. Sally was a ringer on first base. Steve Hawley impressed at shortstop. Kathy could hurl the ball home from deep right field. Anna played catcher with an upbeat attitude, even if the ball did not always land in her mitt. Who was the most resilient? Who could George trust to work well on a team and get the job done? *Who should be first?*[50]

While the New Guys contemplated their prospects, well-established NASA leadership began to change in ways that would have far-reaching implications for the agency and for the Astronaut Office. In June 1981, the Senate confirmed President Reagan's nominee to run NASA, James Beggs. Beggs, a tall, charismatic aerospace executive with a history of working with NASA, believed the agency needed a Kennedyesque goal, a "high challenge manned initiative," to garner Congressional support and public appeal.[51] "The shuttle was never conceived as an end in itself," Beggs argued at a spaceflight symposium in March 1982. "Rather it was conceived as means toward an end."[52] That end, he argued, was the space station, mankind's home in space.[53]

Beggs believed that a space station would revitalize science and technology development at NASA and reinforce the nation's position as an international leader in scientific discovery. The shuttle would play a crucial role in building the station, as a de facto construction truck to the stars ferrying up modules, equipment, and workers to weld the new station together.

Before the space station could become an official NASA project, however, Beggs would have to convince President Reagan to fund the expensive proposition—$8 billion to start—at a time when the economy was sputtering from stagflation and unemployment was over 10 percent. Meanwhile, Reagan had campaigned on a promise to slash wasteful federal spending. Beggs had to prove to Reagan's budget analysts that NASA would not break the bank with this new endeavor. First, he had to show that NASA had the existing technology within its current space program to lay the groundwork for the station. Second, he needed to demonstrate that the shuttle—as the station's midwife—could fly more often and at a lower cost. The shuttle program could meet its own goals *and* build the station.[54]

To tighten the shuttle's budget, Beggs suggested privatizing aspects of the program, like the business of maintaining the shuttle fleet or the on-the-ground operations at Kennedy Space Center.[55] In a showy auction later referred to as the Shuttle Sweepstakes, Beggs's administration awarded the shuttle refurbishment contract to the lowest bidder, aerospace company Lockheed Martin, even though Lockheed as a company had no prior experience with the shuttle.[56]

Beggs clashed over these new policies with the head of Johnson Space

Center, Chris Kraft, who argued that cutting costs would jeopardize safety standards. On the shuttle's second flight in November 1981, one of three fuel cells on *Columbia* failed five hours into the mission. Essential to the shuttle's orbital operations, the fuel cells supplied the vehicle with electricity and the crew with drinking water by combining hydrogen and oxygen. Losing even one fuel cell was cause for concern. When Beggs and Kraft met to discuss shortening the mission from five days to two to ensure the safety of the crew, Beggs argued that they should continue the mission as planned. "You don't know what the hell you're talking about," Kraft snapped at Beggs. Kraft brought the orbiter home early, while Beggs fumed at being dressed down.[57,58]

Kathy, who flew chase on STS-2, remembers seeing tile damage on "about six white squares on the belly of the shuttle."[59] Thankfully, like on STS-1, the damaged tiles were not located in a crucial area of the shuttle's heat shield. Even more concerning, rain at the launch had soaked into the porous ceramic tiles and added hundreds of pounds of weight to the orbiter, requiring more fuel to make it to orbit. The solution for the subsequent flight? *Scotchgard.* The Kennedy technicians bought a "gazillion cans" of the drugstore variety water-repellent fabric spray to spray on the tiles.[60] "That became pretty hilarious," said Kathy. "We were cracking up about it."[61]

The last-minute botch job carried ominous undertones, revealing how rushed NASA was as Beggs and other managers pushed the program to stay on schedule. Within months of Kraft's harsh reprimand, Beggs replaced the legendary Dr. Kraft—the soul of Johnson for a decade—with former Apollo flight director Gerry Griffin. The administrative upheaval was a demonstration of what would happen to those who disagreed with headquarters.

More problems accompanied the shuttle's third flight, STS-3, in March 1982: The toilet malfunctioned the first day, an auxiliary power unit overheated during launch, three communication links were lost, and both astronauts experienced a bout of space sickness.[62] Though frustrating to engineers, controllers, and astronauts, such problems were to be expected from what was essentially a test vehicle—an experimental spacecraft loaded with cutting-edge technology, flying with a very limited base of practical experience. The shuttle program was a startup enterprise still working out the kinks.

Despite these hurdles, when STS-4 landed on July 4, 1982, President Reagan was on hand to pronounce the orbiter "operational."[63] The word struck a nerve with George. In the world of experimental aircraft design, the "operational" label was reserved for vehicles that had undergone hundreds of test flights. Then and only then were these aircraft considered "safe."[64] The shuttle had only flown four times. The operational label gave NASA license to remove the orbiter's ejection seats. If the shuttle was operational, then it was safe enough not to need an evacuation plan. Those who supported ditching the ejection seats argued that at speeds faster than Mach 4, astronauts would be unable to survive an evacuation anyway. *The shuttle flew up to Mach 25 once on orbit.*[65] Installing ejection seats in all the orbiters for all the crew members would be expensive—costing tens of millions of dollars—and time-consuming. The shuttle program could ill afford either.[66]

In the competition to be the first American woman in space, Sally and Judy, dueling masters of the robotic arm, were in a dead heat. TFNG John Fabian, who watched both Sally and Judy put the arm through its paces, thought they were equally talented. Judy, however, was the better hang. "After a long day's work, [Judy was] someone who you could go out with in the evening and have a burger and get a couple of beers and a couple of big cigars," said Fabian.[67]

In her four years at NASA, Judy, who now went by "J.R.," had become someone that everyone wanted to be around, someone comfortable in her own skin. Gone were the long skirts and white button downs she had worn when she arrived at NASA. The frozen orange juice cans her mother used to straighten her hair were a thing of the past. Instead, she let her new best friend, Astronaut Office secretary Sylvia Salinas, make her over. Sylvia introduced her to Zeke, a fashion-forward stylist who layered Judy's hair into rock-and-roll curls. When Judy was unsure about her new look, Sylvia told Judy to "stop fighting it."[68] Sylvia had to get a perm to get those curls. A new leather jacket and aviator sunglasses—Ray-Ban, of course—completed Judy's transformation.[69]

Judy became a magnetic personality to whom men—single and married astronauts alike—flocked. Married Mike Mullane flirted with Judy, playing schoolboy pranks. Once, while she was out for a run, Mullane and his friends snuck into the women's locker room—a recent addition to the astronaut gym, courtesy of Carolyn Huntoon—and put a garden snake into her purse. When she returned, she opened her bag and shrieked, sending Mullane and his friends into fits of laughter.[70] Another New Guy put pink satin sheets on Judy's crew quarters bed.[71] Even NBC's Tom Brokaw could not stop gushing. "What do you say when you meet a guy and he says, 'You're too cute to be an astronaut?'" he asked her on national television. "I just tell them I am an engineer," she responded demurely. Smitten, Brokaw sent her red roses and a six-pack of beer. *Thanks for the beer,* Judy replied, brushing off the obvious flirtation.[72]

Despite her public appeal, Judy's desire to maintain her privacy sometimes made her thorny to the press. Both she and Sally were reserved, but where Sally politely tolerated nosy questions from the press, Judy froze out reporters with answers that left no room to pry.[73] In one Q&A, a journalist asked how the astronauts blew off steam. "We all spend our private time in our own ways," Judy said. The reporter asked what the toughest part of training was. Judy answered, "I don't think any of my training is tough." End of story. "And what do you look forward to?" the poor reporter concluded. "The whole thing." Judy smiled. There would be no embellishment.[74]

George understood that the first American woman to fly would spend the rest of her life in the spotlight. She would represent NASA—and America—to the world. Someone who did not flourish in front of the camera could never be first. Judy's aloof nature with the press was a mark against her. Kathy boiled the whole thing down to a PR contest: NASA needed to decide what "type" of woman they wanted to put out front: "the 'blond surgeon' [Rhea], 'flirtatious single gal' [Judy], 'photogenic . . . married gal' [Anna], 'slender and athletic' [Sally], 'taller, stockier . . . married' [Shannon]," or the woman who was "'not a cover girl type' [herself]."[75]

Unlike Judy, Sally's handicap at NASA—her romantic history with Molly—was a well-kept secret. If Sally ever wondered if being outed would end her career at NASA, her suspicions would have flared a few weeks after STS-1.

On May 1, 1981, tennis star Billie Jean King, Sally's idol, announced that she was being sued for palimony by her female assistant, with whom she had had a romantic relationship. The action made King the first major female professional athlete to come out publicly.[76] Consequences were severe. She lost all her endorsement deals and sponsorships and had to start her career over from scratch.[77] How did the scandal affect Sally? She did not speak about it with anyone, not even her then boyfriend, Steve Hawley. However, King believes the fallout from her own outing "probably scared the hell out of Sally" and "put her more into her shell."[78] Sally seemed to take steps to protect herself from being similarly exposed.

That summer, on July 29, the same day as Princess Diana's wedding to Prince Charles, Sally and Steve quietly moved in together, choosing that date "so that no one would notice."[79] They confessed their love to each other in their signature understated fashion. "I might say that I'm in love with you" is how Steve remembered Sally saying it.[80] He told her, "I love you, too." In many ways, Steve and Sally were a good match. Both were whip-smart with the same sarcastic sense of humor. Sally did not have to apologize when she spent weeks in Toronto working on the robotic arm, nor did she complain when Steve jetted off to Kennedy to work as a Cape Crusader.[81]

That fall, George chose Sally to be the lead communicator in Mission Control for the shuttle's second space voyage. The role of "CapCom," short for Capsule Communicator, was a throwback to an era when astronauts flew in capsule-shaped spacecraft. The CapCom, who spoke directly with the astronauts throughout their journey, was a coveted assignment. As the official voice of Houston, the job required a level head in high-pressure situations and a thorough understanding of the whole mission. STS-2 would debut the robotic arm, in which Sally was becoming an expert. Sally quickly bonded with the STS-2 crew—Dick Truly and commander Joe Engle—by doing hundreds of simulations with them in the months leading up to the mission. As CapCom, she flawlessly coached them through their maneuvers once they were on orbit. For the first time a woman was the voice of "Houston."[82]

The CapCom also got to select the music that woke the crew. Sally arranged for legendary puppeteer Jim Henson to create personalized wake-up calls from *The Muppet Show* sketch "Pigs in Space." The voices of Captain Hogthrob, Dr. Strangepork, and Miss Piggy greeted the astronauts on their first morning in space. Seeing that humans, not pigs, crewed the shuttle, the Muppets lamented, "Well, there goes the neighborhood."[83]

Truly and Engle beamed down eye-popping images of Earth, and the robot arm as it flexed its electromechanical bicep. Sally radioed back, "Super! When do I get my turn?"[84] She was only half joking. Sally did so well that George made her CapCom again on STS-3, a decision that left her fellow New Guys green with envy.

By early spring 1982, the press heard through the grapevine that NASA would send a woman to space in 1983. One reporter speculated it would be Judy, citing the Judy-esque woman astronaut in the mural outside Teague auditorium.[85]

"We are not privy to that information. We just wait our turn," Judy said.[86] Publicly, Judy acted as if she did not care about earning the title of first American woman in space, but she privately tracked her chances, and hoped.

"Judy called me," Marvin Resnik said. "'Daddy, they're going to choose who's going to be the first woman. And it's going to be me or Sally.'"

"Well, you'll make it," said Marvin, always Judy's most ardent supporter. Judy felt less sure.

"We don't know. Nobody knows how they're going to choose—whether it will be the toss of a coin or what."[87]

Building 4, Johnson Space Center. April 19, 1982.

"We've made some crew assignments," George announced standing before a gathering of the entire astronaut corps at the weekly Monday morning All-Astronaut Meeting. George rarely attended these meetings, so his mere presence was enough to rattle everyone. At the head of the conference table, with clipboard in hand, George continued, "The STS-7 crew will be Crippen, Hauck, Fabian, and Ride. STS-8 will have Truly, Brandenstein, Bluford, and Gardner. STS-9 will be Young, Shaw, Garriott, Parker, and two payload specialists."[88] When George finished reading, the room went dead silent. The decision the New Guys had fretted over for nearly two years was now laid at their feet. All delivered with crushing, deadpan brevity, for which George was famous.

George regarded the mixed expressions of those before him: Some smiled gleefully, others tried harder to repress their joy; some sat slack jawed in disbelief, and a few ground their teeth in anger.

"Hopefully we'll get more people assigned soon," he finished.[89] He excused himself from the meeting without extending further pageantry over the earth-shattering news he had dropped like a bomb. The race for first American woman and first African American in space was over. Assuming they did not screw up royally, Sally Ride and Guy Bluford had snatched the titles. "*Poof.* With Abbey's words, TFNG camaraderie vaporized," Mike Mullane said. "As a group wallowing in a common uncertainty and united in a common distrust of our management, it had been easy to share a beer at the Outpost. Now we have been cleaved into haves and have-nots."[90] The unassigned New Guys swallowed their disappointment and shook hands with those lucky enough to be chosen.

"George told us of the assignments a week ago," Sally confessed to her peers, "but he wanted us to keep it quiet until the press release."[91] The admission was salt in the collective class wound.

"I wondered how many times in the past week I had been eating lunch in the cafeteria with Rick Hauck or John Fabian and whining about the delay in flight assignments, and all the while he had been silently celebrating his mission appointment," Mullane recalled. "God, I felt so pathetic."[92]

Of course, Mullane was not the only one battling an intense case of self-pity. As soon as George said "Ride," the other women deflated. "Not getting the nod was a little bit of a wound," Kathy admitted. "I would have loved to go first . . ."[93] Anna was disappointed, but not surprised. A few months prior, Dr. Kraft, her biggest fan, sent her to train on the arm ASAP. She suspected that, behind the scenes, Kraft was pushing for her to be first, but Anna had too much ground to make up. George had assigned Sally to the arm first and given her back-to-back CapCom assignments, putting her leaps and bounds ahead of the others. It seemed George had fast-tracked Sally from the start.[94]

Rhea, who had intuited Sally's early lead, decided to shift her priorities. In February 1981, after a dinner date on Clear Lake's waterfront, Hoot had proposed to Rhea. That May—after they completed their duties on STS-1—the two were married in a small Tennessee ceremony attended by family and a few friends. They danced to "Fly Me to the Moon" and cut into a cake decorated with the astronaut symbol, the same ringed shooting star that graced their silver astronaut pins. Now, one year later, Rhea was pregnant and due in a few months. "I would rather have a child than an early flight assignment," she reasoned.[95,96]

Although Guy had won the title of first African American in space, he had never fixated on the competition. The same was not true for the others. "This is bullshit!" Fred growled under his breath.[97] Meanwhile, Ron, who had only been back at work for a few weeks, could hardly be surprised by the turn of events.

After the morning meeting concluded, Carolyn Huntoon called Sally into her office. "Somebody had to be first, and everyone was adult enough to understand that," Carolyn said. "But these were high achievers, and there was a dip in this office—I felt it."[98] She advised Sally to be gracious. "You got it!" Carolyn said. "Now be nice."[99]

Later that night, the New Guys gathered at the Outpost. The beer flowed. Some celebrated, others drowned their sorrows. "Boy, you work like a slave, and do you get recognized for it? Hell no!" one astronaut grumbled.[100] Hoot called the complainers a bunch of whiners. After a couple of beers, though, even Hoot got saucy with George. He grabbed Steve Hawley and cornered George. *Oh crap,* Steve thought.

"Hey, George," Hoot grinned under his thick mustache. "Stevie and I wanted to tell you that you really screwed up today."

"How so?" George asked, amused.

"You didn't assign us!" Hoot cracked.[101] Steve squirmed underneath Hoot's arm; he wanted to be anywhere else. George merely smiled at the beaming Hoot and blushing Steve.

"Come see me tomorrow," he told Hoot. For the rest of the night, Hoot hoped this meant he was being assigned to STS-10. Instead, it was an offer to be the new deputy for aircraft operations, a desk job.[102]

Hoot's joke and the others' complaints did not surprise George. "They were competitive people. Obviously, everyone wanted to be first," George knew.[103] For a group of type-A people that ran on achievement and accolades, George's opaque management style proved maddening. He did not give them end-of-the-year reviews or feedback for professional growth. He seldom even indulged in an off-handed compliment like "Good job" or "Well done." Without any sense of his plan for their progression through the program, the astronauts were left to stew in the uncertainty. Their frustrations slowly boiled into ire. George may not have realized it at the time, but his lack of transparency would be his undoing.

As for Judy, even if she were heartbroken, she appeared diplomatic. "Firsts are only the means to the end of full equality," she said, "not the end of itself."[104] In part, Judy was relieved not to be first. "Judy was an extremely private person," said Sylvia. "The intrusion into her personal life would have really bothered her."[105] Judy's ex-husband, Michael Oldak, with whom she was still friends, guessed there was another reason she had not been chosen. "She was divorced," he said. Social mores being what they were, NASA probably did not want a divorced woman to be America's first woman in space.[106]

Whatever the reason, George had erased Judy's chance at making history. Someone along the way at NASA Headquarters had contacted Robert McCall, the painter of *Opening the Space Frontier,* to make a change. Now when Judy walked by the Teague Auditorium and glanced up at the mural, she still saw a woman astronaut that resembled her, but the name "J. Resnik" on the astronaut's spacesuit had been softly blurred out.[107]

CHAPTER 9

A FEATHER IN HER CAP

Cape Canaveral, Florida. June 17, 1983.

On the shores of Cocoa Beach is an old house that, for decades, has figuratively sat on the threshold of space. The home is part of a large subdivision NASA purchased in 1963 to clear the area for launches. NASA tore down all the structures in the subdivision—except one. No one knows why that one home was spared, but the fact that it was seems more an act of fate than will.

Through the Gemini and Apollo eras, astronauts used the beach house as a refuge, a last chance to say goodbye to loved ones before they embarked on their dangerous journeys.[1] With red wood siding and an unfinished deck, the beach house was more bachelor pad than ocean cottage. Orange Naugahyde chairs and a tattered brown couch occupied the living room. A pinewood-paneled dining room with mismatched chairs could seat as many as a dozen guests.[2] Glass cabinets showcased consumed bottles of champagne, along with half-finished bottles of liquor left over from previous celebrations.

Outside, among the tall sea grasses, where the sand meets the ocean and the ocean meets the horizon, astronauts contemplated the obscurity or grandness of their lives. Here they battled their fears and embraced their excitement for the day to come. They bid goodbye to their loved ones, said their prayers, confessed their heretofore unshared desires. Here they had one last moment of reflection before they laid it all on the line.

The risks of spaceflight tended to quiet the noise of life and make these travelers confront what was important to them. Some held tighter to what they had. Others reached out for what they had lost. It was the night before the launch that would make Sally Ride the first American woman in space, and she was having just such a moment. She was rattled.

With the start of the shuttle program, George created a new tradition—a barbecue at the beach house a day or two before launch. Since the crew was technically in medical quarantine, they were only allowed one guest—usually a partner, spouse, or parent. Heaping plates of fried chicken, ribs, coleslaw, and corn on the cob lined the table. Guests sipped champagne, drank beer, and nipped exotic liquors they snuck in.

The day before STS-7's launch, the crew and their families celebrated there. Now late in the afternoon, most of the spouses had left. George spotted Sally pacing up and down the hallways. He had known Sally for five years. She was usually as steady as a rock, approaching her work with surgical precision, utterly reliable and famously unflappable. All reasons why George had selected her in the first place. But today, Sally seemed *off.*

Sally had been calm earlier when saying goodbye to friends and family. "I have complete trust in [NASA], and you may not understand that, but you know me," she told her sister, Bear Ride. "I know enough about this stuff that it's worth trusting my life to."[3] As the day wore on and the launch approached, Sally's certainty began to waver. She isolated herself. She paced. George watched her transformation.

The crew had a 3:00 AM wake-up call, and George knew she needed rest. He had to do something to calm her. But what?[4] He thought about contacting Sally's family. Her mother, father, and sister were all staying at a nearby hotel. Steve was also at the Cape. Perhaps some time with one of them would ease her nerves? None of those solutions seemed right to George. Then he got an idea. He picked up the phone and called Molly Tyson.[5]

One Year Earlier. Disneyland, California. April 1982.

HERE YOU LEAVE TODAY AND ENTER THE WORLD OF YESTERDAY, TOMORROW, AND FANTASY. Sally and Steve read the sign above the main gates as they walked into the "happiest place on earth."[6] Yes, Disneyland was just what the doctor ordered for two high-achieving, overworked astronauts. Sally had been burning the midnight oil on back-to-back CapCom assignments and Steve had been ferrying between Houston and the Cape as a Cape Crusader.

However, their holiday was interrupted almost as soon as it began. They returned to their hotel to find an urgent message from George's secretary Mary Lopez: "Come back to Houston by Monday." There were no further details. Thoughts flooded Sally's mind. *Was she in trouble? Was she finally getting a mission assignment? Or was this one of George's special chores?* No matter. UNO had called. She flew back to Houston.

At 7:30 AM, Monday morning, April 19, Sally reported to George's big corner office on the eighth floor of Building 1.[7] The "man of few words" sat at his desk framed by large windows overlooking Johnson's campus.[8] Usually, when George assigned a crew, he met with the whole crew together, but Sally was here alone.

"How do you like the job you've got now?" George began, without preamble.

"Well," Sally answered. "What is my job?"

"We thought that maybe you enjoyed what you were doing so much that maybe you wouldn't want to fly on a crew."[9] Sally had learned to "speak George" by now. He was assigning her.

"I'd be thrilled to fly on a crew!" she told him. Then why was Sally meeting with George alone? George explained that before he dragged her into a full-blown crew assignment, he wanted to make sure she understood the mantle he was bestowing on her: *First American Woman in Space*. Her heart overflowed with emotion.

Then George led Sally up to the ninth floor to see Dr. Kraft. Before she said yes, Kraft had an important message to convey. Wearing a serious expression, Kraft sat her down.

Are you ready for this? You'll be on the cover of magazines, on the nightly news, everyone in America will know your name. Then he spoke of Neil Armstrong. After his historic moonwalk, the humble Armstrong could barely walk down the street without being mobbed by press or fans or both. *Fame took a toll on him.* Kraft warned Sally, *The hardest part is after you land.*[10]

There's no doubt in my mind. I want to do it, Sally said. Her conviction, whether from naïve confidence or raw ambition, might erode with time; she could not yet comprehend what the attention might do to her, but doubts were for another day. Today, she would bask in the glory of being the chosen one.[11]

Over the next year, Sally and her crew spent "twenty-four hours a day, seven days a week together," Sally said. As the robotic arm operator, Sally trained for procedures on deploying satellites. Sally was also tasked with running experiments on the production of pharmaceuticals in microgravity: "The other mission specialists and I spent a lot of time working with the people that built the satellites, working with people that had built the experiments so that we understood what they intended and were able to carry out their plan as they would have if they'd been up there themselves." [12,13] Sally also acted as the mission's flight engineer, assisting commander Bob Crippen and pilot Rick Hauck on launch, reentry, and landing. The job required her to run ascent and reentry simulations daily that lasted anywhere from four to thirty-six grueling hours per sitting.

According to her biographer Lynn Sherr, Sally's life was "organized into stacks of cue cards—ring-bound sets of instructions" and "mountain-sized briefing books" used to study flight modules and contingency plans.[14] She occasionally peeled away from flight prep to catch some hours in her T-38 or jog.[15] Sally *had* to work hard: If she did not perform flawlessly, she would not only disappoint her crew and her female colleagues, but also all the little girls who wanted to break barriers and live their dreams. Every morning and every night, she said her own astronaut prayer: *Dear Lord, please don't let me screw this up.*[16]

Women astronauts had been on campus for five years but flying to space with a woman created new challenges. Norm Thagard, a doctor and mission specialist on the crew, refused to strip down to his skivvies in front of Sally when they had to change into their flight suits. "I'm from the South," he explained. "I couldn't do it."[17] To guard for modesty among the crew, NASA technicians added a privacy screen around the space toilet (a kind of high-tech vacuum cleaner) on the shuttle's mid-deck. The new addition was nicknamed the "Sally Ride Curtain."[18]

Male confusion about women's bodies proved another obstacle. In one preflight meeting, a naive engineer asked, "What if all the mucus that women put out will stop up the [onboard] toilet?"

"We're all sitting there looking at this highly placed engineer who'd been there forever," said Rhea, "and we're saying, 'What mucus?'"[19]

Then there was the "female preference kit"—a toiletry bag for women prepared by NASA engineers. Sure, they had thrown out the British Sterling deodorant and Old Spice shaving cream, but they replaced those items with an offering of makeup: "lipstick, eye shadow, mascara, makeup remover, and blush cream."[20,21] *Gee, do they think I am going to wear makeup in space?* Sally did not even wear makeup on Earth. Kathy looked on, giggling, as Sally pulled out an unwieldy mass of *something* from the kit. "Sally looks up at me with this rolling of the eyes that I had come to know as her 'you have got to be kidding me' look." Kathy said.[22] "She reaches in and picks up this band of pink plastic, and now I can see tampon, tampon, tampon, tampon. Then she reaches the bottom of the string and pulls again, and it was like a bad stage act."[23] The engineers had tried to calculate how many tampons a woman would need for a one-week flight and tied them together so tampons would not—*God forbid*—go floating through the middeck.

"Is one hundred the right number?" one engineer asked.

"No. That would not be the right number," Sally said, hardly able to stifle her laughter.

"Well, we want to be safe," he responded.[24]

Beyond trying to decipher what NASA's men thought occurred in women's bodies, Sally also had to contend with a battery of prying questions from a nearly all-male press corps. *How do you spend your free time when you're in such tight quarters with four men? What does it feel like to be a role model for all American women? Do you plan on becoming a mother after your flight?*[25] Sally answered most of these questions in good humor. She jokingly asked Crip if he was going to give her any free time on orbit. To the "role model" question, she responded seriously, "I've come to realize I will be a role model, even though that's not what I intended to be. What I intend to do is as good a job as I can, and I hope that will serve as a role model."[26] She did not dignify the last question, about her motherhood plans, with a response beyond "You'll notice I didn't answer [that] question."[27]

The questions that reporters directed at Sally's crewmembers were still somehow about Sally. When reporters asked all four male astronauts if she would be an "inconvenience" or if they would have to "defer to her because she's a woman," Crip answered: "Sally's been anything but an inconvenience."[28]

"I haven't felt deferred to in any way. In fact, Crip won't even open doors for me anymore," Sally joked.

When the reporters questioned Sally's competence, Crip ended the discussion: "Sally's on this crew because she's well qualified to be here."[29]

The most offensive question Sally got—one that would continue to irk her years later—came from a *Time* reporter during the crew's preflight press conference: "Dr. Ride, during your training exercises as a member of this group, when there was a problem . . . how did you respond? How do you take it as a human being? Do you weep?"[30] Sally grimaced before he finished the question—it was all she could do not to roll her eyes in front of all those cameras.

Instead, she laughed off the awkwardness, shook her head, and quipped, "Why doesn't anyone ever ask Rick those questions?"[31]

As the launch drew closer, the steady drumbeat of reporters looking for the inside scoop on Sally's life intensified. For the most part, Sally kept her private life private. "Sally Ride is not the sort of person about whom anecdotes cluster," *People* noted.[32] Still, the press could not help but speculate: *Who was the real Sally Ride?* And they could not help but get closer to the parts of Sally that she was trying to hide.

One reporter asked if she was a tomboy growing up. "I don't like that term," Sally rejoined.[33] The article then quoted her sister, Bear's, observation that Sally "never bought into the traditional, female role," noting "she wears no makeup and does not carry around a pocketbook."[34]

"Did you ever wish you were a boy?" another reporter asked at a press conference.

"No, I never thought about that," Sally replied.[35]

Despite her best efforts, some aspects of Sally's nature were seeping through. She simply had a swagger that was not totally straight. One reporter called it the "upright rolling walk of an athlete."[36]

Susan Okie, Sally's childhood friend, was working as a reporter at the *Washington Post* at the time. She heard from another *Post* writer, a woman on the sports beat: "Oh, everybody knows Sally Ride is gay. The tennis players all say."[37] Susan, who had never discussed boyfriends or sexuality with Sally before, decided to ask her about it.

Sally had invited Susan to New York, where Sally would be giving a speech for a scientific organization. Over a quiet dinner after the presentation, Susan worked up the courage to ask Sally point-blank.

People are talking about this . . . Are you gay?

"Molly wanted us to be a couple," Sally admitted. "But I didn't want to, so the answer is no."[38]

"I don't really know if that was a straight answer or not," Susan said. "There was a lot of pressure on her, and having people discover that she was gay would not have been great for her flight."[39]

Although never substantiated, there were plenty of rumors that the higher-ups at NASA disapproved of their unmarried "first woman" cohabitating with her boyfriend.[40] Sally's bosses denied the veracity of these claims. However, a few months after the crew announcement, Sally flew her Grumman Tiger to Salina, Kansas, and, on a warm Saturday afternoon, July 24, 1982, married Steve in his parents' backyard.[41] Both bride and groom wore white jeans and polo shirts.[42] The ceremony was a quiet affair, attended only by family. Even the baker was sworn to secrecy.[43]

There was no honeymoon, no fanfare. A month later, when the *New York Times* ran a small story about the marriage, many of their friends were shocked. Sally had not told anyone. "We didn't want to make a big deal of it," she told the *Times*.[44]

Did Sally marry Steve to appease NASA, or to shield herself from public speculation about her sexuality? Steve would later say that, although Sally was not "a traditional wife type," he believed she entered the marriage with honest intentions.[45] "I think she tried to make it work with Steve," Susan agreed.[46] Although she conceded, "It was a brother-sister relationship. They liked hanging out together, watching *Star Trek* and cracking jokes."[47] Whatever Sally's reason, the marriage seemed to put her at ease. She told Susan that "she felt less protective of her personal life now that she and Steve were married."[48]

As the launch date drew closer, press scrutiny grew more intense, and Sally became more agitated. She tried to deal with her fame, she joked, by flipping "the switch marked 'oblivious.'"[49] Sometimes that worked, and sometimes it did not. "She is going to have a heavy burden to bear," Crip said. "I think Sally can handle it . . . but today, I don't think she fully comprehends what's involved."[50]

Sally may not have wanted to admit just how much her life was about to change, not even to herself. While Sally was visiting Susan in Washington, DC, before her flight, the two made a visit to the National Air and Space Museum. "We're looking at the exhibits and I said, 'Sally, there's going to be a whole exhibit on you. Your statue is going to be here with these other guys.'"[51] Sally "turned completely beet red," Susan said. "She didn't quite believe it; she got so embarrassed."[52]

In her reporting on America's first woman in space, one aspect of Sally that Susan could not quite reconcile was Sally's secrecy. Although they had been best friends since their days at Westlake, Susan wrote, "Sally has always seemed to enjoy being an enigma."[53] However, in her rare, vulnerable moments, the subject that Sally most wanted to talk to Susan about was not her upcoming flight, but Dr. Elizabeth Mommaerts, their old science teacher.

Six years after Sally graduated from Westlake, while she was a graduate student at Stanford, her beloved Dr. Mommaerts, who completed brainteasers in record time and inspired Sally to pursue science, committed suicide.[54] Her death left Sally brokenhearted. When Sally was chosen to be an astronaut, she told Susan that Dr. Mommaerts was "the one person in the world I wanted most to call—even more than my parents. And I can't."[55] Dr. Mommaerts was such a vibrant and brilliant influence on Sally's life that Sally struggled to accept that her cherished teacher had severe struggles with mental health. "Sally still finds it difficult to talk about Elizabeth Mommaerts," Susan wrote, "but she sometimes tries—calling, late at night, when she comes across a book or letter that reminds her. At times of triumph, she said, she forgets for a moment that Elizabeth is dead."[56]

Sally kept her grief, along with almost every other aspect of her personal life, private. In her tougher moments, she would flip her switch to "oblivious" and persevere, trying to smile for the cameras. It was the sort of "restless, smug half smile" a photographer coaxed out of her for her official NASA portrait. During the photo shoot, Sally posed stiffly in front of the American flag in her navy-blue polo while a makeup artist attacked her hair with a curling iron and her face with cosmetic brushes. Sally melted under the bright lights. *Just think,* the photographer said, "Elementary school would-be astronauts will hang this up on their walls." His words were no consolation. It was not a picture she was happy with.[57]

The Night before STS-7's Launch.
Cape Canaveral. June 17, 1983.

Why is NASA's director of flight operations calling me? wondered Molly Tyson. Molly, Sally's former girlfriend from Stanford, was invited to the launch, along with a gaggle of Sally's old friends.

In the years since Molly and Sally's breakup, they had stayed close. Molly visited Sally in Houston and even worked there on a project during Sally's training.[58] Molly was now openly gay. She had a partner. Many of the New Guys had gotten to know Molly over the years, and so had George. George, exceedingly perceptive, noticed how comfortable Sally was around her.[59] George told Molly he was worried about Sally. She was clearly anxious, and it might be nice for her to see a friend. *Would Molly visit Sally?* Molly agreed.

George arranged for an escort to bring Molly through NASA security. Because the astronauts were already in quarantine, only a primary contact could visit them—in most cases a spouse—and only after passing a medical test. No children, extended family, or friends were allowed inside. George had to arrange for a NASA physician to meet Molly, give her a quick physical, and sign her through.

Sylvia Salinas, in charge of signing people in and out of quarantine, helped check Molly through security. George had forewarned Sylvia: *Be discreet, give them their privacy.* Sylvia put two and two together. "That he knew about this person was very insightful," Sylvia said. "But that was George."[60]

When Molly arrived at the beach house, Sally was floored. *Why is Molly here?* She must have experienced an overwhelming mix of gratitude and fear. She must have wondered: *Did George—or others at NASA—know who Molly was to her?* That George chose a woman he suspected might be gay to be the first American woman in space spoke to his character. It was a huge risk for someone in his position, but Sally was qualified, deserving. *What else mattered?*[61]

Sally and Molly talked. Molly kept an eye out for the nerves George had mentioned. Sally seemed amped up, "like an athlete waiting to play on Centre Court at Wimbledon," but that was to be expected, right?[62]

From the porch, they watched the breaking waves. Across the water, *Challenger* sat on Launch Pad 39A.[63] Would the bird carry her safely to the heavens and make her a hero? Or would she be swallowed up by the waves below? Sally hid her fears from Molly until it was time to leave. She still had to complete her final, preflight physical before going to bed. As Molly turned to go, Sally stopped her.

"I'm aware that this is not without risks," Sally said, in an unguarded moment. Any bravado evaporated. "I realize I could die."[64] The last-minute confession rendered Molly speechless.

"It struck me as the most vulnerable thing she'd ever said to me," Molly reflected. "I had her on such a high, unreal pedestal that I was surprised by even this hint that she was afraid."[65]

Sally did not follow up her confession with anything else. Molly thought Sally was reaching out for some connection to their old relationship, but Molly was too shocked to respond. The moment passed. Neither woman would speak of it again.

Sally was already wide awake, showered, and dressed by the time George woke up the crew at 3:20 AM.[66] She donned her flight suit and gathered her keepsakes—banners for the Westlake School for Girls and Stanford University, silver medallions for her family and her five women classmates, a pin for George Abbey, her gold wedding band, a charm for Carolyn Huntoon, and a feather for Molly.[67]

"T minus thirty-five seconds," announced Hugh Harris over the loudspeaker, as Sally and her crew lay strapped in.

"Sally, have a ball," Steve radioed in.[68] She would if she could get over these nerves.

George watched from Kennedy's Launch Control Center. This was not just the flight of the first American woman in space, this was the first flight of his new astronaut class, including John Fabian, Norm Thagard, and pilot Rick Hauck. That morning, George had been like a doting father, escorting the crew out to flashing bulbs and clicking cameras, and riding with them on the Astrovan, a modified Airstream used to transport the crew to and from the shuttle. He bid them farewell at the Launch Control Center before they made the rest of their three-mile journey to the launchpad.[69]

An eight-months-pregnant Anna Fisher, working as the lead Cape Crusader, did the final switch checks and helped strap in the crew, as Fred had done on STS-1.[70] She had spent the night in the darkened orbiter guarding the set configuration. The image could not have been more different from the testosterone-filled era of Mercury astronauts: Here was a pregnant woman helping another woman make history in space.[71]

Dick Scobee joined NBC's Jane Pauley at the Cape.[72] Rhea Seddon, Ron McNair, Fred Gregory, and Guy Bluford all watched from a Johnson Space Center conference room. Guy, who would fly next, contemplated his future. Kathy, who had desperately wanted Sally's place aboard the shuttle, did not attend the launch. Instead, she flew to California to give the commencement address at UC, San Diego. She admitted that her decision to accept the engagement was partly to spare herself "the challenge of keeping up a happy face as I watched the flight I had hoped might be mine leave the pad."[73]

"We're just a few seconds away from switching command to the on-board computers," Harris reported, signaling that the launch managers were relinquishing control to the computers on *Challenger*.[74]

It was then that Sally really started to panic. "I felt totally helpless," she said. "Totally overwhelmed by what was happening. It was just very, very clear that for the next several seconds we had absolutely no control over our fates."[75]

As Harris counted down the final seconds to launch, white steam puffed from *Challenger*'s rocket boosters.

"T minus ten, nine, eight, seven, six, we go for main engine start." On cue, orange and red sparks burst from the main engines.

"We have main engine start . . . ignition . . . and liftoff," Harris pronounced. "Liftoff of STS-7 and America's first woman astronaut."[76] *Challenger* began to climb, leaving the launchpad enveloped in clouds of gray exhaust.

Flashes of light zipped by the orbiter's windows. Rocket fuel exploded beneath her. The experience was "exhilarating, terrifying, and overwhelming all at the same time."[77] The rumble of the engines shook her whole body, causing her teeth to chatter. She relaxed only when they cleared the tower. "All of a sudden, we were going someplace," she said.[78]

At forty-five seconds, *Challenger* was three miles above Earth and supersonic. Pulling free of Earth's gravitational pull, Sally set her checklist in front of her and watched as it floated away. Beside her, John Fabian whooped and hollered. Crip thrusted the *Challenger* to orbit. When they were 184 miles high, Sally radioed down to Mission Control:

"Houston, have you ever been to Disneyland? That was definitely an E ticket."[79]

The five-hundred-thousand-strong crowd—who had camped out in neighboring fields, parked alongside the highways, and docked their boats in KSC's waterways—cheered below, many of them shouting the phrase of the day: "Ride, Sally Ride!"[80] Among them were 1,628 journalists, the fourth largest press turnout in the history of Kennedy Space Center.[81]

Sally's smiling face graced the covers of *Newsweek*, *People*, and *Ms.* In Washington, DC, people danced on the National Mall. Cocoa Beach's hotels were fully booked, and the roads packed with camper vans and cars. RIDE SALLY RIDE T-shirts were being sold on the side of the highway.[82] A local barbeque joint offered Sally's First Ride, a rum cocktail.[83] A local band covered Wilson Pickett's version of the song "Mustang Sally" with corny space lyrics:

All you want to do is ride around, Sally
Ride, Sally Ride . . .
Slow your shuttle down
Plant your moon boots on the ground . . . [84]

"Sallymania" had swept the nation. Over four thousand NASA-invited guests gathered in Kennedy's VIP section. Among them were country singer John Denver, eleven-year-old *E.T.* star Henry Thomas, and Jane Hart, one of the Mercury 13. "This is a day for celebration, for all of us who worked to make this happen," Hart observed with bittersweet pride.[85]

Joyce and Dale Ride, Sally's parents, watched the launch with hope and fear. That was their baby. Even though Dale claimed he was not anxious, his emotions showed in the tears streaming down his face during launch. Joyce kept her tears at bay by joking with the press. "The first one who asks how I feel gets it," she quipped, wielding a cup of water as her weapon.[86] "God bless Gloria Steinem," she added.[87]

The activist and *Ms.* magazine cofounder was nearby. "Millions of little girls are going to sit by their television sets and see they can be astronauts, heroes, explorers, and scientists," Steinem told reporters.[88] Sally had supported Steinem's Equal Rights Amendment (ERA), which failed to pass Congress the year before.[89] Unfortunately, the ERA would never pass, due in large part to the efforts of conservative activist Phyllis Schlafly—who, ironically, was also present at Sally's launch. Schlafly did her best to feign disinterest, saying, "I have no desire to be an astronaut."[90] However, as the countdown got closer to zero, both Schlafly and her eighteen-year-old daughter looked on with the same wonder as the hundreds of thousands of others present.

The most controversial figure at the event was actress, activist, and home-exercise pioneer Jane Fonda, who attended with her husband, California Assemblyman Tom Hayden.[91] Some still called Fonda "Hanoi Jane" after her 1972 trip to North Vietnam, where a photographer snapped an irreverent picture of her in front of an anti-aircraft gun meant to shoot down American planes. "The fact that Sally Ride is on this launch makes me feel a part of the space program," Fonda told the press.[92] Fonda was such a political lightning rod that, the next day, Reagan's deputy chief of staff called NASA administrator James Beggs and reamed him out. "Nancy Reagan is mad, and I am mad, and everybody is mad."[93] Eager to appease the White House, Beggs fired NASA's public affairs director and hoped that would be the end of it. (It was not.)

No matter how anyone else felt, the Rides were thrilled to see Fonda. Bear Ride snapped a photo of her mother linking arms with Steinem and Fonda and announced, "Here's two heroes of the women's movement."[94]

Thousands of miles above Earth, Sally was oblivious to the drama unfolding below. "It sure is fun!" she radioed Mission Control.[95] Sally could not help but be struck by the contrast between the blackness of space and the bright blue orb humanity calls home. "It looked as if someone had taken a royal blue crayon and traced it along Earth's horizon. I realized that the blue line was Earth's atmosphere, and that was all there was of it. It's so clear from that perspective how fragile our existence is."[96]

Hours into their flight, Sally and John Fabian prepared to deploy two satellites. They had trained for hundreds of hours on the simulator, but this was the real robotic arm, all fifty feet and nine hundred pounds of it. "This is real metal that will hit real metal if I miss," Sally thought.[97] Her relentless training paid off. She successfully deployed Canada's $24 million communications satellite, which brought television to millions of North American homes.[98] Then she and Fabian released an Indonesian satellite that provided high-speed data and telecommunication service to the archipelago, earning $20 million for NASA.[99]

Later in the week, Sally and Fabian floated up to the flight deck and used the robotic arm to deploy the first shuttle pallet satellite "SPAS," which was designed by the West Germans and which carried ten experiments to study metal alloys and gasses in microgravity. Sally both deployed the satellite and retrieved it after the experiments were complete.[100] In so doing, Sally and her crew proved an important new shuttle capability—that they could retrieve satellites that needed maintenance or repair. Crip boasted of the maneuver, "Some crews in the past have announced that 'We deliver.' Well, for Flight 7, we pick up and deliver."[101]

When Sally had a free moment, she would drift up to the aft deck to get a look at the spinning Earth beneath her. The view inspired the English major in her:

> Hurtling into darkness, then bursting into daylight . . . The sun's appearance unleashes spectacular blue and orange bands along the horizon, a clockwork miracle that astronauts witness every ninety minutes. We could see smoke rising from fires that dotted the entire east coast of Africa, and in the same orbit only moments later, ice floes jostling for position in the Antarctic . . . One night, the Mississippi River flashed into view, and because of our viewing angle and orbital path, the reflected moonlight seemed to flow downstream—as if Huck Finn had tied a candle to his raft.[102]

Seven days later, bad weather forced the crew to reroute their landing from Kennedy to Edwards Air Force Base in California. The press, Sally's fans, and her family, who were all waiting for Sally's triumphant return at Kennedy, would now have to watch the landing of America's first woman in space on television. At two hundred feet, Rick Hauck lowered the landing gear and the orbiter glided down the Edwards runway.[103] On the other side of the country, inside the wood-paneled walls of the ABC News trailer, Dale and Joyce Ride pumped their fists. "Hot dog!" they screamed.[104]

Aircraft shuttled the crew back to Houston's Ellington Field, where a more elaborate homecoming awaited. Climbing out of the NASA Gulfstream, Sally embraced Steve, deliriously happy.[105] Cameras flashed. Reporters angled their microphones toward Sally, eager for a soundbite. President Reagan called with his congratulations.[106] "You know, I was going to meet you in Florida, but then you decided to land in California," the president joked. "You didn't stop and pick me up off the South Lawn like I asked you to."[107]

Smiling, waving, Sally was ushered to a limo that took her to Johnson. On arrival, a NASA protocol officer handed her a bouquet of white roses, which she graciously accepted, at first. However, when it came time to join her crewmates on the makeshift stage, Sally returned the bouquet to the officer and lined up beside her male crewmates. After all, none of her male crewmates had a bouquet. The crew's spouses were also there; the four wives each held a single red rose, but Steve declined his. Sally and Steve were grinning and had their arms around each other for the ten-minute ceremony. At the end, the officer tried to return the bouquet to Sally, who shook her head no.

This small gesture set off an unexpected firestorm. One headline read "Astronaut Sally Ride Spurns Bouquet of Roses, Carnations."[108] Another read "Sally Ride Turns Down Homecoming Bouquet."[109] Sally had executed a flawless performance on a shuttle mission and proved once and for all that women astronauts were as capable as their male counterparts, but she was criticized because she did not hold on to her bouquet.

The response to "bouquet-gate" was not exclusively limited to pearl-clutching.[110] "That one little action," Sally said, "probably touched off more mail to me than anything I ever did or said as an astronaut. Half of those who wrote were incensed . . . The other half were thrilled."[111] "Good for Sally Ride not accepting a bouquet of roses," one man wrote into his local newspaper. "The jerk who offered them would probably have given Madame Curie an Avon gift certificate."[112]

After shaking off the kerfuffle and spending a few weeks adjusting to life back on Earth, Sally embarked on her US publicity tour. Her first stop was Kennedy Space Center, where Mickey Mouse, tricked out in an astronaut suit and helmet, gave her a kiss and a brand-new Mickey Mouse watch.[113] In August, she traveled to the California state capitol in Sacramento, where she was honored with "Sally Ride Day." In the assembly chamber, she was met with a standing ovation from the joint session of assembly members and senators. Jane Fonda applauded as her assemblyman husband escorted Sally to the speaker's platform, where she was introduced as "a genuine American hero."[114] In New York, Mayor Ed Koch handed her the keys to the city.[115]

Though Sally received most of the recognition, she always invited her crewmates along to these appearances, both for moral support and to deflect attention. Her crewmates were happy to oblige. "I tell people that Sally Ride made me famous," Fabian said.[116] Sometimes it was not so pleasant living in Sally's shadow. At one state dinner in DC, the reporters knocked down astronaut Norm Thagard trying to get to Sally.[117]

Sally's fame was reaching a fever pitch. "Fans" found her home phone number and left voicemails for her at all hours. The local constable had to send deputies to patrol the Ride-Hawley neighborhood to keep gawkers at bay.[118] Police apprehended one stalker at UC, San Diego, where Sally was scheduled to make a speech.[119] Events like these put Sally on edge. "She was a huge target," Fabian noted.[120] At one point, NASA received twenty-three press requests for her an hour.[121]

Even though she had been warned about attention coming her way, Sally could never have anticipated the true burden of fame or its effects on her mental health.[122] "She felt like she could never be alone except if she was at a podium giving a speech." Susan said. After those speeches, Susan noted, crowds would descend, making her claustrophobic. "She had no privacy."[123]

"Luckily I have a very forgettable face and people tend not to recognize me in the grocery store," Sally joked. Nevertheless, the reality was that she was struggling. Sally started seeing a psychologist, who encouraged her to take some quiet time to preserve her sanity.[124] Sally then asked her mentor Carolyn Huntoon to screen press requests and to attend future events with her as moral support. Before embarking on her international tour, Sally set some new ground rules. No appointments longer than fifty minutes. Mandatory bathroom breaks every two to three hours. At least fifteen minutes of downtime if she was doing a meet-and-greet for more than four hours. A day off every week. NASA agreed to every request.[125]

That September, Sally, Steve, Rick Hauck, and Rick's wife, Dolly, jetted off to Europe. They visited eight capital cities over three weeks, starting in London and ending in Bonn, West Germany. Sally and Rick Hauck presented a gold commemorative medal they had taken into space to London's Royal Society. They met with Holland's Queen Beatrix and Prince Claus and the King of Norway.[126] They dined with Russian ballerina Natalia Makarova.

The highlight of the trip was Budapest, where Sally and Rick Hauck were invited to speak to over 650 scientists and engineers from thirty-two countries, including the Soviet Union. Prior to their trip, NASA had warned its astronauts not to fraternize with the Russians—Cold War tensions were running high. A month before their visit to Hungary, a Soviet military plane shot down a Korean Air passenger jet traveling from New York to Seoul. All 269 passengers onboard, including sixty Americans and US congressman Larry McDonald, were killed.[127]

Sally and the rest of the crew did their best to keep their distance from the Russians at the event until, during the reception, Sally felt a light tap on her elbow. It was Svetlana Savitskaya, the Soviet cosmonaut and second woman in space. Svetlana congratulated Sally on her flight. Sally thanked her, looking around to make sure the press was not watching. After they parted ways, Sally wondered if she had missed an opportunity to speak with one of the only women in the world who might understand her. At an event the next morning, Sally cornered their Hungarian translator, Tamas Gombosi.

"You know, I'd really like to get a chance to talk to Svetlana," she whispered.[128] Gombosi understood immediately. At an embassy reception that night, he pulled Sally aside, inviting her and Steve to Hungarian cosmonaut Bertalan Farkas's apartment. "And there will be *other people.* Are you interested?"[129] Sally agreed to meet him in the hotel lobby at 9:00 PM.

When Sally told Steve, he balked: He would not disobey headquarters. She then consulted with Rick Hauck, the highest-ranking member of the *Challenger* crew. "Sounds like a great opportunity," he said with a glint in his eye. "Don't pass it up."[130] Sally resolved to go alone. At nine o'clock on the dot, she walked through the lobby, doing her best to appear casual. Gombosi ushered her into a chauffeured car. As they wove through dark streets, Sally's doubts multiplied. Were the secret police following her? Had she been bugged? What if she caused an international incident?

When they arrived at the apartment building, they snuck in through the back. Sally relaxed when she saw that the cosmonaut's apartment was decorated with spaceflight photos and paraphernalia, like the astronaut home she shared with Steve. Svetlana spoke passable English, and Gombosi filled in the blanks with his translations. Soon, Sally and Svetlana were chatting like long-lost friends—talking about their crews, sleeping in zero gravity, and what it was like to barrel back through Earth's atmosphere. They were tunneling underneath the geopolitical barrier between them.

"There was a code between the two women. From the first minute they just showed so much affection for each other," said Gombosi.[131] Svetlana gave Sally Russian books, dolls, and a scarf. Sally gave Svetlana an STS-7 charm and promised to send her the TFNG T-shirt she had worn on her flight. They signed autographs for each other and linked arms, taking photos. After six hours of stories and toasts of vodka for the men and juice for the women, it was nearly three in the morning. Sally hitched a ride back to the hotel with the Russians. "Don't get into a wreck!" the giddy passengers joked with the chauffeur.

When they arrived back at the hotel, Sally got out first so she and Svetlana would not be seen together. They hugged goodbye. "I felt closer to her than I'd felt to anyone in a very long time," Sally said.[132] Though all of Sally's future written communications to Svetlana would be vetted by the government, Sally had made a friend for life. Somewhere, halfway around the world, there was a woman cosmonaut wearing an STS-7 charm who was just like her.

At the height of her national press tour, Sally was invited to go on Bob Hope's NBC special *Blast Off with Bob,* the comedian's tribute to NASA. Guests would include Alan Shepard, Neil Armstrong, Bob Crippen, and Guy Bluford. NASA asked Sally to join. She declined, objecting to Hope's brand of comedy. Hope famously surrounded himself with scantily clad women on his shows and was known to make misogynistic comments, not the least of which referred to women as cattle.[133]

Gerry Griffin, the new director of Johnson Space Center, got involved, inviting Sally for a beer one night. Griffin told her that Hope promised to focus on her achievements: "Nothing funny where she's involved."

"No, I'm not going to do that," she told him.

"Really. Why not?" Griffin said.

"Because I don't like the way he exploits women," Sally replied.

"He's kind of a national icon," Griffin implored.

"No," Sally said. "I'm not going to do that."

"I knew it was final," Griffin said."[134] Bob Hope was a longtime friend of NASA, who often featured astronauts on his television specials.[135] Griffin felt he owed Hope for his support, but he could not fault Sally.

"She was a person of principle, stuck to her guns," he observed.[136] Everyone who knew Sally agreed: Her sense of right and wrong was unwavering.

Shortly after her conversation with Griffin, Sally disappeared.

"No one knew where she was," Fabian said. "No one knew when she was coming back, but one thing we were pretty sure of was that she was not going to appear with Bob Hope."[137] Even Steve did not know where Sally was. She simply ran away. He understood why she had refused to go on Hope's special, but not why she left.

"Her reaction was a little inappropriate—going AWOL and not telling anyone, in particular not telling me, and not taking responsibility for her decision. Particularly given that she accepted the [STS-7] assignment with the knowledge . . . that this was going to happen," Steve said. "And it was almost like, now that she's had her good deal, she doesn't want to go through with her part of the bargain."[138]

Bob Hope's *Salute to NASA: 25 Years of Reaching for the Stars* aired on NBC on September 19, 1983, without Sally.[139]

Sally had fled to her "hideout," Molly's home on a quiet street in Menlo Park, California.[140] After months of living in the public eye, with every aspect of her life under a microscope, she needed to escape. She spent time with friends who knew her before NASA, tagging along to their softball practices, margarita parties, and long runs. Her old boyfriend from Stanford, Bill Colson, even dropped by to grill salmon for dinner. It was nice, so very nice, to relax, to be herself—not just with her friends but with Molly, the person who perhaps knew her best.

Even though Sally's hideout was the home Molly shared with her new partner, Sally was happy there. She was safe. No stalkers. No cameras. No annoying reporters. Just people who saw her for who she really was.

Sally stayed in Menlo Park for two weeks. Despite her protests, Molly threw a party in her honor. Sally endured it for a while, but soon needed to get some air. She grabbed Ann Lebedeff, a mutual friend from their tennis days, and they went for a twilight walk around Molly's neighborhood. As they distanced themselves from Molly's house, the sounds of the party faded away. In the silence, Sally sighed and confessed to Ann, "I miss the days when no one knew who I was."[141]

CHAPTER 10

ROCKET DAWN

Launch Pad 39A, Kennedy Space Center.
August 30, 1983.

Suited up in his sky-blue flight suit and white helmet, Guy Bluford crawled into his seat in the shuttle's crew cabin, behind commander Dick Truly and pilot Dan Brandenstein, another New Guy.[1] Like Sally on STS-7, Guy was the flight engineer: Sally had occupied his seat two months prior.

Rain scattered outside the orbiter's windows. Jolts of lighting streaked through the night sky, striking terrifyingly close to *Challenger* and its half-million gallons of liquid fuel. Just after midnight, the shuttle was poised for its first nighttime liftoff.

Cape Crusader and fellow New Guy Shannon Lucid straddled Guy, tightening his straps and double-checking his communications connections and oxygen hoses.

"Not too tight," he teased Shannon as he awkwardly lay on his back beneath her.[2]

Guy heard the thunder, saw the lightning flash

through the orbiter's windows, and thought, *Why are we even up here right now?* The answer was obvious. This was the first time that an African American was blasting off into space. Not even a thunderstorm could make NASA cancel this trip. The closeout crew cleared out once mission specialists Dale Gardner and William E. Thornton were strapped in, sealing the door shut behind them. The crew sat for another two hours as thunderstorms swept the Florida coastline. Outside, Crip cruised in the Shuttle Training Aircraft (STA), a modified Gulfstream II jet that was used to assess the weather conditions prelaunch. Their prospects were improving.[3]

In the stands outside the Launch Control Center, invited guests waited under towels, tarps, and umbrellas with bated breath.[4] Among the two hundred-plus VIPs was a Who's Who of Black America, from Washington dignitaries to talented NASA scientists to TV stars like Bill Cosby and Nichelle Nichols.[5] Thousands of spectators packed into buses along NASA Parkway to watch the liftoff in person.[6] A million more people stayed up all night, crowding around their TV sets to watch the liftoff. During a brief lull in the rain, launch controllers made the call. *Go for launch.*[7]

What did this mission mean to America, a nation that had long dragged its feet when it came to acknowledging the humanity—much less the contributions—of its Black citizens?

The 1980s were a time of undeniable progress for the Black middle class, and yet the Black working class and poor suffered setbacks. As Guy prepared for his historic launch, the city of Chicago and his hometown of Philadelphia elected their first Black mayors.[8] In 1983, the same year as Guy's launch, Jesse Jackson became the first Black man to run in the presidential primaries as a major party candidate. When Jackson announced his campaign, he proclaimed, "Our time has come!"[9]

Jackson was both right, and wrong. Thanks to new federal laws ensuring civil rights, as well as the affirmative action movement, the Black middle class in America was expanding. Between 1970 and 1990, the number of Black doctors tripled, and the number of Black engineers quadrupled. Over the same period, the country gained six times more Black attorneys and twice the number of Black university professors.[10]

Black culture was, slowly, becoming mainstream. *The Cosby Show*, which premiered in 1984, portrayed a Black middle-class family with a father who was a doctor and mother who was a lawyer. The '80s kicked off with Tina Turner's spectacular comeback and Eddie Murphy's debut on SNL. In 1983, Michael Jackson's *Thriller* topped the charts, Alice Walker became the first Black woman to win the Pulitzer Prize for Fiction for *The Color Purple*, and Vanessa Williams was crowned the first Black Miss America.

Despite the advancements for the Black upper and middle classes, the Black unemployment rate was still double that of whites.[11] One-third of Black Americans still lived at or below the poverty line, compared to 10 percent of white Americans. Reagan's policies perpetuated the inequality gap.[12] The administration oversaw cuts to social safety net programs like Medicaid and food stamps, which supported poor Americans of every race but were especially vital for Black Americans. His advisors labeled affirmative action a "new racism" and slashed funding for the Equal Employment Opportunity and Civil Rights Commissions.[13] The Reagan administration, for all its symbolic gestures—like making Martin Luther King Jr. Day a federal holiday—did little to address the systemic economic disparity between white and Black Americans.[14]

Amidst this inequity, Guy's flight to space assumed even greater significance. Twenty years had ticked by since Ed Dwight Jr. was rejected by NASA, and sixteen years had passed since Bobby Lawrence was killed in a tragic flying accident. Lawrence's family sat in the stands, and for them, and so many others, Guy's triumph meant more than a trip to the stars. His mission was a vindication, a beacon of hope for the future, a stride toward a more just America.[15]

At 2:32 AM, Guy and his crew were cleared for liftoff. With the force of the main engines, the upper fuel tank pulled back several feet before rebounding—the "twang" they had trained for. Then the solids ignited, and things really got going. Within each booster, a stack of four thick donuts of solid propellant—a chunky candle with its center hollowed

out—waited for its turn.[16] Small rockets at the top of each solid rocket booster sent flames roaring down the center of each candle. Their glow illuminated the darkness, turning night into day. Oohing and aahing rippled through the crowd.[17] Some spectators wept; others started to cheer.[18] The space shuttle illuminated the night in a "rocket dawn."[19] From as far away as North Carolina, people saw the shuttle light up the night sky with their naked eyes.[20]

"This has to be one of the most spectacular things I've ever seen!" exclaimed National Urban League president John Jacob.[21]

"Awesome is the word!" echoed basketball star Wilt Chamberlain.[22]

Onboard the cabin shook and rattled like a wooden roller coaster. The windows glowed with the rocket's red glare.[23] As *Challenger* rose, the candle burned from the inside out, releasing over five million pounds of force.[24] Guy slammed back into his seat. Thirty seconds in, everything was going as planned.

What Guy did not know—what no one knew—was that with each passing second, the crew drew closer to a full-scale disaster.

The solid rocket boosters were carefully controlled bombs whose nozzles directed thrust downward. Hydraulic actuators gimbaled the nozzles to point that thrust, in effect steering the shuttle. Each nozzle was lined with three inches of resin, an ablative material designed to absorb the extreme heat around the nozzle by burning away, but the resin was burning too fast. One minute in—halfway through the boosters' burn time—only one and a half inches of material remained. Inside the cabin, Guy tensed his muscles to counteract the g-forces and push blood out of his legs and into his brain.[25]

Two minutes in, and over twenty-seven miles above the ground, the nozzle's resin burned down to 0.19 inch—half a finger's width. The resin should have eroded down to one inch, at most.[26] In mere seconds, the 4,000°F flames could have breached the nozzle's metal skin, sending the shuttle into a deadly tailspin.[27] At that precise moment, *Challenger*'s computers fired sixteen tiny rockets, eight attached to each booster. The boosters fell away from the orbiter and plummeted harmlessly into the Atlantic.[28] The crisis, of which the astronauts had not a clue, was averted.

After jettisoning the boosters, the ride to orbit was liquid-smooth, exclusively powered by the shuttle's main engines, which used liquid fuel from the external tank. Six and a half minutes later, the main engines cut off and the shuttle was in orbit.

Traveling at twenty-five times the speed of sound, Guy joyfully unstrapped from his seat and let his body float above the cockpit.[29] "Oh my goodness," an awestruck Guy whispered to himself, "zero-G."[30] It was almost three in the morning, and he had become the first African American to break free from Earth's gravitational pull. He looked outside the window to watch the sun rising over the eastern horizon, bathing the dunes of the Sahara Desert in amber light.

"Wow." Guy whistled. "What a view."[31]

As a fighter pilot flying his Phantom F-4C over Vietnam, Guy had been far enough away to dehumanize the enemy and do his darn job. Now, with Earth unfolded beneath him "like a *National Geographic* map," he could see there were no such things as borders. The "enemy" was a fiction.[32] They were all just humans, living on the same blue ball.

Guy thought of his parents, regretful that he would never get to share this success with them.[33] Guy's father had died right after he returned from Vietnam, and his mom had passed shortly after he was selected to the astronaut corps. He wished they could see him now. Could they have guessed their shy son was now riding on top of the world?

Slender, with a goofy smile below a professional 1970s coif, Dr. Guion "Guy" Bluford had been underestimated his whole life.

The Blufords were a family of erudite and talented intellectuals: musicians, educators, and writers.[34] They were a middle-class, African American family living in a comfortable row house in desegregated Philadelphia. Neither Guy's mother, Lolita, nor his father, Guy Sr., would ever allow their children to use race as an excuse not to succeed. The Blufords felt their "ancestors had achieved without joining or bucking the white world," one reporter noted.[35]

Guion Sr. and Lolita Bluford practiced "cultural conservatism." They were "Eisenhower Republicans" who voted for Nixon over Kennedy in

1960.[36] They did not discuss the civil rights movement at the dinner table. "If it came up in a political way, it was put down quickly . . . and that was the end of that," said Guy's youngest brother, Kenneth. After all, hadn't previous generations of Blufords succeeded, despite racial discrimination?

In a family of achievers, the shy Guy, whom his mother nicknamed Bunny, was often overlooked. Lolita thought him the "least likely to succeed" of her three sons.[37] "Bunny just had to work harder than the rest of us," said Kenneth. "[Guy] was always a little behind and trying to catch up."[38] A high school guidance counselor once told Guy's parents that "Guy was not really college material and might be better off as a carpenter or a mechanic."[39] Lolita and Guy Sr., who had generations of college graduates on both sides of the family tree, found her advice unacceptable.

By Guy's own acknowledgment, he had trouble reading, but everyone missed that he excelled in math and science. Instead of letting his teachers and family discourage him, Guy focused on his strengths, solving math problems and puzzles. He joined the Boy Scouts and taught himself how to play chess. For fun, he built model airplanes and spent hours thinking about why planes fly.

Guy deeply admired his father, his namesake. When Guy Jr. was eight, Guy Sr. developed epilepsy, a grossly misunderstood disorder in the 1950s. Epilepsy cost Guy's father a career as a mechanical engineer and sent him through a revolving door of mental institutions. Lolita became the sole breadwinner, left to raise three boys on her own. She took on a second job, teaching at night.[40]

"I don't know if I noticed that my dad was gone as much as I noticed that my mom had become angry all the time," Guy said. He would downplay the effect that his father's epilepsy had on his childhood, but his brother Kenneth saw the change: "I think the decline of my father, and the decline of his self-respect as he got sicker, affected Bunny more than anything in his life . . . Bunny was always quiet and serious. But he got more so after that."[41,42] Guy had already dreamed of becoming an engineer, but after witnessing his father lose his career because of his illness, he may have become even more resolved to succeed.

Guy went on to study aerospace engineering and join the then-mandatory ROTC program at Penn State.[43] One of his professors remembered Guy, "as a quiet fellow and an average student, not the sort you would expect to be interviewed about twenty years later."[44] That same professor also admitted he doubted he would have remembered Guy if he had not been the only Black student in the engineering school. Guy graduated, married his longtime girlfriend Linda Tull, received his commission, and moved west to Arizona to attend pilot training at Williams Air Force Base.[45] There, he attended a lecture from a Tuskegee Airman.

"I had never seen a Black pilot!" Guy marveled. He walked out of the talk with a new dream: "I want to be a fighter pilot."[46]

Guy would get his chance, high in the skies over Vietnam. In 1966, Guy was deployed to Cam Ranh Bay with the 557th squadron—known by their call sign as "the Sharkbaits."[47] The once-shy aerospace engineer was now in a world dominated by bone-rattling g-force blackouts and daily brushes with death. During his nine-month deployment, Guy flew 5,200 hours in 144 combat missions across North and South Vietnam.[48]

"Don't you realize you are bombing peasants and children?" Kenneth once asked his big brother. Guy just stared at him blankly.[49] "To Guy that just didn't compute," Kenneth said. "He was doing his duty, doing a job he was good at."[50] Guy did not get upset by his brother's comment, but Kenneth took the hint and dropped the issue.

Despite Kenneth's political objections, he would later say that Guy "found himself" in Vietnam.[51] The boy who had grown up building model airplanes now had his chance to be a real pilot.[52] Two out of every three days, Guy was stationed on the "alert pad," living in a trailer next to the runway, where his Phantom F-4C sat poised for action. To Guy, the Phantom was a thing of beauty: Sleek and aerodynamic, camo-painted, and boasting thirty-five thousand pounds of thrust, but when packed with napalm and a five-hundred-pound bomb, it looked more like a pregnant goldfish.[53] During those anxiety-filled forty-eight hours, Guy was on high alert, waiting for the base klaxon to sound. When it did, he shot out to his plane in a mad scramble to get up into the sky and provide air support to fellow soldiers under attack.[54]

When he finished his tour, Guy left the bloodshed behind to serve as a flight instructor at Sheppard Air Force Base in Wichita Falls, Texas. However, death followed him home. Within two months of Guy Jr.'s return to the States, after years of being cycled in and out of institutions, Guy Sr. passed away. The loss called to mind Guy's own dream to one day become an aerospace engineer.

Guy sought engineering opportunities within the Air Force, but the corps was reluctant to sacrifice a talented pilot while the war still raged. Guy's superiors suggested he first get a graduate degree to improve his career opportunities. Afterward, they would revisit his request. In 1972, Guy applied and was accepted to the Air Force Institute of Technology at Wright-Patterson Air Force Base in Dayton, Ohio.

Guy was only a year out from earning his doctorate, when he heard rumors that the Air Force planned to send him back to training command as a flight instructor. He thought he had proven his intellectual competence, but here he was, underestimated again. "It wasn't a step up the ladder, but a kind of stagnation," he said. "Like running in place on a closed track."[55]

Then Guy saw NASA's ad for astronauts—and his way out. Astronauts were both pilots and engineers. This was the job for him.

Convinced that NASA would not accept him, Guy did not bother telling anyone he was applying. He could not believe his luck when he made it to the interview round. He observed all the other extremely impressive applicants. "These were smart people in good shape," he said, "and I barely got through Penn State alive."[56] One morning on his drive into work, he heard the news over the radio: "NASA selects new astronaut class."[57] *Guess I wasn't selected*, Guy thought. Shortly after arriving at the office, he got the good news from George.

Later, Guy would describe January 1978 as a "bittersweet month."[58] His mother, Lolita—who still had no idea that Guy applied to be an astronaut—learned of her son's selection when the news broke on national TV. She was lying on her deathbed in the hospital. Guy did not even know that his mother was sick. "I have nine months to live," she calmly informed him two days later.[59] She would die two weeks into Guy's AsCan training, that July.[60]

"I'm just an average guy," he told the reporters who swarmed him after the announcement.[61] In fact, he was so shocked by his selection that he kept asking NASA administrators why they had chosen him.[62] Once he cornered Dr. Joseph Atkinson, who had been on the selection board, and quizzed him. *Why me? Why not anyone else?* Atkinson looked at Guy with disbelief. "You were good," he said. "You have a PhD in aerospace engineering. You were a flier. A combat veteran. You were running an engineering branch. You had supervisory experience. And you could pass the physical. You were easy!"[63]

Guy did not concern himself with jockeying for an early flight assignment. He figured he would fly when he flew. He liked to do his job and go home to his family, in a neighborhood far from Johnson, "where astronauts weren't a big thing."[64] He did not schmooze with George or party with the other TFNGs.[65] He was not particularly close with the other members of his class. His one TFNG activity involved playing racquetball with his officemate Kathy Sullivan, who usually beat him.[66]

Guy thought charismatic Ron McNair or tough-as-nails Fred Gregory would make better public figures. "It might be a bad thing to be first," Guy said, presciently. "It might be better to be second or third, because then you can enjoy it and disappear."[67]

He assumed he was in trouble when, in March 1982, John Young pulled him aside after an astronaut all-hands and instructed him to be in George Abbey's office at 10:30 AM.[68] *Wonder why Abbey wants me?* Guy puzzled on his walk over to Building 1. Then he noticed Dan Brandenstein and Dale Gardner heading in the same direction. When the three of them awkwardly walked through the doors of Building 1, they realized they all had the same destination. As the trio waited nervously outside George's office, they whispered among themselves. *What was going on?* After a few tense minutes of waiting, George's secretary Mary Lopez summoned them in.

"We need a crew to fly STS-8," George told the assembled astronauts, in typical Georgian fashion. "Are you interested?"

After years of overanalyzing each technical assignment and wondering, maddeningly, how they were stacking up in Mr. Abbey's eyes, the men experienced a flood of joy and relief. Guy may not have wanted to be first, but after having this once-in-a-lifetime opportunity laid at his feet, he was over the moon. Then Dick Truly piped up from the corner. "Can I fly with these guys?" He grinned.[69]

Guy waited until dinner that evening to share the news with his wife and his teenage boys, who, as far as he could tell, could not have cared less.[70] Or perhaps the boys were taking their father's lead. "I'm more excited about flying than I am about being the first Black to fly," Guy told reporters. "Whatever significance there is, won't last long."[71]

On the evening after launch, the start of the crew's workday, Guy positioned himself in the commander's seat to deploy their payload, the Indian weather and communications satellite INSAT-1B. The deployment of INSAT-1B was the "primary objective" of STS-8.[72] As one of the first commercial satellites that would be deployed from the shuttle, not only would INSAT-1B earn NASA over eight million dollars, but it would also display the commercial potential of the fledgling shuttle program.[73]

Guy sent commands to the spring-loaded communications satellite and its propulsion module, spinning the spacecraft up before releasing it from the open payload bay. The rotation would keep the spacecraft's mass symmetrical, making it less susceptible to forces that might push it off course. He watched the satellite travel in a straight and predictable line, glinting in the sunlight.[74] Objective complete.

For someone who had flown napalm sorties over Vietnam, it felt low stakes, but that was exactly why George had selected him. The only issue Guy encountered on his mission came from Earth—his wife used the shuttle's onboard teleprinter to inform him their house had termites.[75] The teleprinter also kept Guy abreast of Penn State's performance in football and the Phillies' in baseball.[76]

Before the crew returned home, they fielded a call from President Reagan. "You are paving the way for many others," Reagan waxed. "You are making it plain that we are in an era of brotherhood."[77] To some, the paean to "brotherhood" rang hollow from a president who was not seen as an advocate of civil rights.[78] No matter what Guy might have thought of Reagan, he was gracious.

The eighth shuttle mission was, as Guy put it, "a fabulous adventure" and had achieved every flight goal.[79] With melancholy, the crew configured *Challenger* for the flight back to Edwards Air Force Base on their fifth day. "The mission seemed to go faster than we had wanted it to," Guy said, "and all of us were hoping that we would have the chance to fly again."[80]

After touchdown, Guy, like Sally, went on a national publicity tour, but was met with far less press interest than the first American woman in space. While Sally graced the covers of *People, Newsweek,* and the *New York Times,* Guy donned only the covers of Black publications like *Jet* and *Ebony. Newsweek* focused on the novelty of the shuttle's "first night launch," rather than the first African American riding aboard. The *New York Times* ran Guy's story on page six.[81]

Guy reasoned that "the press was more interested in the unique difference of male versus female in space, than in black versus white on orbit."[82] The introverted Guy had never sought the limelight. He saw the title as a threat to his privacy and was wary of his new fame.[83]

Even with these explanations, it is hard not to think racism was also at play. A 1983 *Jet* survey of magazine editors indicated many mainstream publications had quotas for how many Blacks were permitted to appear in their pages. "It's conventional wisdom in the magazine business that Black faces don't sell as well as white faces," one editor explained. "Whether that is from research, or hunches, or racism, or a combination of the three, I don't know."[84] When *Jet* asked why they had these quotas, one media critic answered, with shocking sincerity, "Magazines sell escapist entertainment. People don't want to read about nuclear holocausts. Blacks represent images of violence, crime, and reality."[85]

Despite the more muted response to his flight, Guy, much to his surprise as a self-proclaimed introvert, enjoyed his PR tour. He joined Bob Crippen to do Bob Hope's NASA special in Hollywood—the same one Sally refused to attend. He received an NAACP Image Award. He met Billy Dee Williams, who was in the middle of filming *Star Wars* and seemed "as excited to see me as I was to meet him."[86] He hobnobbed with celebrities like Johnny Carson, Redd Foxx, and Miss America herself, Vanessa Williams.

Guy appreciated this chance to thank his mentors and the American people, whom he credited with giving him the opportunity to fly in space. In every speaking engagement, he made a point to thank his alma mater, Penn State; the city of Philadelphia; the US Air Force; and the Tuskegee Airmen.[87]

Guy may have once been considered the "least likely to succeed," but now he was an American hero. For a few days in November, he returned home to Philadelphia. He rode in the Thanksgiving Day Parade, met the mayor of the city and the governor of Pennsylvania, and spoke at several schools, including his former high school, Overbrook Senior High, where a counselor had once suggested he was not "college material."[88]

During his stop in DC, Guy received his Air Force astronaut wings and dined at Blair House with Secretary of Defense Frank Carlucci, General Colin Powell, and other high-ranking military officials. "When are you coming back to the Air Force?" Carlucci asked. Guy remembered how, only five years ago, the Air Force had been happy to keep him in flight instructor purgatory. Now they were begging for his return.

Vindication must have felt sweet, but Guy never gloated. He said he "felt very privileged to have been a role model for many youngsters, including African American kids, who aspired to be scientists, engineers, and astronauts in this country."[89] He may not have asked to be first, but he wore the mantle well.

Years later, Guy would meet Ed Dwight Jr. on a NASA-commissioned art project. Dwight spent four hours recounting his experiences to Guy and Fred Gregory. Guy thanked him when he finished.

"I couldn't have done it," Guy said, shaking his head. "I couldn't have done it, Dwight."

"You guys [are] already here," Dwight pointed out. They had *already* done it.

"Yeah," Guy replied. "Because of you."[90]

Guy was well into his press tour when engineers at the Cape disassembled the motors in the solid rocket boosters and discovered how much resin had burned. The carbon-phenolic rings located in the forward section of the nozzle of the left-hand booster were so badly eroded, the crew came within eight seconds of losing their lives.[91]

The incident could not have come at a worse time for contractor Morton Thiokol, which was defending its decade-long monopoly in producing the booster rockets. On their heels was a younger, cheaper, Utah-based company called Hercules Inc., which had the support of Senator Jake Garn from Utah, the chairman of the Senate subcommittee that set NASA's budget. Hercules had financially supported the senator's political bids and hailed from his hometown of Salt Lake City.[92] Publicly, Thiokol tried to downplay the incident, but behind closed doors, managers were frantic. They dug into their supply chain and discovered the left-hand nozzle had been fabricated by a different supplier than the right. They traced the cause back to one batch of faulty resin that had slipped through quality control. The supplier confessed that the same batch had been used to line the nozzle of the right rocket booster attached to *Columbia*, which sat waiting upright for the ninth flight of the shuttle. Stunned crews at Kennedy pulled *Columbia* off the pad and replaced its right booster with a new rocket.[93]

Morton Thiokol, having now identified the source of the problem, managed to secure its monopoly on producing the booster rockets. As for the STS-8 crew, Guy said, the nozzle issue "wasn't a high priority" for them. "They were already concentrating on the next mission."[94]

CHAPTER 11

WE DELIVER

Low Earth Orbit. February 3, 1984.

As the payload bay doors opened, Ron watched as Westar 6 spun up on its turntable, like a metallic ballerina emerging from a music box. The rotation would give the satellite enough angular momentum to keep its trajectory stable when he released it from the spring-loaded platform. Westar 6 was a communications satellite for Western Union.[1] The payload was the sixth communications satellite of its kind, but the very first to be launched aboard the shuttle. Ron's mission was part of a broader strategy proving the shuttle's superiority over expendable rockets as a launch vehicle for commercial customers, whose satellites would receive a boost to space, plus astronauts specially trained to deploy their payloads and troubleshoot problems.[2]

NASA had begun lining up customers even before the shuttle's maiden voyage. Big-name clients like Dow Chemical, Johnson & Johnson, General Electric, and AT&T booked flights years in advance.[3] The shuttle's promise as a space delivery truck had even inspired the New Guys' class slogan, "We Deliver."[4] Judy helped design class shirts around the slogan, drawing thirty-five cartoon astronauts working around the shuttle.[5] Now Ron needed to prove that NASA could keep its promise.

"All systems just look perfect," Ron said as he tracked Westar with a 16mm camera.[6] As the satellite drifted into a higher orbit, it looked like a silvery speck against the pitch black of space. The deployment seemed to go off without a hitch, but forty-five minutes later, the Westar's upper stage rocket failed to boost the satellite to geosynchronous orbit, twenty-two thousand miles above the shuttle's current orbit.[7]

The telemetry data showed the spacecraft was orbiting miles off course. *Had the rocket failed to ignite?*[8] Ron's heart sank. Their seventy-five-million-dollar satellite was floating uselessly in low Earth orbit.[9] *What went wrong?*

In a back room adjoining Mission Control, engineers from Hughes, makers of the Westar, frantically triaged with Western Union managers. The Westar's payload assist module, an upper stage rocket that should have fired for eighty seconds, only lit for a paltry fifteen.[10] Even though the fault lay with McDonnell Douglas, the manufacturer of the payload assist module, Ron and his crew still felt the sting of disappointment.[11]

Ron had higher hopes for his first shuttle mission. His family and friends from Lake City, South Carolina turned out en masse for STS-41B, the tenth shuttle flight. Ron's mother, Pearl, doted on Cheryl, who showed a baby bump: Reggie, now two years old, was going to be a big brother.[12] Hoot Gibson was the mission's pilot, so Rhea and her son Paul, the same age as Reggie, joined Cheryl and Reggie on the roof of the Launch Control Center. The friendship that Rhea and Ron had begun as swim buddies now extended to their families.

On day three, the crew had an opportunity to redeem the mission by deploying an identical communications satellite, the Palapa, made by Hughes for an Indonesian telecommunications company. Since Palapa would be boosted to geosynchronous orbit by an identical rocket, the crew worried the spacecraft would suffer the same fate. Despite these warnings, the telecommunications company pressed on. Painfully, the crew watched as their second satellite failed to fire. "We just didn't want to let ourselves believe that it had happened to us twice in a row," confessed mission specialist Bob Stewart as the satellite joined Westar in limbo.[13]

The NBC Evening News' graphic read "Chunks of Junk" as Connie Chung reported on the embarrassing malfunctions.[14] In Washington, NASA administrator Jim Beggs sulked in his office. Even if NASA's subcontractor McDonnell Douglas was to blame, NASA was taking the heat, and perception mattered.

In 1978, NASA advertised that a satellite deployment aboard the shuttle would cost anywhere between $5 and $6 million, but NASA managers underestimated costs and timing.[15] By 1985, the price tag for a deployment had climbed to as much as $27 million per satellite.[16] Technical debt, like the tile problem that plagued the first shuttle flight, continued to come due, resulting in delays and even cancellations of missions. When a satellite deployment failed altogether, as two had on Ron's mission, customers got antsy. Meanwhile, European companies were developing more affordable single-use expendable rockets to compete with the shuttle. Chester Lee, director of shuttle customer services, cautioned, "If we would lose all our commercial customers from the shuttle, NASA's budget would go up at least $250 million each year."[17] Such a dramatic increase would threaten Beggs's dreams for the space station.[18]

One week before Ron's flight, on January 25, 1984, President Reagan made the space station an explicit goal of the US space program during his State of the Union address. Standing before Congress and the nation, with rhetoric decidedly reminiscent of President Kennedy's moonshot directive, Reagan officially announced his administration's commitment to build the space station, declaring, "America has always been greatest when we dared to be great. We can reach for greatness again. We can follow our dreams to distant stars, living and working in space for peaceful economic and scientific gain. Tonight, I am directing NASA to develop a permanently manned space station, and to do it within a decade."[19] Reagan's decree came after months of concerted lobbying of the president from Beggs and other high-ranking space and DoD advisors—lobbying that had to overcome equally impassioned pushback from cabinet members close to Reagan.

George Keyworth, a presidential science advisor and former weapons designer, had attempted to block Reagan's support, dismissing the space station as a "motel in the sky for astronauts."[20] Keyworth believed that the entire shuttle program should be handed over to the Air Force, which he thought had superior operational expertise. White House domestic adviser Edwin Meese, counselor to President Reagan and a member of his National Security Council, simply thought the station was too expensive.[21] Eventually, Beggs found a way to work around Keyworth, Meese, and the other naysayers. He arranged a personal meeting with the president a few weeks before Sally's history-making flight, without the knowledge of Meese or Keyworth.[22] Beggs came prepared with a pitch tailored made for the former movie star. "Knowing [Reagan] was an actor," Beggs bragged, "I quoted from [Shakespeare's] Julius Caesar."[23]

> There is a tide in the affairs of men
> Which, taken at the flood, leads on to fortune . . .
> Omitted, all the voyage of their life
> Is bound in shallows and in miseries.
> On such a full sea are we now afloat,
> And we must take the current when it serves,
> Or lose our ventures.[24]

"[It was] not an apt quote," Beggs later laughed at his reference to Brutus and Cassius's conversation, "because both of the men who participated in the conversation were dead the next morning."[25] Still, Beggs's ploy worked. The president liked Beggs's vision of using the station as a waypoint to Mars, a plan resurrected from NASA's Nixon-era spaceflight dreams. Reagan's favorite books growing up had been Edgar Rice Burroughs's Martian fiction (such as *A Princess of Mars*), and Reagan brightened at the thought of opening a new planetary frontier. Adding urgency to the matter, in July 1983, Reagan's advisors briefed him on Soviet efforts to build a "continually manned space station," predicting the Soviet space station would be in orbit by 1986.

Beggs won over the Gipper. His victory, however, came at a steep cost. In end-running both Keyworth and Meese, he had earned two powerful enemies.[26]

For the time being, though, Beggs and his space station project had gained valuable political momentum. To maintain the tide of support, he now needed to demonstrate that the space shuttle was a capable workhorse, able to ferry space station components to orbit and meet the satellite launch demands of private industry. The twin failures on Ron's flight did not help prove either point. Beggs had to shore up customers' concerns and fast.

Hoping to recover from the failed satellite deployments, Ron's crew turned their attention to their final goal: flying the Manned Maneuvering Unit. Bruce McCandless had been working on the jetpack technology for years and now he would finally get to use it in space. If all went well, he would perform the world's first untethered EVA, an essential test for future construction and maintenance of the space station.[27]

On day four of the mission, McCandless stripped down and slid into adult diapers and sweat-wicking long-johns. He squeezed into a form-fitting spandex mesh jumpsuit lined with tubes, through which cold water would run to prevent him from overheating under the full glare of the sun, then floated into the pants and "upper torso" of the EVA suit, a rigid shell with bellows-like accordions at the shoulders and elbows.

As Ron watched from the flight deck with his camera, McCandless fearlessly flew out into the payload bay, maneuvering with his jetpack until he was hundreds of feet away from the shuttle. One false move and he could spin off into oblivion.[28] "It may have been one small step for Neil, but it's a heck of a big leap for me," radioed McCandless.[29] As an official photographer for the mission, Ron captured that big leap. The photo would become a defining image of the shuttle era.[30]

After McCandless returned, Ron moved on to his passion project. He broke out a small soprano sax, wet the reed with his tongue, and smoothly played Jackie DeShannon's "What the World Needs Now Is Love" and Diana Ross's "Reach Out and Touch Somebody's Hand."[31] Ron had hundreds of songs committed to memory, but he chose these two songs in particular, Cheryl told reporters, "to give a message, a universal message—a solution to malice that exists among us."[32]

Since his tenor saxophone was too big for the limited space on the shuttle, Ron borrowed this soprano sax for the flight. He practiced blowing with more force to account for the low pressure of the crew cabin. What he had not accounted for was the effect of weightlessness. Wind instruments rely on gravity to pull a player's saliva down the instrument. In microgravity, the moisture got stuck in the mouthpiece. After fifteen minutes, Ron's sax-playing sounded no different than someone blowing bubbles into a glass of milk.[33]

Alas, his two-hundred-mile-high flourish took place during a media blackout, and his recording was accidentally taped over.[34] "The project is not yet complete," Ron said. "Having a picture of a sax in space and having the world hear one being played from space are two entirely different events with different impacts."[35] Ron set his sights on a grander plan for his next space flight.

In the few quiet moments of the flight, Ron gazed out the shuttle's windows, noting the stark differences between Earth and its neighbors. "The moon ain't nothing but one big ol' rock!" he would later tell friends back on Earth. "And there's Mars over there, and Jupiter way over there—you can look at them and tell that there's no life. But when you turn to look at Earth, it's blue with a white mist around it, and any alien could look at it and tell that there's life somewhere on that rock."[36] Looking down on his blue home, he felt his spirituality intensify. Ron had a religious conviction some thought odd for a scientist. For Ron, faith and scientific query were equally valid responses to the natural wonder of the universe. His commitment to one only deepened his commitment to the other. With all of planet Earth spinning before him, protected by its thin atmospheric shroud and safe from the meteorites that collided so frequently with the moon and Mars, religion and science were in harmony. "There's a god out there somewhere," Ron thought.[37]

With its objectives cleared, *Challenger* prepared for a landing at Kennedy—the first time the shuttle would land where it launched. Kennedy's runway, unlike Rogers Dry Lake at the Edwards base, was short and unforgiving. "The runway didn't look near as big from [the air] . . . as it had looked from on the ground," said mission pilot Hoot Gibson.[38] With its classic double sonic booms, *Challenger* signaled its arrival home.[39]

Commander Vance Brand maneuvered the orbiter like a skier, carving a slalom at two hundred miles per hour. "Flies like a brick," he joked.

Ron landed to a cheering crowd of four thousand.[40] "There are those who said after the first shuttle or two, no one would care," Ron told the spectators. "Well, you have proved them wrong again."[41] He went on to tell the press assembled on the tarmac, "I feel like this job and I—we're sort of made for each other."[42] He could not wait to fly again. He wanted a do-over.

His family and community bore none of the disappointment Ron had internalized. Lake City exploded with pride, throwing Ron a massive parade, and renaming its main thoroughfare Ronald E. McNair Boulevard. He visited his alma mater, MIT, and stopped by the Massachusetts legislature to lobby for better resources for inner-city schools. "Black minds and talent have skills to control a spacecraft or scalpel," he impressed upon the state senators, "With the same finesse and dexterity with which they control a basketball."[43]

Honorary doctorates were bestowed upon him, and he fielded countless speech requests. With each one, Ron had an ask of his own: a chance to speak at a local school. There, he echoed the message he had first heard from his guidance counselor Ruth Gore at North Carolina A&T years before. Ron reassured the schoolkids that they were each "more than good enough." He told the students about an eagle who had been raised by chickens and believed it was a chicken, too. "Then, one day, the eagle saw a flock of eagles fly by. The eagle ran across the barnyard, flapped his wings, and left the chickens on the ground, soaring over trees and mountaintops. Black students, minority students," he cried. "You're not chickens! You're eagles! Stretch your wings!"[44]

NASA could have used a dose of Ron's inspirational zeal. As Ron toured the country, the press continued to examine the twin failures of Westar and Palapa, keeping NASA's embarrassment fresh in newsprint and under scrutinizing public glare.

Beggs hoped that presenting a new vision for the future of spaceflight—the space station—would help galvanize support for the agency. That March, he embarked on an international tour to pitch the station to America's allies. The French, Italians, Germans, Japanese, and

Canadians gave their enthusiastic support for his wildly ambitious proposal, which included building a hangar on the space station in order to construct spaceships that could travel farther to Mars, a garage for repairing satellites, and four orbiting laboratories.[45] Only Prime Minister Margaret Thatcher and the English declined Beggs's offer.

Armed with international backing, Beggs now made his appeal to Congress.[46] The House Appropriations Committee agreed to help pass a bill allocating $150 million for the space station. The House bill would pass on July 11.[47] However, there was a slight hiccup at the hearing before the Senate appropriations subcommittee. Utah Republican Senator Jake Garn, the subcommittee chairman, spent most of the time expressing his desire to fly aboard the shuttle, touting his experience as a Navy pilot. *What do I have to do to get to space?* Garn asked.[48] The unspoken answer, of course, was to vote for the space station. Garn agreed and got his ticket—Beggs authorized the senator to fly on a science mission in 1985.[49]

To support these promises and earn back the trust of commercial customers, NASA needed to demonstrate its shuttle was a reliable spacefaring vehicle. Its next flight had to be a winner.

CHAPTER 12

YELLOW DEATH

Kennedy Space Center. June 26, 1984.

Judy Resnik and the crew of STS 41-D lay strapped on their backs aboard *Discovery*, NASA's fresh-off-the-assembly-line shuttle. The cabin still had that "new car" smell, and the fuselage, unmarred by fiery ascents or reentries, gleamed white on the launchpad. Judy and *Discovery* were making their first flight to space together.

On the rooftop of the Launch Control Center, Judy's aunts, uncles, cousins, father Marvin, and even her estranged mother, Sarah, waited for the chance to see their studious, intense little girl fly into space. Judy fled her conflict-filled childhood only to end up on top of over two million pounds of highly explosive propellant—and was far happier for it.[1]

The countdown hit T minus six seconds, and the main engines ignited.[2] Flame and exhaust burst out from the tail of the orbiter and plumed out over the pad on cue. Spectators counted down with the launch announcer, watching the seconds pass on the giant clock towering above the stands. Suddenly, the clock froze. A terrifying metallic moan echoed across the Cape as the shuttle stack rocked against its bolts.[3]

A chill went through Marvin Resnik's spine.

The crowd's view of the launchpad was obscured by the morning haze, but the bright flash of flame and loud roar, all without a liftoff, gave the impression that the shuttle had exploded. Silence fell over the crowd. Mike Mullane's wife, Donna, burst into tears.[4] Judy's mother, Sarah, looked on in wordless shock.[5]

In the windowless middeck, adrenaline pumped through Judy's body. With four seconds to go before liftoff, an onboard computer had automatically aborted the launch, but no one knew why.[6] The engines needed to be shut down immediately. If they kept firing while the shuttle remained on the pad, *Discovery* and its crew would be engulfed in flames.

"We've had engine shutdown," said launch control.

"I thought we'd be higher at engine cutoff," Steve Hawley said, trying to lighten the mood in the cabin. No one laughed.

Judy saw two of the three engine shutdown lights glow. *But what about the third engine? Surely the third engine isn't still running?* Pilot Mike Coats poked the dark shutdown button, but the light would not illuminate.[7]

"Break break, break break, GLS [Ground Launch Sequencer] shows engine one *not* shut down," an engineer said over the comm.[8] Commands came quick and muffled, but Judy heard one word loud and clear: *Fire.*

One hundred forty feet below, fire suppression systems kicked in, blasting the shuttle stack with jets of water, though no one could see any fire.[9] Judy climbed over to the side hatch window and peered out. The fire suppression systems were spraying water over the arm, but she did not see any smoke or flames.

Judy gamed escape scenarios with her commander, Henry "Hank" Hartsfield. *We're lucky that the solid boosters haven't lit yet. What if they do? What if the liquid fuel in the external tank explodes? Should we bail out?*

"Henry, do you want me to open the hatch?" Judy asked.[10] Outside the hatch was a sidewire escape system with baskets to zipline the astronauts twelve hundred feet down to a bunker, but the system was so new, it had not been tested yet. The crew could be injured or killed if it malfunctioned.[11]

Hartsfield gave the order: *Keep the hatch shut.* As he and Judy conferred, one of the three main engines was leaking hydrogen, which burns invisibly and floats. The flames crept up the left side of the orbiter unseen. If the crew had exited, they would have burned alive.

Thirty-eight long minutes later, the fire was extinguished. Shaken, Judy made her way down the elevator and through sheets of pouring water. Hair matted and flight suit soaked, she climbed into the Astrovan and shivered.[12] At least she was alive.

A pad abort was bad news for NASA, but a main engine fire on a brand spanking new orbiter? Catastrophe. Big customers signaled their flagging confidence in the shuttle. The Air Force announced it would add two expendable rocket launches to its annual manifest, paring back its original promise to use the shuttle exclusively.[13] AT&T and Satellite Business Systems pulled their shuttle flights, banking instead on new French expendables.[14]

To make up for lost time, NASA managers placed a record-setting three satellites—twenty-four tons of cargo—on Judy's rescheduled flight, while Kennedy engineers raced to replace *Discovery*'s faulty main engine and update its software.[15] "The future of the floundering US space transportation system may depend on the success of *Discovery*'s seven days in space," a *Miami Herald* reporter wrote.[16]

On August 30, 1984, Judy was back on the pad. This time, liftoff was a go.

Judy's signature curls offered her first indication that they had entered zero-gravity: She could feel her hair levitating. Fellow mission specialist Mike Mullane compared the floating tendrils to "Medusa's snakes."[17] As Judy watched Earth shrink beneath her, she noticed a few stray nuts, a couple bolts, and a roll of duct tape floating by, evidence of the techs' rush job to get *Discovery* off the pad. When Hartsfield opened the payload bay doors, a Coke can drifted past the high-tech satellites into space.[18]

Within the first few hours, Judy and Mullane had their first big success, deploying a communications satellite for a consortium of businesses.[19,20,21] *Discovery* shuddered as the heavy satellite and its payload assist module (together weighing more than ten thousand pounds) emerged from the payload bay.[22] Unlike on Ron's flight, the McDonnell Douglas module fired correctly and boosted the spinning SBS-4 to its designated orbit.[23] *Check one for* Discovery*!*

On day two, the team prepared for the release of Syncom, their second commercial satellite, while Commander Hartsfield maneuvered a multimillion dollar, seventy-five-pound IMAX camera into position to film the event. Based on the suggestion of Michael Collins himself, astronauts were compiling footage for *The Dream Is Alive,* an upcoming documentary about the shuttle, to be narrated by Walter Cronkite.[24] As Hank rolled film, the unthinkable happened: The belt drive of the unwieldy camera snagged a lock of Judy's hair and began drawing her in.[25]

Shit! Judy's scream rang out through the shuttle, as the camera's motor pulled so hard it felt like her hair would rip right out of her scalp. Finally, the motor jammed, and a fuse blew. Dangling uncomfortably from the jaws of the IMAX, Judy could practically hear her mom nagging in the background. *Maybe if you'd straightened your hair like I taught you . . .*

Unable to untangle her, Mullane began to cut Judy free, snipping off chunks of her mane, which then floated into the crew's eyes and mouths.[26] *What's taking so long; what are you all doing up there?* Mission Control wanted to know. Hartsfield was about to radio an answer when Judy stopped her commander. "If you so much as breathe a word to [Mission Control] about my hair jamming the camera, I'll cut your heart out with a spoon."[27]

As the second American woman in space, Judy did not want to make it harder for women astronauts to be taken seriously. The incident would have been a field day for the press, which was already obsessed with her appearance. One journalist called her "frail" and pointed out how Judy, unlike Sally, wore makeup and had painted nails. Another commented that she was "wearing short pants" on orbit.[28] On launching, the Associated Press noted her "Brillo pad look" and a slew of other journalists pointed out the halo effect of her curls standing up and out in weightlessness. The same articles mentioned that Judy was the first Jewish person in space.[29] *Imagine what the press would say if they found out that America's first Jewish woman astronaut had got her "Brillo pad" hair caught in the IMAX?* Hartsfield told Mission Control about the IMAX jamming, but he and the rest of the crew dutifully kept their mouths shut about what had caused it.[30]

The next day Judy and the crew launched their third communications satellite for AT&T without incident. Maneuvering the camera on the robotic arm to get a look, Judy watched in wonder as the satellite fired into its correct orbit, with the Orion constellation serving as a glittering backdrop.[31] At the end of the third day, a reporter asked flight director Randy Stone how the mission was going on a scale of one to ten. "Twelve," he shot back.[32]

A few days into the mission, flight controllers in Houston picked up low temperature readings on the orbiter's waste spouts. The two external valves that dumped excess water and urine had not disposed of their waste at all, but instead built giant icicles that clung to the side of *Discovery.* To make the shuttle lighter, engineers had replaced the tile that surrounded the dump valves of older orbiters with insulating blankets on *Discovery.*[33] The blankets were not up to the job.[34]

Using the camera at the end of the robotic arm, the crew estimated that the "peecicle" was about a foot long, weighing between eight and twenty pounds—certainly large enough to cause serious damage if it dislodged and struck the orbiter on reentry.[35]

The toilet was promptly decommissioned. The crew would instead whiz into the old Apollo bags, buried deep in a locker in the middeck storage area. Urine bounced right back up as soon as it hit the bottom of the Apollo bags, scattering droplets throughout the cabin. The men found they could minimize splashing by putting absorbent socks in the bottom of the bags. Socks became the "coin of the realm," said Steve Hawley. "If you had socks, you were king."[36] Mullane became a highway robber who attacked his crewmates for fresh laundry.

"Help, I'm being socked!" Judy screamed as Mullane grabbed her and yanked the socks right off her feet.[37]

Anxious to find a solution quickly, Judy worked with *Discovery*'s flight control team: *Could they knock off the peecicle with the robotic arm?* In Houston, as the crew slept, arm expert Sally Ride tested out such a maneuver on a simulator. If the arm misfired in space, it could damage the orbiter's heat tiles, or worse, rip a hole in its aluminum skin. She worked out precise joint movements to shimmy the arm around the side of the orbiter and up next to the peecicle without touching anything else.

The next day, commander Hartsfield oriented the peecicle into the sun, hoping the solar warmth would melt the ice. He then fired the thrusters to shake the ice loose.

"No joy," Judy reported when she peered out at the recalcitrant stalactite.[38] Time to try the arm. As the commander, Hartsfield insisted on operating the arm himself. If anything went wrong, he wanted to be the one to blame. A barefoot Judy floated over to the commander's seat where she received instructions, codenamed "Ice Cube Ops," from Sally in Mission Control. Then Judy monitored the black-and-white camera feed as Hartsfield guided the arm along the side of the orbiter. When Judy lost sight of the arm, he pressed forward at a snail's pace before tapping the icicle. Miraculously, an ice chunk floated up past the camera's view.

"There it goes!" shouted Mullane. Judy watched the murderous shard drift into the abyss. *Good riddance.*

On the morning of the fourth day, 160 nautical miles above Earth, Judy put on a pair of aviators and looked through *Discovery*'s aft windows down at the Pacific Ocean, dotted by cirrus clouds. She was about to deploy one of the largest structures ever put in space: a thirteen-foot-wide, ten-story-high solar array.

With her checklist Velcro'd to the control panel, Judy issued a command to pull a coiled fishing line taut and stretch the solar array to its full height. The line acted like a sailing mast, unfolding the delicate, paper-thin folds of Kapton fabric. Made up of thousands of gold solar cells that folded into a four-inch-tall stack, the array expanded like a giant accordion. As the mast climbed ever higher, Judy had to crane her neck to see the top of the golden panels through the overhead windows. Like a translucent, golden skyscraper, the solar array loomed above them.[39]

If successful, the array's ability to convert solar energy into electricity would be used to power future spacecraft, like the space station. This lone array could generate enough energy to power an entire shuttle mission while on orbit.[40]

"It's up and it's *big*," Judy chuckled at her own ribald commentary.[41]

"Up and big, copy that," CapCom replied, stifling a laugh.[42]

On the way home, the urine smell lingered, so much so that it eclipsed *Discovery*'s new car smell. The odor would cling to the cabin for months after landing, earning it the nickname "Yellow Death."[43] When the crew returned to Edwards Air Force Base, the receiving team conspicuously wore coordinated yellow socks, and in Houston, Henry Hartsfield arrived home to a handful of white porcelain toilets in his front yard.[44,45]

Despite the insider ribbing, dozens of newspapers across the country proclaimed the mission a resounding success, and the shuttle's reputation was restored.[46] They had deployed three satellites and the solar array, a signpost that might one day power space stations and sustain long trips to deep space.[47]

A few weeks after 41D returned, President Reagan informed Congress that he was "pleased" with the progress of the shuttle program.[48] Of the twenty-seven satellites launched the previous year, 1983, Reagan boasted that five had been launched aboard the shuttle, and two of those with the help of "first American woman astronaut in space" Sally Ride. The nation could be proud of its twenty-five years of space exploration achievements, the president declared. In another month, the country would celebrate not one, but two new space firsts: the first time two women traveled together to space and the first spacewalk by an American woman.

CHAPTER 13

MUCH HAVE I TRAVELED

Kennedy Space Center. October 5, 1984.

Two days after Kathy's thirty-third birthday, she was poised to make history as the first American woman to perform a spacewalk. First, though, would be "the walkout," a Kennedy Space Center launch tradition that had remained almost unchanged since Apollo 7. The suited crew would emerge from a pair of double doors at the Operations and Checkout Building, give a quick wave goodbye to the press, and jump into the Astrovan, which would whisk them away to the launchpad.

When George indicated it was "go time," Kathy felt a distinct shove from behind. Not from George but from Sally, pushing Kathy up to walk by Crip's side. *The showcase spot.* Was Sally really giving up the limelight? Sally nudged her again. It was a moment of generosity from her greatest rival. *It's your turn,* Sally said.

Kathy stepped up next to Crip and walked out to the

flashing cameras. She proudly swung her arms and gave the slightest of smiles back at Sally. Here were two women, traveling to space together, for the first time in history.[1]

Early on, Kathy suspected that Sally was George's favorite and therefore her main competition. On paper, Kathy and Sally were both athletic, smart, twenty-six-year-old PhDs from southern California. They had even attended the same grade school. In their approaches to life, though, they took a different tack. Where Kathy wore backpacking slacks, Sally wore designer jeans. Where Kathy was rugged and strong, Sally was lithe and compact. While Kathy followed all the rules, Sally was not afraid to break them.[2] Sally seemed to understand the politics of the job. She went to the mixers, drank beer with the guys, and understood and respected the power structure. Whereas Kathy was, in her own words, clueless. "I'd never been in that large of a bureaucracy before." Kathy said. "I was a real babe in the woods, a naive kid . . . I didn't know how to get traction."[3] Sally's native understanding of office politics paid off. Kathy watched Sally receive all the best technical assignments, culminating in her "first American woman" flight.

A couple of months after Sally's historic launch, Kathy was backpacking in the Rocky Mountains when she got an unexpected call from Sally. *Just so you know,* Sally said, *George is going to call you and assign you to a flight. It's going to include a spacewalk.*[4] Kathy rejoiced.[5] This particular spacewalk would be key in demonstrating that satellites could be refueled and serviced in space, extending their lives and expanding the shuttle's commercial capabilities. Hearing the news from Sally, however, not George, stung. *Of course, Sally would be the first to know.* Sally would also be Kathy's crewmate, flying again before Rhea, Anna, or Shannon even got their first chance. It was a cruel, cosmic irony that Kathy was sharing her first flight with her archrival.[6]

To highlight the pecking order, commander Crip named Sally interim leader while he trained for another mission, making her Kathy's boss.[7] "I wrestled a lot with second-class citizen feelings," Kathy said. "The media didn't help any. The only interest the media had in me was that I was not Sally but might know where she is."[8] Kathy coped with humor. For press events leading up to the flight, she wore a nametag that read SALLY with a line drawn above it, a mathematical notation that meant "not Sally."[9] Sally was not amused by the prank, "which made it even better," Kathy chuckled.[10]

Kathy shuffled back and forth in the White Room outside *Challenger*'s flight deck, as she and Sally waited to board their flight. "What do you think the news anchors are saying about us right now?" Kathy asked.

Maybe we should give them something to talk about? Sally wondered.

The two pretended to synchronize their watches.[11] *That's what people do in the movies before a mission, right?* Luckily, there were not any microphones in the White Room to catch them muttering, "Do you think we've stretched it out long enough?"[12] Moments later, Sally and Kathy climbed into *Challenger,* which launched without a hitch.

On orbit, one of Kathy's first tasks was to open the leaflike antenna of the shuttle's imaging radar.[13] The thirty-five-foot-long by seven-foot-wide antenna would send millions of microwave radar pulses down to Earth and receive the pulses that the surface reflected. Since different terrain reflects radar in its own unique way, imaging experts could use the reflected radar data to piece together remarkably detailed maps.[14] First Kathy would have to assemble the giant device, which sat in a stack of three pieces in the payload bay.[15]

From the flight deck, Kathy released the latches holding the three pieces of antenna together. She sent a command to unfold the first leaf. Moments later, the whole stack of leaves began swinging wildly. *That never happened in the simulation.* She felt blood rushing to her head, flushing her cheeks with the red-hot stain of failure. She typed away wildly at the shuttle's cockpit display, commanding the next leaf of the antenna to unfold. Brilliantly, that stopped the seesawing.[16]

Before Kathy could relax, another antenna issue cropped up. Every shuttle was outfitted with a small satellite dish mounted on the starboard side of the cargo bay. About the size of an average television satellite dish, the antenna bounced information from orbit to a geostationary satellite that then transmitted that data to the ground. Once on orbit, the payload bay doors opened and released the antenna for use. However, the dish needed to be stored for launch and reentry.

The antenna dish could pivot and tilt to orient itself to continually face the geostationary relay satellite, like a flower growing toward the sun.[17] Thirteen hours into 41G's first day in flight, a circuit failure sent the antenna dish into a spin.[18] Mission Control cut power to the affected circuit, but now Kathy had to realign the dish with the antenna's arm and manually stow it for reentry while she was out on her spacewalk. Her EVA, originally scheduled for their fifth day on orbit, was pushed to the last full day in orbit of the mission.[19] NASA generally banned spacewalks on the day before reentry because of their complexity and risk, but here they had no choice. The errant receiver needed to be stowed before landing.[20]

Low Earth Orbit. October 11, 1984.

Kathy listened to the hum of her spacesuit's oxygen pumps filling her ears. Despite the importance of their mission, she worried that she would not get her chance to spacewalk, her shot at being "first."[21] "The EVA experts in mission control followed every step as Dave [Leestma] and I worked through our preparation, their scrutiny becoming harsher and harsher as we approached the critical moment," Kathy said. "Any glitch they saw or concern that arose in mission control could bring everything to a screeching halt, and possibly even cancel the EVA altogether . . . I was itching to get outside."[22]

Kathy's EVA would demonstrate to NASA's commercial and military clients that satellites could be maintained on orbit and refueled successfully, even with the danger of using the "highly explosive and extremely toxic" hydrazine fuel the satellites required.[23]

On the seventh day of STS-41G's mission, the staticky radio came to life: "*Challenger,* Houston: You are GO for EVA."[24] They were the sweetest words Kathy had ever heard. Kathy had been preparing for this mission for the past five years.[25] She had spent all of flight day six "prebreathing" air at a lower atmospheric pressure to remove the nitrogen from her system. Otherwise, in the low pressure of space, nitrogen would automatically be released by her bloodstream, escaping as bubbles and giving her a nasty case of the bends.[26]

The hatch opened. Leestma exited the airlock first, but Kathy was close behind. She sailed out into the payload bay, claiming her slice of history as the first American woman to walk in space. She hung on to a handrail as Earth beneath her flew by at 17,500 miles per hour. Commander Crip's voice came in over her headset, interrupting her rush. "Look around, guys," Crip reminded them. "Take a moment to look at the Earth and appreciate where you are."[27]

Kathy obeyed, turning her back on the orbiter so she could gaze down at her home planet. "A burst of clashing thoughts flashed through my mind. I was in this absolutely extraordinary place, keenly aware of how deadly the environment outside my suit was, yet being there seemed perfectly natural," Kathy said. "I felt utterly comfortable."[28]

After a minute of Earth-gazing, Kathy turned her attention to her work, securing her safety tethers and grabbing her tools from the cargo bay. Kathy and Leestma used the official EVA handrails to make their way to the refueling experiment at the port side of the cargo bay. Kathy attached herself to the fabric fingerholds nearby so she could monitor Leestma, as he used their specialized kit to install a valve and transfer the hydrazine from one tank to another.

Kathy and Leestma had designed their equipment to allow them to refuel the satellites, without touching the hydrazine itself. Kathy compared the task to filling "the gas tank on your car without ever touching the gas cap," all while being dressed like the Michelin Man.[29] It was tedious work.

"Is it lunchtime yet?" Leestma grumbled.

"I bet they ate our lunch," Kathy joked, referencing their crewmates safe in the crew cabin.

"You'd have loved it," Sally said devilishly. In the end, the transfer rig operated "flawlessly," and "Kathy and Dave [Leestma] made this task look easy," said Crip.[30,31]

Next, Kathy had to travel across the cargo bay, from the port to the starboard side, to store the errant antenna. The orbiter's EVA handrails, which she had used to journey over to the refueling experiment, would not allow her to reach the antenna dish. Instead, she would have to make her way to the middle of the bay, cross there, and inch along the starboard sill to the dish. That also meant clambering over a large pallet containing earth science experiments while making sure not to damage anything.[32] Only her safety tethers would keep her from drifting into the endless void of deep space.[33]

Kathy's unconventional route across the cargo bay would give her crewmates an excellent opportunity to use the IMAX to film an unforgettable scene for the upcoming *The Dream Is Alive* documentary. Pilot Jon McBride had the IMAX ready for the shot. Kathy used her hands to creep her way across while McBride filmed the scene.[34] "As soon as I lowered my gaze from my hands, I felt like I was hanging from a tree limb and looking down at the ground," Kathy said.[35] Only the ground was 140 miles beneath her. From the maps she loved as a child, she recognized where the northern tip of South America met with the Caribbean Sea. She watched, awestruck, as Venezuela's Maracaibo peninsula glided beneath her boots.[36]

As Kathy locked away the troublesome antenna, she took her last look at space from her extraordinary vantage point. She thought of her mother, who had passed away two years after Kathy had been accepted to the program. *Was her spirit out here among the stars?* Kathy had called her mother after NASA accepted her. In place of congratulations, Barbara joked that Kathy should pursue a career on solid ground. Her mother called back the next day to apologize. *It would have meant the world to me to have my mother's support if I had been given such an opportunity,* Barbara told her. *You have mine.* In a mother-daughter relationship that had been filled with pain, this was a rare healing moment.[37] Barbara lived to see Kathy become an astronaut, but never saw her daughter in space. Kathy knew, if her mother could see her now, she would approve. Kathy had kept her promise: *She never stopped short of her full potential.*

CHAPTER 14

SEND ME IN, COACH

Johnson Space Center. July 1983.

I'm thinking of sending Anna," George said, eyeing Bill. "How do you feel about that?"[1] Anna, who sat right next to Bill, wondered why George was asking her husband the question and not her. When George invited both Anna and Bill into his office that morning, Anna had not expected a crew assignment. George never met with couples, and she did not need Bill's blessing to join a mission. Yet George seemed to be asking for Bill's permission. Anna cradled her eight-months-pregnant belly. Perhaps that was why George felt the need to include her husband.

Bill had fulfilled his lifelong dream of becoming an astronaut when he won a spot in the Astronaut Class of 1980. Twice rejected from NASA, both as a boy and as a potential TFNG, Bill had tenaciously honed his resume after moving to Houston with Anna, throwing himself into biomedical engineering classes while earning his private pilot's license and continuing his emergency medical practice.[2] Now, reporters heralded NASA's first "Mr. and Mrs. Astronaut," and the nation fell in love with the picture-perfect couple.

With both Fishers at NASA, in theory, they would spend more time together. In reality the couple found their marriage sidelined in favor of work.

"We don't have a private life like most other people," Bill admitted. "Sometimes days or weeks go by before we see each other."[3] It was not ideal, but Anna and Bill knew their relationship rested on a solid foundation of shared trauma. "If a couple can survive medical school together," Bill joked, "they can survive anything."[4]

Now George seemed to be asking if they could survive as a family if Anna were to take a flight, leaving their infant in Bill's hands. If Anna accepted the flight assignment, she would be the fourth American woman to space, and the first mother. *Maybe the world wasn't ready to send the mother of an infant to space? Maybe George wasn't?* But Anna was ready. Before Bill had a chance to answer George's question, Anna interrupted the conversation between the two men.

"Send me in, coach," she said with an irrepressible grin.[5]

Two weeks later, Anna was training with her crew: TFNGs Rick Hauck, Dale Gardner, and Dave Walker, plus Joe Allen, one of the scientist astronauts recruited in the Apollo days. Their job was straightforward: deploy two spacecraft. Or so they thought.

While wrapping up a mind-numbing day of reviewing their mission's cargo, Anna felt a dull cramp in her lower abdomen.[6] *This is it,* she thought. *I'm going into labor.*[7]

The very next day, a Friday, Anna gave birth to daughter Kristin. Anna beamed at her little bundle and thought of how lucky she was. She was on the precipice of her life-long dream of becoming an astronaut and she had created a family. By Sunday, though, she was dead tired. In 1978, Congress passed the Pregnancy Discrimination Act, which forbade employers from showing bias against women who were pregnant, but there was no federal law that mandated parental leave. Some states enacted their own protections, but Texas was not one of them.[8] Not wanting to give up her career or her chance to fly, Anna garnered all her energy and enthusiasm and breezed into the Monday morning all-hands meeting, donut pillow in hand. "It was worth it just to see the looks on their faces," she said with a grin.[9]

Over the next year, Anna juggled work, simultaneously serving as CapCom for STS-9 and training for her assigned flight, plus the immense responsibilities of new motherhood.[10] As CapCom, Anna developed a strategy: As soon as the shuttle lost comms with the ground, she ran to the ladies' room to pump milk in one of the stalls.[11] She called Kristin—who was often with her mother, Elfriede, or nanny, Susie—to hear the baby coo. She did not sleep much. When Hauck questioned her demanding schedule and suggested she rethink her CapCom assignment, Anna insisted that the experience in Mission Control would make her a better crew member. He relented.[12]

What she had not expected was the skepticism and ire of the public. "Anna Fisher is a good astronaut, a good doctor, and a good citizen. Is she a good mom?" *Boston Globe* reporter Aimee Lee Ball asked before her flight.[13] During one press conference, another reporter asked Anna, "Do you believe that the astronaut job is compatible with a mother's affection?"[14] After her flight announcement, she received harsh letters, insisting that she was a bad mom. Bill angrily tore them up. "Don't read these things," he responded protectively. "No one says, 'Omigod, how could you leave your family?' to a male astronaut."[15]

In the public eye, Anna spoke with authority as a doctor, a mother, and an astronaut. "Everyone puts so much stock in the mother being more important to the early development of the child, but I think both parents are equally important," Anna explained to the press. "I made my commitment to NASA before I decided to have a family, and I can't back out of that commitment."[16]

Privately, Anna harbored her own fears. She had witnessed her share of near misses and understood the risks of going to space aboard the shuttle better than most.[17] Right after landing her crew assignment, Anna bought a video camera. If she did not return from her mission, she wanted

Kristin to have something by which to remember her. She recorded dozens of home videos. "Mommy is going to go on the space shuttle, this is my last night home, and I love you more than anything in the world," she said softly to the baby while Bill filmed.[18] Her entire life, Anna had searched for a home. She had found it in Bill and Kristin. *Was she about to throw it all away for a chance to ride a rocket?*

At the same time, her once-straightforward mission grew more complex. After Ron's flight left two satellites stranded in a useless orbit, Lloyd's of London, an insurance market, was forced to pay out $75 million to Indonesia and $105 million to the Western Union.[19] *Today*'s Jane Pauley grilled Anna in front of millions of viewers. "Why did NASA deploy the second satellite when the first failed?" Pauley asked. "Do you think NASA will go back and get those satellites?"

"No one's ever done anything like that," Anna explained. "Those satellites were not designed to be retrieved."[20] It was true. The satellites had no handrails or hooks to grapple. Anna had no idea that behind the scenes, NASA administrators were nevertheless already dreaming up ways astronauts might recover the wayward spacecraft—or that *her* mission would attempt the rescue.[21] NASA was in the midst of negotiating a deal with Lloyd's syndicates in which the insurer would pay $5.5 million to NASA to attempt the salvage mission and $5 million to Hughes to create strategies to maneuver the satellites into an orbit where they could rendezvous with the shuttle and build the hardware to capture them.[22]

When Anna and the crew learned about the rescue, they threw themselves into overdrive.[23] A salvage and retrieval mission like this had never been attempted before. That said, NASA had already begun to lay the groundwork for such a procedure. On STS-7, Sally used the robotic arm to deploy and then retrieve the first shuttle pallet satellite (SPAS-1), which carried several small experiments. On STS-41C, Commander Bob Crippen and his crew performed the first on-orbit satellite repair when they captured and triaged Solar Max, a $250-million astronomy satellite that photographed solar phenomena (sunspots and solar flares) to study their effect on Earth's weather. The repair had been harrowing. First, the crew maneuvered *Challenger* to within two hundred feet of the Solar Max. Then TFNG Pinky Nelson jetted over to the satellite in the MMU and attempted to wrangle the satellite back into the shuttle's payload bay. Pinky found the task impossible and returned to the shuttle empty handed. Eventually, the crew was able to grapple the SolarMax with the robotic arm, bringing it into the payload bay where Pinky and fellow TFNG James "Ox" van Hoften completed the repair.

Anna's crew, along with mission planners and the satellite's hardware engineers, carefully studied the ways that STS-7 and STS-41C used the robotic arm and the MMU as they prepared for their own flight.[24,25] Not only would the mission be an incredibly complicated procedure with many variables, but due to the shuttle's tight flight schedule, their mission would not get any extra time to prep. In the press, NASA was optimistic about the mission's chances at success, but at a meeting with an associate administrator and the new head of NASA's public affairs office, Commander Rick Hauck summed up the crew's feelings: "If we successfully capture one satellite, it will be remarkable, and if we get both satellites, it will be a fucking miracle." Then Hauck added: "You can quote me on that."[26]

With Bill busy at work, Anna regularly rushed home to scarf down dinner before driving back to Johnson in the evenings with baby Kristin in tow. As the mission specialist assigned to the robotic arm, Anna crammed in extra time on the center's simulator to develop procedures for retrieving the satellites while little Kristin sat quietly in a blue carrier on the simulator floor. Every so often, she would let out a gleeful babble as she took in her futuristic surroundings and watched her astronaut mother focus on her controls.[27]

Kennedy Space Center. November 8, 1984.

On the morning of her launch, Anna woke at 4:30 AM, put on khakis and a navy polo, and slid a comb through her feathered perm. She fastened a gold necklace Bill had given her, in the custom of astronaut spouses. Swinging from the chain was a shining miniature Earth, etched with two stars: one over Houston and one in space.

At breakfast with her crew and George Abbey, Anna swallowed ScopeDex, an antinausea medication, with a sip of black coffee.[28] She could not stomach a bite of food. There was only one thing on her mind.

Outside crew quarters, Bill parked his rental car behind the press. He was supposed to be with the other crew families at the Launch Control Center, but as an astronaut himself, he had taken advantage of his access privileges at the Cape. In the back seat, his daughter Kristin sat with Anna's mother, Elfriede. Bill warned Elfriede that he was skirting the rules. *Stay in the car.*[29]

Elfriede's searching eyes, behind big gold glasses, found her daughter. She knew at that moment what Anna needed. Throwing caution to the wind, Elfriede flung the door open and thrust Kristin aloft like Simba in *The Lion King*. Dressed in a flight suit of the same blue material as Anna's, Kristin giggled as Elfriede held her arm up in a wave. *Mama!*[30] Anna spotted her strawberry blonde child in the crowd and their eyes met. Instantly, immense joy and then peace filled her. She nodded. *I can do this. I'm good to go.*[31] She boarded the Astrovan, and with one last wave to Kristin, Elfriede, and Bill, turned to face her destiny.

On the flight deck, Anna's eyes stayed glued to *Discovery*'s instrument panel as they rocketed up. Finally, eight minutes into flight, the main engines cut off and Anna looked out the window as her gold necklace floated up to graze her cheek.[32]

After deploying two communications satellites, Anna rose early on their fifth day on orbit. Looking out the flight deck windows, she recognized the familiar nighttime sprawl of Houston two hundred miles below. "Tallyho!" she radioed down to Mission Control. "We see you really clearly."[33] In Houston, Bill, out on their deck, spotted two bright objects streaking overhead: the shuttle and the Palapa. Elfriede, holding baby Kristin, shrieked. Her Anna was in that little dot of light. Grandma's scream would forever be burned into Kristin's mind as her earliest memory.

Meanwhile, Anna and the *Discovery* were now in hot pursuit of the Palapa, charging toward it at a speed that was over 230 miles per hour faster than the satellite's own. Twenty-four hours later the shuttle caught up with its target.[34] As *Discovery* made its final approach towards the Palapa, Commander Hauck, with Anna's help, fired the orbiter's rear rockets to slow them down. A crucial burn—an operation in which the crew fired small propulsion rockets to adjust the shuttle's orbit—was coming.

"I have to go to the bathroom," Hauck turned to Anna.

"Rick!" she cried. Anna did not need to remind him that NASA required two crew members to supervise every burn.

"I gotta go," he shrugged, sliding onto the middeck and out of view. "Check it twice."[35]

Alone on the flight deck, Anna fired the rockets and locked *Discovery's* orbit with the Palapa's. Out the aft windows toward the payload bay, thirty feet above the orbiter, the satellite glittered.[36] She had successfully put them into position.

Joe Allen and Dale Gardner prepped for three long hours in the airlock, completing their pre-breathe and carefully getting into their EVA suits. Now it was go-time.[37] As David Walker was putting on Allen's helmet, Allen stopped him. "I'm so hungry," Allen cried. "I really need a cookie."

"Oh, Joe, how could you?"

Allen urgently cut him off. "David, I *need* a butter cookie."

Walker muttered to himself as he searched for a butter cookie in the pantry. Returning with Allen's prize, he jammed the cookie into the spacewalker's mouth.

"Eat it," he said. "But don't choke, you little rodent."[38]

Butter cookie consumed, Allen floated into the airlock, donned the MMU, and exited through the payload bay. He was flying free in "an overstuffed rocket chair." No, it was the shuttle's "special dinghy."[39] No, it was "a silent and magical carpet." He could not quite find the perfect words for the MMU. Eventually, he landed on: "EVAs have to be the most fun ever invented."[40]

As the shuttle crossed the terminator line that divided the night from the day, Allen buzzed his way toward the Palapa. Waiting in the payload bay, Gardner grew impatient as his hands warmed up with the rising sun. "It's hot!" Gardner radioed Walker. Their cooling garments did not extend to their hands. "Our gloves are 120 degrees inside right now!" he moaned. With the shuttle seeing a sunrise and sunset every ninety minutes, Gardner and Allen would have to deal with wildly fluctuating temperatures as they did this precise, laborious work.

As Allen was about to dock with the Palapa, the sun blinded him.

Hauck acted quickly, shifting the orbiter to shade Allen and the six-foot stinger he was about to insert into the Palapa to capture the satellite. Allen looked like "a space-age medieval knight entering a jousting contest."[41] To perform the procedure, Hughes had slowed the spacecraft's spin from fifty rotations per minute to two, the slowest rotation possible without destabilizing the spacecraft. Still, the satellite spiraled briskly. After grappling the Palapa, Allen began to rotate, too. Using the MMU thrusters to stabilize the satellite, he drove the Palapa back to the spacecraft.

Anna snagged the astronaut-satellite pair with the robotic arm and drew them back into the payload bay, where Gardner waited to secure the satellite. However, Gardner encountered an unexpected piece of hardware that was not in the contractor's blueprints. A waveguide, a small tube of metal used to direct radio signals, prevented him from clamping on. *Seriously?*

Onboard the shuttle, Walker scanned his list and read the backup plan should the clamp fail: "Improvise."[42] *Well,* thought Walker, *this is our chance to prove the value of human improvisation in space.* Walker suggested Allen hold the Palapa in place while Gardner secured it to the payload bay *with his hands.* It had never been tried before. Mission Control gave the go-ahead.

Allen disconnected himself from the MMU while Anna used the arm to hold the satellite via the grapple fixture on the stinger. After Allen retethered himself to a work platform on the rail of the cargo bay, Anna used the arm to gingerly hand the satellite over to Allen. The satellite was fifteen thousand pounds, and, although weightless in space, was not massless. Allen called on his background as a state wrestling champion to wrangle the unwieldy spacecraft.[43]

Allen hung on to Palapa as Gardner connected himself to a work platform on the robotic arm and attached an adapter that would allow them to secure the satellite to the payload bay. Since neither astronaut could see fully around the massive satellite, Anna watched carefully through the back windows of the orbiter and guided the work, cautioning Allen when the satellite moved too close to the orbiter's walls.

The maneuver took another ninety minutes, a full orbit around

Earth, but finally, five and a half hours into their EVA with just a half hour of oxygen left in their tanks, Gardner and Allen locked the Palapa into place. Part one of their two-part retrieval mission was complete.

Anna and the crew celebrated their immense success with rehydrated steak, the shuttle equivalent of luxury, and the satisfied crew floated into bed. As Anna dreamed, the Westar still loomed 690 miles behind them.[44]

The next morning, they awoke to attempt the second and final retrieval of their mission. If recapturing the Palapa had been difficult, recovering the Westar 6 was a high-wire act. Floating into position behind the controls of the robotic arm, Anna felt disoriented. "The shuttle is moving . . . I'm moving the arm. You're going by the Earth. The clouds are moving," she explained.[45] The Westar rotated ever so slowly. Anna hovered at the arm's control panel while peering out the aft windows. Balanced on the tip of her robot arm was Allen himself in his full EVA suit. Gardner, floating nearby in the MMU, had the Westar in the stinger's grip.[46]

Anna maneuvered Allen over to the Westar, his toes the only thing stopping him from drifting off, lost and alone into the great expanse. The thought terrified both him and Anna. Allen trusted her skill and finesse with the arm, but Allan's EVA helmet restricted his field of view so severely that he had no idea if his feet were staying in their restraints. He had never had a fear of heights, but now he felt as if he might fall off "the world's highest diving board." Allen's reptilian brain told him that if he slipped, he would plummet into the payload bay, or worse, down toward Earth.[47] Anna, for her part, could not think too much about the fact that she held a life on the other end of the robotic arm, lest she freeze up entirely.

Three sunsets and three sunrises later, the Westar was secure in the bay. "Doggie's on the ground, we're putting the rope around the feet," the space wranglers reported. [48,49]

Having salvaged Westar and Palapa, their insurers could now resell them, up to $35 million each, to recoup their losses. Dale Gardner and Joe Allen indulged in some "Madison Avenue pizzazz," holding up a FOR SALE sign over one of the recaptured satellites before heading back to the bay.[50] The image became iconic, affirming that NASA was in the space *business.*

When Hauck passed the news to the ground, Mission Control erupted into robust cheers. Though Anna had never been one to savor attagirls, she swelled with pride. She had been disappointed when George passed her over for first woman in space, but she knew now that she would not have had it any other way.[51]

Anna unwound that night, playing a mixtape Bill had curated for her. As David Bowie struck the opening chords to "Space Oddity" (*good choice, Bill*), she marveled at the precarious, precious blue marble that held nearly all human life, save herself and her crew.

For here
Am I sitting in a tin can
Far above the world
Planet Earth is blue
And there's nothing I can do.

As Bill's tape wound down, Anna heard, "I luh, I luh"—a little voice sounding out, "I love you."[52] It was Kristin. As fulfilling as her work had been, as much as it had been a dream come true, at that moment, Anna could not wait to get home to her little girl.

Landing at the Cape on November 16, 1984, Anna exited the orbiter and ran into Bill's arms, whispering, "It was all worth it."[53]

At the post-flight press conference, reporter Thornton Page asked Anna, "Would you recommend a mechanical arm for taking care of children?"

"I think when Kristin is a little older, I'll have some interesting stories to tell her," Anna laughed. At the same time, she could not resist the opportunity to gently remind Page, "Every crewmember here is a parent."[54]

Press conference disposed of, Anna raced home to Houston. At Ellington Field, she saw her mom holding Kristin. She raced down the airstairs, grabbed her baby, and covered her in kisses. "Mommy's so glad to be back. I missed you so much!" Anna said over and over, hugging Kristin tightly.[55]

Anna's flight was the triumphant last act of 1984. "The mission was among the most spectacular in the twenty-six-year history of the American space program," a *Time* reporter declared, noting the fact that this mission marked the first time that satellites had been salvaged from space.[56] The mission's loudest cheerleader was President Ronald Reagan. "You demonstrated that we can work in space in ways that we never imagined were possible," Reagan had radioed the crew.[57] Most meaningful to Anna, Dr. Kraft wrote the crew a letter likening their heroic feat to Alan Shepard's historic launch and the Apollo 11 moon landing. Anna's childhood dream of spaceflight had come full circle.[58]

After stumbling out of the gate, the New Guys managed to close the year on a high note, ensuring President Reagan's continued support for the shuttle program and the space station. The age of the commercial shuttle was in full swing. As for George, he could not have been more pleased with his first class of astronauts. His hard work, keen judgment, unorthodox management style, and strategic pairings had paid off. However, behind all the welcome-homes, presidential congratulations, and pats on the back lay a more uncertain future for the shuttle. Every time the program reached for a new goal or pushed the bounds of its still experimental technology, the risk quotient mounted.

CHAPTER 15

BLOOD MOON

Sunnyvale, California. 1984.

Trust no one. Ellison Onizuka was on high alert as he drove down the unfamiliar streets of Sunnyvale, California. He kept a watchful eye on his rearview mirror. A KGB agent might well be following them. After all, he was preparing for an assignment of the highest national security, the first shuttle mission entirely dedicated to the US Department of Defense.

El passed fast-food joints, gas stations, strip malls, and seedy motor courts, but no sign of the motel where he and his crew would be staying during their visit to the Air Force Satellite Control Facility. There, they would finally meet their highly classified payload: a national security asset that might well tip the scales in the Cold War.

Tensions between the US and the Soviet Union were running high. Months before, Soviets had misinterpreted a US war games exercise as the real thing, and the world—unbeknownst to the public at large—came perilously close to nuclear catastrophe. Disaster had been avoided, but in a game where surveillance was king, the US needed better intel on Soviet chess moves.[1] El's payload would help the US intercept important communications between the Soviets and their allies.

Due to the nature of the mission, El could not say a word about his launch date, the manifest, or the payload to his colleagues, friends, wife Lorna, or two daughters—Janelle, fifteen, and Darien, nine. When the crew left Houston in their T-38s, they filed plans to land in Denver, then rerouted midflight to Moffett Airfield outside of Sunnyvale. The slick redirection created the illusion that the sojourn was an unexpected detour, not the crew's true destination.

Now the crew was crammed into a tiny rental car as it rumbled through the Bay Area sprawl. Next to El sat mission commander Thomas "T.K." Mattingly, who orbited the moon on Apollo 16. In the back seat sat TFNG pilot Loren Shriver and TFNG mission specialist James Buchli. They were like *The A-Team*. Except instead of being on the run from the military police, they were working for the Air Force and instead of the iconic GMC van, they had this rusted-out rental. Sometimes real life did not live up to the high-octane adventures of Tuesday evening television. Sometimes even undercover astronauts had to find their own motel.[2]

"Stop here!" Buchli yelled from the back seat.

Uh-oh. What's wrong?

"We made extra stops to make sure that we wouldn't come here directly, and they can't trace our flight plan. We didn't tell our families. We didn't tell anybody where we were," he reasoned. Then he pointed out the window to a motel marquee. "What does that say?"

In big block letters, the sign read WELCOME STS-51C ASTRONAUTS. Below that, each crewmember's name was proudly displayed. Checking in at the front desk, they saw their own photos hanging on the wall.[3] So much for a low profile.

El chuckled at the absurdity of the moment. Even though the security breach ended up being inconsequential, the event showed the challenges NASA and the DoD had to surmount when they worked together. NASA sought publicity to win support and funding; the DoD operated in the shadows.

El had a foot in both worlds. An Air Force major, he was professional, mission-oriented, and well acquainted with classified defense operations. He had also fallen easily into the role of gregarious ambassador for NASA.[4] As the first Asian American astronaut, El curried enormous support from his home state of Hawaii and from Asian American groups nationwide.[5]

El was one in a long line of Onizukas—one of whom had even ruled Japan as shogun.[6] His grandparents immigrated to Hawaii from Japan in the late 1890s, laboring in Oahu's sugar fields before planting the roots of a coffee farm on the Big Island. There, the Onizukas reestablished the familiar patterns of home, organizing into neighborhood associations called *kumiai,* building Buddhist temples, and establishing Japanese language schools.

Their son, El's father, opened a general store on the Kona coast with his wife. That store saw Kona's farmers and ranch hands through the worst of the Great Depression. "One did not starve, no matter how poor the times, when the Onizuka store was doing business," El's biographer Dennis Ogawa wrote.[7] El saw his parents give generously to their community in hard times, and he carried that spirit with him always.

El grew up in a tropical paradise. The island's active volcanoes piqued young El's curiosity about how the planet worked. He watched 1,500-foot-tall geysers of molten lava at Kīlauea Iki crater and wondered if—and how—all that energy could be controlled.[8] He loved taking things apart to see how they worked—a doorknob, a bicycle, a chair hinge. Though his parents were sometimes vexed by his handiwork, they appreciated his native curiosity.[9]

His neighbors knew him as a charismatic, outgoing *mohan seinen*—Japanese for "model youth." He did it all: Eagle Scout, National Honor Society, class treasurer, star poultryman at 4-H, and center fielder for his school's champion baseball team. El also had a need for speed. He once raced his friend's car on foot down a stretch of the Mamalahoa, a highway that encircled the Big Island. He lost. When he got his driver's license, he upped the ante to racing with his dad's Jeep.[10] "When I grow up, I will drive an airplane," he told his grandmother. Like fellow TFNG and aerospace engineer Guy Bluford, young El was obsessed with building and flying model planes. After NASA introduced the Mercury 7 to the world, then-thirteen-year-old El refined his vision: He wanted to be an astronaut.[11] El enrolled at the University of Colorado Boulder, where he joined the Air Force ROTC and studied aerospace engineering.

In college, El met and married Lorna Yoshida, a student at nearby University of Northern Colorado and—like himself—a third-generation Japanese American from the Big Island. The two bonded over their shared heritage and a love of one-liners. Yet as proud as El was of his Japanese background, he saw himself first as an American.[12] "American ends with two words: I Can," he once told a rapt Rotary Club audience. "How does it feel to be one of the minorities selected?" a reporter once asked El. "I didn't know I was a minority," he joked.[13]

He had reason to mistrust labels like "minority." They had caused immeasurable harm during the Second World War, when selective internment to Oahu robbed the Japanese-Hawaiian community of its leaders.[14,15] No, El was happy to fly under the radar, and this, his first mission, granted his wish.

NASA's relationship with the DoD hung in the balance. George had handpicked El for the all-important STS-51C mission because he was qualified, trustworthy, and unwavering. "When [El] started something, he wanted to make sure that thing was completed and everything done right," his eldest sister, Shirley, said.[16] El also happened to be George's number one bubba. George, now a divorcé, often ended his day with a drink at the Onizukas' home. El was an affable, willing sidekick to George and "the very best judge of chili" at George's beloved cook-offs.[17] Quite simply, George trusted El to get the job done. The stakes for El's flight were high. In 1979 the DoD had agreed to phase out its use of expendable launch vehicles and use the shuttle exclusively. Within NASA, the DoD established its own group of military astronauts who would only deploy top-secret payloads. The Air Force established a group of officers in Johnson's Mission Control to help coordinate military shuttle flights and built an $85 million secure room on the third floor of Mission Control.[18]

With this kind of investment, the DoD had expected its payloads to be launched on schedule.[19,20] On early shuttle flights, the agency grew increasingly frustrated by technical mishaps that stood in the way of their plans. The shuttle's maiden voyage flew two and a half years behind schedule, and subsequent missions were rife with engineering problems. Tile troubles, main engine breakdowns, fuel line leaks, and computer failures caused the DoD to question the reliability of the vehicle on which they had staked their entire satellite surveillance program—*in the middle of the Cold War.*

So far, the shuttle had flown only one classified DoD mission, STS-4. The military equipment onboard was meant to test the viability of Reagan's Strategic Defense Initiative (SDI). Popularly known as Star Wars, SDI promised to launch a network of lasers that could detect and disarm nuclear missiles in space, neutralizing the threat of nuclear attack. Many in the scientific community called SDI pure science fiction, but Reagan was not dissuaded. He secured $2.5 billion for Star Wars from Congress in 1985.[21] STS-4 was not an auspicious beginning to the program: The classified payload's two ultraviolet and infrared imaging sensors, designed to detect missiles from space, failed to operate after deployment.[22,23]

In 1983, the shuttle flew its sixth mission carrying an inertial upper stage rocket (IUS) developed by the Pentagon. The rocket that should have boosted the satellites to geosynchronous orbit failed on deployment and had to undergo a complete redesign. The redesign pushed El's flight, which would also use the technology, for fifteen months. Other upper-stage rockets developed by the Pentagon faced similar delays. Meanwhile, cost overruns bedeviled the construction of the space shuttle facilities at Vandenberg Air Force Base that would allow for launches into polar orbit, flying directly north from Southern California, over the north pole, and to Russia for surveillance, and then back to America's West Coast.[24] The shuttle's first launch from Vandenberg had been pushed from October 15, 1985, to late January 1986, at the earliest.[25] The military worried the shuttle program would not be able to meet its promised flight schedule, thirty-four flights between 1983 and 1985. Administrators at the DoD ground their teeth as cutting-edge surveillance satellites collected dust in their top-secret Sunnyvale warehouse.

By 1984, the DoD was searching for a way out of their exclusive partnership with NASA. After the disastrous pad abort on Judy's flight, the DoD had announced it would renew its expendable rocket launch program to ferry satellites to space.[26] Now the DoD was lobbying Congress to pay for ten new launch vehicles for this program. Without the DoD as a customer, Beggs saw his budget evaporate. As El readied for his flight, Beggs desperately tried to stall the DoD's efforts in Congress.[27,28] El's mission would either point the way forward or prove that NASA's partnership with the DoD was bound for failure.

Launch Pad 39A, Kennedy Space Center. January 24, 1985.
A cold front whipped through the Southeast. The chill in Washington, DC, forced President Reagan to cancel his second inaugural parade. In Florida, the days prior to liftoff set record low temperatures. The night before dipped into the teens, bringing clear skies.[29] "Ironically," a *New York Times* reporter wrote, "the visibility for the secret mission was better than it had ever been for the 'open' missions."[30] By the time the shuttle launched, right before 3:00 PM, the temperature on the pad had risen to 53°F. Still, El's was NASA's coldest launch to date.[31]

Soon after *Discovery* reached orbit, Mission Control's massive screens flickered off. A red background replaced displays of the shuttle's path around the globe and the public affairs officer, who had announced each step of the ascent, fell silent. The DoD took over *Discovery*'s operation; henceforth, all communication between crew and ground was shielded from public record.[32]

In orbit, El and his two fellow mission specialists—Buchli and DoD astronaut Gary Payton—readied the payload and the upper stage rocket that would propel the satellite to orbit. This was a second chance for the retooled IUS. If it failed again, El's mission would be a bust. With that in mind, the crew ran the IUS through its full range of motion, wiggling the conical rocket nozzle on all axes to ensure it could properly guide its cargo into position. Then they released the payload and waited. A fiery tail burst from the nozzle and cheers rang out across the flight deck as their payload zoomed to its designated orbit.[33]

While 51C's payload technically remains classified, reports indicate the asset was almost certainly a state-of-the-art signals intelligence satellite called Magnum.[34] The IUS deployed Magnum 22,300 miles above Earth, in geosynchronous orbit, allowing it to stay locked over the western half of the Soviet Union. With a dish that stretched one hundred yards wide, the $300 million satellite could intercept waveform communications like phone calls and telex cables sent within the Soviet Union and China. In the years to come, Magnum would be an important tool in helping ensure the Soviets upheld their nuclear nonproliferation agreements.[35]

Their primary objective complete, the crew now ticked off their short list of experiments and indulged in the joys of space travel. El posed for a photo wearing a headband that said KAMIKAZE and was adorned with the rising red sun of the Japanese flag.[36]

It was a subversive image—a Japanese American Air Force officer floating on the middeck of a US spaceship, donning the symbol of suicide pilots who attacked Pearl Harbor forty-three years earlier, drawing America into World War II. Was the photo one of El's jokes? Or political commentary? The events of Pearl Harbor had also spawned Japanese internment camps in the United States and the imprisonment of over 120,000 people of Japanese descent, many of whom were American citizens.[37] How far society had come: His prank was regarded as a harmless joke, and now a Japanese American was flying a highly classified US defense mission.

On January 27, 1985, *Discovery* touched down at Kennedy. A cheer went up from the viewing area. George welcomed the crew home and congratulated them on their success. The next day, back in Houston, El's guests from Hawaii celebrated their hometown hero at one of his favorite spots—the Villa Capri, sister restaurant of Frenchie's. With El's astronaut friends, they toasted the flight's success, swilling Chianti late into the night. George serenaded everyone in an Italian duet with the Villa Capri's owner, Frankie Camera.[38] "*Luna rossa, chi mme sarrá sincera?*" they warbled. No one knew George spoke Italian. "Oh, luna rossa, you're out tonight, a moon of red in a sky of white, because I'm telling a lie tonight . . ."

Luna rossa translates to "red moon" or "blood moon," when a full moon is eclipsed by Earth and takes on a reddish tinge. In ancient times, the blood moon was a harbinger of trouble, or the coming wrath of God.

The crew of 51-C, however, was blissfully unaware of any danger that surrounded them or that might be on the horizon. Indeed, none of them knew they had come close to death a few days earlier.

STS-51C's Launch Day. Fifteen Miles off Cape Canaveral. January 24, 1985.

On the deck of the *Freedom Star*, Jim Devlin watched *Discovery* rise and become a speck on the horizon. Devlin counted in his head. *One minute. Two minutes.* As NASA's retrieval operations manager, Devlin and his 176-foot ship were to retrieve the shuttle boosters after launch.

The boosters separated from the orbiter, right on time, continuing upward to an altitude of forty-five miles before gravity arched them back to the sea, beginning a descent that would last almost four minutes.[39] The captain guided the *Freedom Star* close enough to where the boosters might land, but not so close that the booster's tremendous impact might rock the ship. Devlin saw a splash in the distance. The familiar white cap of a booster bobbed to the surface. He signaled his team: *Time to get to work.*[40]

The reusable boosters should have saved NASA money, but no one had predicted how difficult retrieving the boosters from the Atlantic Ocean would be, especially in the middle of winter.[41] For the job, NASA built two specialized ships, the *Freedom Star* and the *Liberty Star.* Each was responsible for retrieving one booster—right and left. Once they got close enough, highly skilled divers sped the rest of the way in a dinghy and performed an elaborate series of dives to pump the booster with air. When the 149-foot-long structure floated to the surface, they hitched it to the tugboat and hauled it back to the Cape.

At dockside, crews lifted the boosters onto dollies, disarmed their pyrotechnic ignition systems, and depressurized their fuel systems. An assessment team documented any anomalies that may have occurred during flight. Then they washed the boosters with detergent, disassembled them piece by piece, and shipped them back to Morton Thiokol in Utah by railcar.[42] The segments' steel casings would later be rejoined together with the clever tang-and-clevis connection and sealed with two rubber O-rings at each joint.

Roger Boisjoly, an engineer from Morton Thiokol, was overseeing the booster rocket refurbishment process when 51C's boosters arrived back at the Cape. Upon examining the O-rings, Boisjoly grew concerned. The primary O-ring in the left booster's forward field joint had eroded.[43] Even more troubling, soot appeared in the joint—clear evidence that hot gas had escaped past the rubber ring, a phenomenon called "blowby." The right booster incurred damage in the center field joint—an unsettling first—and the secondary O-rings had been degraded by heat.[44] If any more hot gas had leaked, fire spilling from the boosters might have reached the external tank—and its explosive fuels.[45]

Thiokol had seen O-ring damage before. After STS-2, Thiokol engineers found that—despite their careful calculations—flames had burned through a fifth of the primary O-ring. On the tenth shuttle flight—Ron's—both the primary and secondary O-rings showed unexpected erosion.[46] Engineers at Marshall Space Center recorded the events but did not surface the issue to their colleagues at Johnson. Then it happened again on the eleventh flight . . . and the twelfth. After that twelfth flight—Judy's—Thiokol engineers found black soot between the primary and secondary O-rings, meaning the primary O-ring had failed to keep the hot gasses from leaking past the seal.[47] However, the secondary O-ring had sealed appropriately, and kept those hot gasses from escaping the joint.[48] Because of the effectiveness of the secondary O-ring, Thiokol deemed the blowby a "self-limiting" issue.[49] On El's flight, though, something looked different.

"The grease between the O-rings was blackened, just like coal," Boisjoly observed in his thick Massachusetts accent.[50] The hydrocarbon grease coated the interior surfaces of the booster rockets, to protect the metal from saltwater corrosion after they splashed into the Atlantic.[51] He furrowed his brow, making note that it was the worst blowby he had ever seen. What had been different on El's flight? Boisjoly personally inspected the joints, ordered the technicians to take photographs, and sent grease samples back to Thiokol for chemical analysis.[52] The technicians at the plant informed him that the seal temperature during launch was 53°F—the coldest temperature yet. Boisjoly suspected that the cold weather during launch made the O-rings inflexible, unable to expand and contract to form a tight seal.

Boisjoly raised the issue with his bosses at a flight readiness review for the next shuttle launch. He specifically pointed out that cold temperatures appeared to greatly reduce the O-rings' ability to seal. "That presentation was a very pointed discussion," Boisjoly remembered. "I felt very strong about that fact because I spoke from conviction, and I was challenged by just about everybody in the room about what I had reported."[53]

Thiokol concluded that although the temperature enhanced the probability of blowby, the secondary seal would provide enough of a barrier to keep hot gasses from causing any real damage. In an internal memo Thiokol concluded that while the condition was "not desirable," it was "acceptable."[54] After all, none of these anomalies had ever prevented the shuttle from reaching orbit.

Although El's return home received limited fanfare—no *Newsweek* covers or press tours awaited the first Asian American, as they had for Sally—as far as Beggs and the DoD were concerned, El's mission was an unqualified success. On February 14, 1985—nineteen days after 51C landed at Kennedy—Beggs met with Air Force Secretary Pete Aldridge and came to an agreement that recommitted both institutions to the partnership. Over the next few years, NASA officially agreed to ramp up the shuttle roster to twenty-four missions per year, with a third of all missions reserved for the military.[55] By comparison, NASA had only flown five missions in 1984. By the end of 1985, Beggs promised, NASA would more than double that number with a dozen flights on the manifest. It was a flight schedule that almost no one at NASA thought possible.

CHAPTER 16

THE PRINCE AND THE POLITICIAN

Low Earth Orbit. April 15, 1985.

Rhea hovered on the middeck of *Discovery*, a pair of scissors clenched in her teeth, brow furrowed. She held a large sailmaker's needle steady for a moment then plunged it through several layers of duct tape and rigid plastic. After pulling a long thread through the eye of the needle, Rhea sewed through the mess of duct tape and plastic. The thread pulled taut through her stitchwork, securing an eight-by-ten-inch sheet of plastic to the metal pole around which it was wrapped. It felt, Rhea thought, a bit like stitching a patient back together. When Rhea finished, the makeshift tool managed to look both futuristic and primitive, with the large rectangle of plastic and duct tape hanging off one end of the metal pole. Her STS-51D crew started calling her invention the "flyswatter."[1]

The crew hoped to stick the flyswatter to the end of the robotic arm and use it to toggle a stubborn "separation switch" from the off to the on position. The glitchy switch was on the underside of Syncom, a Navy communications satellite built by Hughes.[2] When the crew had launched Syncom days prior, they had been dismayed to see that the satellite's booster rocket had failed to ignite. Much like the two satellites on Ron's flight, Syncom was stranded in low Earth orbit. However, this time ground control believed they had devised a way to rescue the satellite. Rhea's crewmates, TFNGs David Griggs and Jeffrey Hoffman, would perform the shuttle program's first *unscheduled* spacewalk to manually attach the slapdash flyswatter to the end of the arm. Then Rhea would maneuver the arm so that the swatter could flip the switch.

Rhea regarded her finished work for a moment before taking a few pictures and downlinking them to Houston, so Mission Control could confirm that the flyswatter looked as they intended. Controllers agreed: It looked even better. CapCom Dave Hilmers commended the work, calling Rhea "a fine seamstress." Sally Ride, watching from Mission Control, corrected him. "That was a surgeon."[3]

A surgeon who had finally earned her gold wings. Rhea's flight had been delayed over five times, her payloads reshuffled, and her mission patch changed, but doggonit, now she was getting her turn. When *Discovery* launched (three days earlier), Rhea became the fifth American woman in space.[4] To get there, Rhea and her crew had to accommodate a highly political request, flying with Senator Jake Garn, who had leveraged his appropriations subcommittee chairmanship for a ride on the shuttle. Garn was tasked with conducting a study on motion sickness in space, but he became victim number one, forcing Rhea to assume the work.[5]

On April 19, after two extra days on orbit to tug Syncom's lever in the correct position, it was time to come home. As the shuttle screeched to a stop on Kennedy's notoriously short and narrow runway, and one of the tires blew out, Rhea thought twice about her career as a spacefarer. Although technicians replaced the brakes and tires after every shuttle flight, the landing gear was a systemic problem.[6] The orbiter, which weighed anywhere from ninety-five to one hundred ten tons on reentry, only had four main landing gear tires that, as Rhea said, were "pretty dinky for that large a vehicle."[7] The brake system was first designed in the early 1970s, during the shuttle's developmental phase, before the engineers could fully comprehend all the stress that the vehicle would experience on landing. As the orbiter developed, and grew heavier, the landing gear had never been adjusted. "The brakes and tires were pretty chewed up after each flight," Rhea said.[8] The landing gear had sustained varying degrees of damage in sixteen of the shuttle's first seventeen flights, but there was never time to fix the design problem.[9]

NASA workers struggled to refurbish the shuttles under impossibly tight turnarounds. Beggs's decision to award the shuttle refurbishment contract to Lockheed as part of his 1983 cost-cutting Shuttle Sweepstakes took its toll as the process slowed to a crawl at the Cape. Technicians, engineers, and project managers found themselves with increasingly long shifts, weekends and holidays be damned.[10] There were predictable effects on workers: fatigue, exhaustion, accidents, and the breakdown of safety standards.[11] Since 1983, NASA had also reduced its number of inspection personnel involved in the daily oversight and refurbishment of the shuttles, instead relying on Lockheed to provide quality control.[12]

The accelerated flight rate exacerbated the already strained system. NASA's goal of twelve shuttle flights in 1986 required a larger inventory of spare parts to outfit the orbiters. In October 1985, however, management at Johnson reduced the program's $285.3 million logistical budget by $83.3 million in order to shift resources to higher-priority engineering issues.[13] The nearly 30 percent budget cut postponed the purchase of spare parts and forced ground crews to regularly cannibalize returning shuttles to install parts on departing ones, a practice that increasingly exposed the process to human error. Sometimes the sophisticated choreography faltered, producing further holdups and, ironically, increasing costs.[14]

To minimize delays, NASA, along with its contractors, increased the practice of signing waivers to accept risk for potentially compromised parts.[15] The waivers allowed flights to continue on schedule, despite the safety threats. Hans Mark, then NASA's deputy administrator, became increasingly concerned about shuttle safety issues by the mid-1980s. His concerns were continually ignored, and he resigned in mid-1984.[16,17]

Beggs pushed the ambitious schedule not only to maintain his renewed DoD relationship, but also to win support for the space station. In March 1985, space station planners officially adopted a blueprint for the orbiting laboratory. They settled on a dual keel design, a massive, rectangular frame of scaffolding with a long beam running through it crosswise, like a toothpick spearing an olive. Space station compartments, solar energy panels, and radar dishes for data transmission would attach to the dual keel frame. The design would be more stable than a single support beam, but its construction would also be far more complex. To develop the dual keel, NASA contracted $144 million of work from six aerospace companies, awarding $27 million each to McDonnell Douglas and Rockwell.[18,19]

By May, international partners including Canada, Japan, and the European Space Association (ESA) all joined NASA in signing a memorandum of understanding and drafting preliminary studies that outlined design objectives for the space station project along with technological commitments from each country's space program. Canada agreed to supply a robotic arm like the one that it had built for the shuttle. Japan and the ESA would each add a lab module.[20] The space station was well on its way to becoming a reality.

Kennedy Space Center. April 29, 1985.

As *Challenger*'s boosters ignited, Fred Gregory, strapped into the shuttle's cockpit, felt the incredible power of boosters as they sped him to orbit.[21] Fred may not have been the first African American in space, or even the second, but he was "behind the wheel" as the first African American shuttle pilot. Now that *Challenger* had leapt from the pad, he felt positively giddy. Fred's mission, the second fully operational Spacelab flight, served as proof of concept for the space station and its ability to conduct larger science experiments in space. The shuttle's first nonhuman mammals—two squirrel monkeys and twenty-four rats—joined the crew for their journey.

The flight appeared successful, but, weeks after Fred's return, Thiokol again discovered O-ring damage.[22] The first seal on the left booster's nozzle was totally destroyed and the second seal had 24 percent of its diameter burned away. Fred had come within three-tenths of a second of death, a fact he would not learn until years later.[23]

Larry Mulloy, head of the solid rocket booster program at Marshall Space Center, requested a report on the O-ring damage from Thiokol. As the chief engineer on both the inertial upper-stage satellite booster project and the external tank project, Mulloy had proved himself to be an exceptional mechanical engineer and an apt administrator. In 1982, center director William Lucas promoted him to head the solid rocket booster project.[24] Known as the "smiling Cajun," Mulloy also had an affable side, preferring to spend his weekends shucking oysters with his friends on the Tennessee River.[25] Mulloy understood the politics of his job, having spent the lion's share of his career under the hard-nosed directorship of Lucas. He knew that his boss detested failures and launch delays, especially those emanating from Marshall.

Mulloy assigned Allan McDonald, Thiokol's director of the solid rocket motor program, to investigate the issue and head up an "in-house O-ring Seal Task Force." Where Mulloy was a tactician, McDonald was an earnest chemist who said what he thought. McDonald had completed his graduate work at the University of Utah when James Fletcher was president of the school. At forty-eight, he had a full head of gray hair and worried eyes accustomed to carefully parsing data as he tackled abstruse engineering questions that often had life-or-death consequences.[26]

McDonald's conclusion was simple, but disturbing: The primary O-ring on Fred's flight never sealed, and most likely had even leaked.[27] The engineers had not noticed the leak during factory checks because the test was run at a low enough pressure that the zinc chromate vacuum putty, which cushioned the seals, acted as a pressure seal on its own. McDonald realized that if the check had happened at a higher pressure, the leaky O-ring would have been spotted. He and the engineers doubled the maximum amount of pressure exerted during those checks to prevent a similar mistake from happening again.[28]

Fortunately for Fred and the crew, the second O-ring had prevented the gas from leaking through the booster joint. Thiokol's team pointed out that, once again, the gas pressure was not strong enough to break the second seal. They also believed they could stop potential issues by running tougher tests on the booster joints before the rockets went out to the pad. McDonald recommended that NASA proceed with the next launch as scheduled.[29]

Roger Boisjoly disagreed and wrote a scathing memo to his bosses at Morton Thiokol warning them that if they did not address the O-ring issue: "The result would be a catastrophe of the highest order—loss of human life."[30] His warnings fell on deaf ears. Boisjoly later claimed the culture at Marshall—to which Morton Thiokol reported—was to blame, saying he had been "personally chastised" and "crucified" in flight readiness reviews when he voiced doubt. He believed the man who helped to foster that culture was Mulloy's boss, Dr. William Lucas, the head of the Marshall Space Center.[31]

Boisjoly was not the only one concerned. Paul Wetzel from NASA Headquarters learned about the issue on Fred's flight and called McDonald directly to ask about the O-ring trouble.[32] That is when McDonald realized that Mulloy never passed his initial report on to headquarters.[33] McDonald told Wetzel what he knew and Wetzel invited him and his team to Washington, to discuss what went wrong on 51B.[34]

Somewhere over the Atlantic. June 1985.

Sitting in a 747 bound for Riyadh, the capital of Saudi Arabia, Shannon Lucid fumed. The Oklahoman mother of three, the oldest of the six TFNG women, usually approached life with optimistic exuberance and polite heartland charm. A chemist by trade, she especially enjoyed moments when her work in the Astronaut Office overlapped with NASA's science division. A few weeks earlier, Shannon had become the last of the six TFNG women to travel to space when her flight, STS-51G, launched from the Cape. Now she had to assume the title "honorary man" to accept an invitation to visit the king of Saudi Arabia, an invitation that she did not want to accept in the first place.

Like Rhea, Shannon found her first flight influenced by politics. Sultan bin Salman Al Saud, a prince and nephew of Saudi Arabia's King Fahd, joined her crew to support the launch of the Arabsat-1B satellite, becoming the first Muslim, the first Arab, and the first royal to go to space.[35] At twenty-eight years old, he was also the youngest shuttle astronaut ever.

The Saudi satellite promised to deliver telecommunications to the Arab states at a time when relations between the US and Saudi Arabia were especially congenial. Throughout the 1980s, the Saudis promised to defend US interests in cheap Middle Eastern oil, while the US returned the favor by providing state-of-the-art military weaponry. Together, the two nations battled Soviet communism in Afghanistan and Shiite extremism in the Middle East. The Saudi prince's ride aboard the shuttle was a diplomatic cherry on top.

Perhaps that was why, when the Saudi satellite flunked all its safety tests, NASA higher-ups ignored the objections of everyone in the Astronaut Office and greenlit the satellite anyway.[36] A previous iteration of the Arabsat had lost power, attitude control, and the function of its orbit gyros, which helped control the satellite in orbit. Not only did the malfunctions render Arabsat 1A useless, but had it launched from the shuttle, instead of an expendable rocket, deployment may have been dangerous to a crew.[37] Since Arabsat 1B failed its safety reviews, Shannon's crew might face the same malfunctions.

"It was very disappointing to a lot of people at the agency," said John Fabian, a fellow TFNG and Shannon's crewmate on STS-51G. "If they hadn't flown the satellite, you see, political embarrassment. What are we going to do with the Saudi prince? What about the French astronaut [payload specialist Patrick Baudry]; what's the French government going to have to say about us [not] flying their satellite on the shuttle? What will be the impact downstream of other commercial ventures that we want to do with the shuttle?"[38]

Despite the politics, the crew enjoyed a successful flight, launching Arabsat and a high-precision tracking experiment for Reagan's Star Wars initiative. Shannon also deployed and retrieved SPARTAN-1, a three-hundred-pound pallet of astronomy experiments, using the robotic arm.[39,40]

After touchdown, Saudi King Fahd bin Abdulaziz Al Saud invited the astronauts to Riyadh. Shannon's invite came with a catch—she would need a male escort to be allowed into the country. Loath to implicate herself in the country's treatment of women, she declined the invitation. The king and prince called President Reagan to express their dismay; President Reagan called NASA administrator James Beggs; Beggs called Johnson director Gerry Griffin, and soon enough a reluctant Shannon hopped on a flight to Saudi Arabia.[41] Although she did not bring a male escort, the Saudis allowed her to visit as either the "honorary daughter" of commander Dan Brandenstein, the "honorary sister" of mission specialist John Fabian, or most infuriating of all, an "honorary man."[42] Shannon reluctantly accepted the dubious "honorary man" distinction, as the queen of England had during her visit to Saudi Arabia in 1979, and made it a quick trip.

In spite of the mission's success, Shannon's crewmate John Fabian tendered his resignation after landing. His wife, a mission controller at Johnson, warned him that decision-making at headquarters now favored schedule over safety.[43] "This was an unhealthy environment within the agency," Fabian said.[44] As a former Air Force commander who had flown combat missions in Vietnam, Fabian was no stranger to peril. He was, however, a firm believer in controlling and eliminating any known risk. To him, NASA was courting danger. Fabian, beloved by his peers, was a loss for the office.[45]

As NASA managers accelerated the shuttle's flight rate, technicians had little time to address outstanding engineering issues. With no way to repair the shuttle's delicate silica tiles in flight, tile damage was still a concern. Astronauts and engineers alike worried the main engines might be a ticking time bomb. Rocketdyne workers were still finding cracked turbopump blades in post-firing autopsies of the space shuttle's main engines.[46] That, at least, was a serious enough issue for NASA to regularly switch out the engines.[47] On the shuttle's ninth flight, commanded by John Young, two of the shuttle's three auxiliary power units—responsible for lowering the landing gear and powering the brakes and nose-wheel steering—leaked flammable hydrazine fuel on orbit, caught fire on landing, and exploded only twenty minutes after wheels stop.[48] Some said the event caused John Young to hang up his wings.[49]

Then there were the O-rings. By the end of 1985, the boosters' O-rings sustained damage on thirteen of twenty-three flights. Among those flights were not only Fred's and El's, but also Ron's and Judy's first trips to space.[50] The Thiokol engineers closest to the problem, like Roger Boisjoly and Allan McDonald, assumed that the astronauts were informed of the O-ring issues, as they were supposed to be aware of all the engineering risks that affected them. In fact, none of the astronauts knew.

In August 1985, McDonald finally made his presentation to NASA Headquarters, with two notable changes from Marshall. First, Marshall officials asked McDonald to include Thiokol's recommendation that it was safe to continue flying the boosters, so long as "certain inspections and leak-check procedures were verified at the time of assembly."[51] Second, according to McDonald, Larry Mulloy asked him to remove a statement that colder temperatures might contribute to O-ring failure—the issue Boisjoly had identified on El's flight. Not understanding how critical that detail would become, McDonald followed orders.[52]

The day after the presentation, Thiokol formalized its extant O-Ring Seal Task Force. McDonald would lead the full-time task force, with Boisjoly as the main engineer.[53] In September, the task force suggested each field joint be outfitted with a "capture feature" to help the O-rings to seal under any environment.[54] The team at Marshall, concerned this recommendation would only increase the cost of an already out-of-control shuttle budget, requested a formal proposal, complete with cost-benefit analysis.

McDonald and the task force struggled to translate improved safety into a dollar amount. They ultimately suggested the revised design would improve the boosters' reusability, thus defraying the cost of the capture features over the lifetime of each booster.[55] They sent the new proposal and waited to hear from Marshall.

Kennedy Space Center. August 1985.

Anna Fisher squinted through heavy rain toward pad 39A, where her husband, Bill, sat strapped into *Discovery*. It was still raining when the boosters ignited. *You've got to be kidding me? They're taking off in this?!* To Anna's shock, *Discovery* launched, vanishing into the stormy cloud cover.[56] Bill's flight, STS-51I, also happened to be pilot Dick Covey's first flight. With Covey in orbit, all thirty-five New Guys had made it. Every member of George's first class had succeeded in becoming an astronaut—the gold pin–wearing kind. No one had failed. No one had been left behind.

For the New Guys, the journey—rattling violently on the trip up, stepping into the vacuum of space, returning home in a ball of plasma—forced a direct confrontation with the extreme risks of leaving this planet, but the true extent of the danger was hidden among burnt tires, missing tiles, O-ring remains, and contractor waivers. "We had lulled ourselves into thinking that since we had gotten away with it before, we could get away with it again," said then flight controller Paul Dye. "We felt comfortable because nothing terrible had happened. But what we didn't realize was that the terrible thing just hadn't happened . . . yet."[57]

CHAPTER 17

BEAUTIFUL, LIKE AMERICA

Kennedy Space Center. January 10, 1986.

Are they going to kill my husband?

Rhea tried to push the thought from her mind as she clutched an umbrella on the roof of the Launch Control Center. NASA had already scrubbed the launch of STS-61C four times. Even though she and the other crew families had already been hunkered down in the Cape for over a week, Rhea predicted they would scrub again. Torrential rain poured around her. Lightning flashed in the distance. Three miles away, mission commander Hoot Gibson lay strapped inside *Columbia* readying for the launch. The two had been married for five years now; their son, Paul, was nearly four. In Hoot, Rhea had found her true love and her home. She did not want NASA blotting out her husband in the prime of their lives.

You know where you don't want to be when lightning strikes? Sitting on almost two million pounds of liquid fuel,

Steve Hawley thought as he lay strapped in behind Hoot. As a mission specialist and one of two astronomers on the mission, he would be responsible for observing Halley's comet, which would be visible to earthlings for the first time since 1910. As excited as Steve was about this job, he did not want to die doing it.

Steve's wife, Sally, dutifully stood on the roof of Launch Control with Rhea, but her mind was elsewhere. She and Steve were growing apart. Their rigorous work schedules meant they did not spend much time together. These days, they were more like roommates than husband and wife. More to the point, Sally was changing. She bought a new red Pontiac Fiero, a jaunty, angular midengine sports car that soon developed a reputation for catching on fire.[1] Although she previously shied away from politics, now she openly campaigned for Geraldine Ferraro, the first woman vice presidential candidate, who ran with Walter Mondale against Ronald Reagan. Sally took the stage at the National Women's Political Caucus, where she defended women's rights to abortion and equal pay.[2]

In her personal life, Sally was taking an even bigger risk. She had reconnected with a tennis pal from her childhood—Tam O'Shaughnessy, an elegant woman with high cheekbones, expressive blue eyes, and short brown hair.[3] In their youth, Tam had been the better tennis player. Tam's childhood tennis coach was Sally's idol, Billie Jean King, and Tam had played at Wimbledon and the US Open.[4] When Sally and Tam were kids playing on the junior tennis circuit, their coaches would lament that they preferred to sit on the bench, twirling their rackets and talking instead of playing the match.[5]

Sally and Tam stayed in touch through their twenties. While Sally was getting her graduate degree at Stanford, the two worked at rival magazines, *Sportswoman* and *womenSports*, and occasionally played tennis. They also spent time together off the court. Once, Tam had Sally over to her apartment overlooking San Francisco Bay for a steak and wine dinner. Sally thought the dinner felt borderline romantic, but they both were dating other people.[6]

While Sally toured the country for NASA public relations, she repeatedly found excuses to get to Atlanta, where Tam taught eighth-grade biology. There, Sally and Tam would go on long walks. "We'd talk about the old tennis days," Tam recalled. "And I'd talk about biology, and we'd talk about what we wanted to do in the future."[7] Without a tennis match to distract them, the two could talk for hours.

After one such walk, they returned to Tam's place where her old cocker spaniel Annie waited for them. When Tam leaned over to pet her dog, Sally gently placed her hand on Tam's lower back. "It gave me the chills," Tam said.[8] Tam knew then that she was in love with Sally, and Sally was in love with her.

"We are in trouble," Tam said, turning to Sally.[9]

The trouble was not just that Sally was married. She was famous—a household name, eminently recognizable by her puffed curls and guarded smile. She had thrown out the first pitch at the 1985 World Series, been inducted into the Women's Hall of Fame, and had graced the covers of *Time, Newsweek*, and *People*.[10] If the press ever learned about Sally's affair with Tam, her career would likely be over.[11]

Despite the huge risks, the two fell into an intense romance. Sally flew to Atlanta with increasing frequency. "We were both just madly in love with each other," Tam said. "We hardly ate."[12]

Steve, busy training for his mission, had no idea about the affair, but he recognized that his wife seemed less and less interested in their relationship.[13] Even when Sally was present, as for this launch, "she pretty much made it clear that she wasn't enjoying it," he said.[14]

As Steve sat inside *Columbia,* watching lightning fork through the dark Floridian sky, he was not enjoying life very much either. Rain pooled in the shuttle's windows and soaked through the porous tiles.

"How do you read?" the first communications officer keyed the mic from Launch Control. Hoot was tempted to radio back: "This is the Submarine *Columbia,* how do you read?" Deciding that was not professional, he stopped himself.[15] The crew endured the relentless storm for hours. Finally, when a clap of thunder shook the shuttle, and a bright flash of lightning illuminated the cabin, Hoot's frustration boiled over. "You can come and get us anytime, guys!" he yelled into his mic. Sure enough, the launch control team scrubbed the launch. On the ground, Rhea exhaled a sigh of relief.

George was furious that the launch controllers had sent Hoot and company to the pad in the first place. "You launch when you are ready and when the weather is acceptable," the flight operations director privately fumed. "No one remembers the day you launched, but they do remember if you have a problem."[16]

Because of the delays and cancellations of other flights, STS-61C was bursting at the seams, chock-full of science experiments, odds and ends that made it seem "almost like a year-end clearance sale," said the mission's pilot, Charlie Bolden.[17] Mission specialist Pinky Nelson called their payload "trivial," adding that, "Our main cargo was [Congressman] Bill Nelson."[18] Space enthusiast Nelson, the US House Democrat representing Florida's 11th district, and chairman of the House Space Subcommittee, had been lobbying to fly aboard the shuttle for years.[19] NASA Administrator James Beggs finally wrote him an invitation:

> Dear Mr. Chairman:
>
> As you know, the Space Shuttle has long enjoyed the support of Congress and, in particular, your support . . .
>
> The maturing of the Space Transportation System to the point where we can fly private citizens, as well as engineers from private companies and representatives of foreign governments, is, of course, the beginning of a broadened participation in our space program by many in the United States and abroad. It will culminate with the development of the permanently manned space station in the 1990s. It is therefore appropriate for those with congressional oversight to have flight opportunities to gain personal awareness and familiarity, and because of your well-known interest, we are pleased to invite you to fly.[20]

The crew—nicknamed "Delta House" for their close fraternal bond and affinity for the film *Animal House*—welcomed Nelson as one of their own. Nelson, however, soon discovered that a "maturing" shuttle program did not necessarily mean risk-free flying. In fact, technical issues plagued Delta House and almost yielded disaster.[21]

During the January 6 launch attempt, a valve on a fuel line leading to the external tank did not close properly. Over four thousand pounds of liquid oxygen fuel drained from the tank.[22] Controllers, tired and overworked, did not notice until unusual sensor readings started flashing across screens in Launch Control. There was nitrogen gas in the external tank, where it should not be, engines were undertemping, and the tank pressurization was taking twice as long as normal.

Despite the anomalies, controllers wanted to press on. Faulty sensor readings were common, given the complexity of the shuttle, and the fact that the engines were colder than usual was not a problem in and of itself. The space shuttle main engine controller announced to his team, "Okay, the engines are too cold, but we're just going to mask that parameter, and we can continue."[23]

Thankfully, cooler heads prevailed and Horace Lamberth, director of engineering at the Cape, called off the launch. "Okay, everybody, this countdown has been a disaster. I don't know what's wrong. But something's wrong. Engineering is no go."[24] The decision likely saved the crew. If they had launched, they would not have had enough fuel in the external tank to make it to orbit and would have been forced to attempt a risky emergency landing at a launch abort site on the other side of the Atlantic Ocean.

The next day, Delta House strapped in for another attempt, but bad weather scrubbed that launch. They tried again on January 9, but controllers halted the countdown at T-31 seconds when a liquid oxygen sensor broke off and lodged itself in one of the main engines. If Delta House had launched, the shuttle's engine would have exploded on the pad.[25,26] When they scrubbed on January 10, due to the cataclysmic lightning storm, it marked six failed launch attempts.[27]

A record. Hoot, tongue planted firmly in cheek, identified the reason for their problems: Steve Hawley was bad luck. Steve's last flight, with Judy Resnik, had endured a similar series of delays. For 61C's sixth launch attempt, Steve donned a Groucho Marx disguise: horn-rimmed glasses, a big plastic nose, bushy eyebrows, and a caterpillar black mustache. *If* Columbia *doesn't know it's me,* Steve told Hoot. *Then maybe we'll launch.*[28] Hoot approved the extra cargo.

The disguise fooled the launch gods. With no delay, on January 12, 1986, *Columbia finally* lifted off the pad just before seven in the morning.

"Houston, *Columbia*'s with you in the roll!" Hoot chirped into the com as the bird arced smoothly onto its back.

"Roger roll, *Columbia*," Fred Gregory, CapCom, replied.[29] Mission Control erupted in boisterous applause.

Delta House was not in the clear yet. Seconds after completing the roll program, an alarm rang through the cockpit. Bolden noticed an alert blinking to life on the shuttle's dashboard, indicating a helium leak in the right-side main engine.[30] Helium, an inert gas, acted as a buffer between the highly volatile oxygen and hydrogen fuels. Without enough helium, the rockets would become bombs. The alternative was not much better: If they had to shut down the engine, they would never make it to orbit. With the cabin shaking violently as the shuttle approached max q, Bolden and Hoot began triaging.

Bolden recalled "Hoot's Law," a warning Hoot had offered earlier during training: *No matter how bad things get, you can always make them worse.* Cautiously, Bolden and Hoot searched for the leak. They checked the helium levels, which held stable, and discovered an erratic sensor triggered the alert. Shutting down the faulty sensor, they eased back into their seats, and rocketed to orbit. *Another crisis averted.*

Almost as soon as *Columbia* slipped free of Earth's atmosphere, the crew learned their mission would be cut short by a day.[31] Engineers at Kennedy needed to scavenge parts from *Columbia* to turn *Challenger* around for its upcoming launch. However, Mother Nature had other plans: Persistent bad weather at the Cape forced Hoot to land two days late, and on the other side of the country at Edwards. During descent, an alarm signaled that an auxiliary power unit was overcooling and pulling in extra water. Without a moment's hesitation, Bolden shut off water to conserve the hydraulic system until touchdown.

Is it just me, Steve thought, *or is everything falling apart?*[32]

Columbia touched down at Edwards under the bright glow of xenon lights. TFNG Dan Brandenstein led the ground team to the parked shuttle, ferrying Burger King meals to the tired crew.[33]

Steve Hawley called his dad to tell him he had returned. "This just proves to me that you guys are really operational and that the space shuttle is just like an airline," his dad said.

"Why do you say that?" Steve asked.

"Because you're in California and all your luggage is in Florida."

The year 1986 promised fifteen scheduled flights, NASA's most ambitious year yet. Three classified DoD payloads, seven commercial communications satellites costing their customers an average of $115 million per flight, and five astronomy missions filled the schedule. The year would also mark completion of the Air Force's $3 billion launch site at Vandenberg and see the first shuttle launch from the West Coast.[34] George tapped Crip to command the first flight from Vandenberg, which would launch into polar orbit. Crip and his crewmates would be able to view nearly every square foot of Earth as the shuttle made dozens of longitudinal orbits around the rotating planet each day, allowing them to photograph and study any number of terrestrial targets for the Defense Department.

After Crip's mission, in October 1986, NASA would launch its scientific crown jewel, the billion-dollar Hubble Space Telescope, which would peer into the dark corners of the cosmos, gazing fourteen billion light-years away and illuminating the origins of the universe.[35,36]

STS-51L, the first of the astronomy missions slated for 1986, and STS-61E would include observations of Halley's comet, which only appeared once every seventy-five years. STS-61E, the more ambitious of the two missions, would scoop the Soviets, whose probe was due to snap photos of the comet on March 9, 1986. The other two exploratory shuttle flights in May 1986 would ferry two deep-space probes, Galileo and Ulysses, to orbit. They were scheduled to launch within six days of each other—first on *Challenger*, then on *Discovery*—taking advantage of the alignment of the planets. One probe would slingshot around Jupiter to study the sun, and the other would research the striped gas giant itself. Both probes would rely on a powerful Centaur second stage rocket, a Pentagon-funded propulsion system that John Young nicknamed the "Death Star," a dark acknowledgement that "this was going to be the riskiest mission the shuttle would have flown up to that point," according to mission commander Rick Hauck.[37,38] Debuting on Hauck's mission, the Centaur upper stage rocket was more powerful than previous upper stage rockets and would be able to ferry heavy Department of Defense payloads into geosynchronous orbit on future flights.

Hauck worried about the rocket technology and approached higher-ups in early January 1986, requesting they improve the safety of the Centaur before its first flight. When a shuttle program manager denied his request, Hauck went back to his crew and candidly told them, "NASA is doing business different from the way it has in the past. Safety is being compromised, and if any of you want to take yourself off this flight, I will support you."[39]

Equally infuriated, George went through a backchannel to raise the issue with his friend, Congressman Don Fuqua, chairman of the House Science and Technology Committee, which had oversight over the shuttle program's budget. George had been friends with Fuqua's administrative assistant Tom Tate since the 1970s and kept a close friendship with the congressman through Tate. Hoping that pressure from outside the organization might spur change, George had Fuqua briefed about the dangers of the Centaur rocket in a closed-door meeting. Although Fuqua sympathized with George's plight, he was not sure he could move the needle internally at NASA.[40]

Grand ambitions require wise leadership, but by the mid-1980s, the agency's management was tottering. Beggs was still persona non grata with conservatives for allowing Jane Fonda at Sally Ride's historic launch. He had further ostracized himself by orchestrating a private meeting with the Gipper to secure funding for the space station, bypassing key Reagan officials like science advisor George Keyworth and trusted advisor Ed Meese.

Beggs's career detonated the Monday after Thanksgiving in 1985, when a federal grand jury indicted him for graft during his private industry days. Beggs claimed the charges were politically motivated and refused to leave NASA unless President Reagan personally asked. His deputy William Graham took over his station, but Beggs continued to lurk in a back office, undermining Graham's leadership and assuring

NASA employees he would be back soon.[41] A year later the Department of Justice would dismiss the charge, with US Attorney General Edward Meese issuing Beggs a "profound apology" for any embarrassment.[42]

The damage had already been done. In the weeks that followed the November 1985 indictment, NASA's chain of command fragmented. Some officials—acting either from loyalty or fear of reprisal—reported to Beggs. Others to Graham. As the agency faced administrative disorder, Johnson's director Gerry Griffin, who had been at NASA for more than twenty years, resigned, leaving the agency without a continuity of leadership.

On January 25, 1986, Jesse Moore, the associate administrator for spaceflight was installed as Griffin's replacement. Some, including George, thought Moore did not have the chops to run human spaceflight.[43] Moore had worked in the Space Science and Spacelab program at NASA Headquarters only two years, and had little experience with Johnson. As George said, "Jesse was a nice guy, but he had no business being put in charge of human spaceflight."[44] The flurry of management changes did little to engender confidence in the program's leadership. "This place is like a ship without a captain," one engineer reflected.[45]

Despite the dysfunction under the hood, NASA's disaster-free shuttle record seemed proof of its promise to make space travel routine and ordinary. In the press, riders of the shuttle touted that it was safer than driving in city traffic.[46] Ironically, NASA's success was also its undoing. The regularity of spaceflight dulled public interest. Major networks stopped broadcasting launches, due to their low ratings. Astronauts were no longer household names. The loss in public interest undermined congressional support.

NASA needed something new and splashy to attract public interest. Beggs, leaning into the shuttle's image of reliability, suggested an "ordinary person" do the extraordinary: Go to space.[47] *But who?* Should they send a journalist, an explorer, an entertainer, or a poet? NASA announced the idea to the public, and applications flooded in.[48] Famed newscasters Walter Cronkite and Tom Brokaw as well as Tom Wolfe, author of *The Right Stuff*, all applied.[49] NASA administrators even considered sending *Sesame Street*'s Big Bird. They quickly scrapped the plan when they realized the costume would not fit through the orbiter's hatch.[50]

Then Reagan, up for reelection that year and seeking support from the teachers' union, parried opponent Walter Mondale's barbs about education with an announcement in August 1984: "Today I'm directing NASA to begin a search . . . [for] the first citizen passenger in the history of our space program, one of America's finest—a teacher."[51]

Over 11,400 applications poured in from schoolteachers around the country. On July 19, 1985, in the Roosevelt Room at the White House, Beggs and Vice President George H. W. Bush gathered with the ten finalists. "When the shuttle lifts off," Beggs opened, "so will our winning candidate, right into the hearts and minds of young people around the country." Then Bush announced the winner: "Christa McAuliffe."[52]

A thirty-six-year-old high school social studies teacher from Concord, New Hampshire, Christa would take a year-long sabbatical from teaching—as would her backup, Barbara Morgan—to prepare for the mission. Onboard the shuttle, Christa planned to broadcast elementary science experiments live to classrooms across the country. Wearing an oversized, butterscotch blazer decorated with a red rose corsage, Christa stepped up to the podium next to Vice President Bush.

"It's not often that a teacher is at a loss for words. I know my students wouldn't think so," she began, smiling and emotional. "I've made nine wonderful friends over the last two weeks. And when that shuttle goes," she said, fighting back tears, "there might be one body, but there's going to be ten souls that I'm taking with me."[53] The nine other finalists standing behind Christa, who would travel with her in spirit only, warmly congratulated their new friend.

After the announcement, Christa's life changed overnight from garden-variety teacher to NASA celebrity. Much beloved by her students, Christa embodied a real-life Ms. Frizzle with a heavy New England accent. Undeniably cheerful, blessed with curiosity, intelligence, and an ability to make complex ideas accessible, she kindled in her students a love of learning and a belief in themselves.

"Students will say 'this ordinary person is contributing to history,' and if they can make that connection, then they're going to get excited about the future," she said.[54] Christa told her students that she was "reaching for the stars, and so should they."[55] Her motto was "I touch the future. I teach."[56]

Some at NASA privately disparaged the mission as a press stunt and warned that flying a civilian courted an unacceptable risk. However, Beggs figured, "The biggest receptive audience we have in this country are the kids. Kids love space. A teacher could give you an introduction to those kids that no one else could."[57]

Christa was assigned the role of payload specialist on STS-51L, one of the astronomy missions set to launch on *Challenger* in January 1986. With her addition, the 51L crew would be the most diverse yet.[58] Ron, El, and Judy had already been assigned to the flight in January 1985. TFNG Dick Scobee would be *Challenger*'s commander. The New Guys had come to know him by his "if I can do it, anyone can" attitude.[59] Scobee had also recently been tapped to lead the military's Vandenberg launch site. He had a long career at NASA ahead of him.[60] Mike Smith, a naval aviator who had flown in the Vietnam War and joined the Astronaut Office with Bill Fisher in 1980, would pilot *Challenger*. Though the launch would be his first, he won respect for his unimpeachable flying skills and his diplomatic touch, which would be particularly helpful on such a highly publicized mission.[61]

The final seat on the flight went to payload specialist Greg Jarvis, an employee of Hughes Aircraft Company one of NASA's most important aerospace contractors. Jarvis had already been bumped from two previous flights in lieu of more politically-connected passengers, Senator Garn and Congressman Nelson.[62] Now his turn had arrived.

Christa, a consummate professional, worked hard to prove herself an equal part of the team. She kept a full schedule—flying in the Vomit Comet, training for EVAs in the WETF pool, practicing her experiments for broadcast, and running through simulations with the crew.[63] However, press requests ate into her schedule and she only had four months total to train. In the end, Christa's preparation did not extend beyond the basics—how to go to the bathroom in space, how to prepare food, and what switches not to touch.[64]

Technicians at the Cape worked through the Thanksgiving and

Christmas holidays to ready *Challenger* before the scheduled launch date of January 22.[65] Despite their best efforts, the launch pushed, when weather delayed "Submarine *Columbia*'s" return. Kennedy technicians met *Columbia* on the runway at Edwards—harvesting a propulsion system, a temperature sensor, a nose-wheel steering box, an air sensor, and one of its five computers—before flying back to Kennedy to install them on *Challenger.*[66]

NASA managers reset *Challenger*'s launch for January 26, 1986—Super Bowl Sunday, a day when millions of Americans would be glued to their screens. On January 23, the crew left Houston for the Cape. They landed their T-38s in close formation—Dick Scobee with El Onizuka, Michael Smith with Judy Resnik, and John Young with Ron McNair—on the same runway where *Challenger* was to land at mission's end. They jogged over to meet up with Christa McAuliffe and Greg Jarvis, whom George escorted on NASA's Gulfstream, and marched in loose succession to the waiting press.[67]

"Looks like a war-movie propaganda platoon," one reporter said, referencing World War II movies like Frank Capra's *Why We Fight*, which put America's diversity on center stage. "Got one of everything."[68] Another reporter remarked that the crew "exemplified the sweeping social changes that had gripped the country during the fifteen years since the shuttle project began"—from Judy, a divorcée engineer and "new American woman," to El, a flight engineer and the first Asian American in space, to Ron, an African American who defied the expectations of his upbringing in the segregated South, to Christa, who represented the promise of America's future.[69]

Judy removed her helmet to speak, her dark curls falling around her shoulders. She joked to the press that she hoped Steve Hawley's "affliction"—his bad luck with delays—had not rubbed off on her. She handed the mic to El, who expressed his gratitude and said he was "ready to go fly."[70] Ron, wearing tinted aviators, spoke next, recalling his first shuttle landing at Kennedy and affirming his joy to be doing it again. He then brought up "the person you came to see," Christa McAuliffe.[71] Christa's baggy flight suit and tennis shoes—as opposed to black boots—made clear that she was the non-astronaut. "Well, I am so excited to be here," Christa beamed. "I just hope everybody tunes in . . . to watch the teacher teaching from space."[72]

After the press conference, their families filled in around the crew for photos. Because of intense training schedules and quarantine, the crew had not seen much of their loved ones over the past few weeks. It was a real reunion—gussied up for the press to wring hearts—but real nonetheless.[73]

As expected, the press loved it. The crew's enthusiasm was infectious. "They were beautiful," one twenty-something education writer covering her first shuttle mission gushed. "They're just like . . . well, the best America can offer."[74]

CHAPTER 18

GODSPEED

Clear Lake, Texas. January 1986.

"Bring it to me screaming," Judy told the waiter.[1] It was Saturday night and Judy was out at her favorite steakhouse, the Longhorn, with her best friend Sylvia Salinas.

Sylvia had been Judy's first assistant at NASA, but now the two were more like sisters.[2] The daughter of a Mexican American grocer, Sylvia joined NASA at the tender age of eighteen as a stenographer, after the Apollo 13 mission. NASA fell far out of her family's experience, but she performed excellently on a civil service typing test and NASA recruited her to the steno pool. Support staff like Sylvia did not socialize with the more seasoned Apollo astronauts, but when the New Guys arrived—women and minorities like her—the astronauts and administrative staff became fast friends. They played on each other's softball teams; they ate barbecue together; they partied together.[3] The Astronaut Office became Sylvia's family and Judy became her best friend. Judy saw that Sylvia was bright and encouraged her to take night classes, sometimes even tutoring the "math-averse" Sylvia. Now Sylvia was well on her way to earning an undergraduate degree.[4]

The Longhorn was dimly lit. Large, private, brown leather booths lined the room and a bull's head hung over the fireplace. Sylvia and Judy stood out from the restaurant's typical clientele. "Two trippy young girls . . . '80s hair, sitting there with these big steaks in front of us," Sylvia recalled. "It was hard to miss the two of us."[5] Indeed they were not missed by the men sending kamikaze shots to their booth. The women were drinking wine but did not want to reject a good-natured gesture.

Who sent them?

It did not matter, their dance cards were full. Sylvia had recently married, with Judy by her side as maid of honor, and Judy was dating astronaut Frank Culbertson, a Southern charmer with a kind heart. Like Judy, Frank was divorced. When he asked Judy to meet his three daughters at Christmastime in 1985, Judy brought Sylvia as emotional support. "I cannot do this by myself," she told Sylvia.[6]

Frank had brought up marriage, but Judy was not sure she was ready to be a stepmom.[7] Sylvia advised her not to string Frank along. "It's not fair to him; it's not fair to you."[8]

Buying the women's drinks was as far as any admirers would get that night, but the kamikazes were strong. Despite the ribeyes lining their stomachs, Judy and Sylvia were flushed.[9]

"I think you better come to my place," Judy told Sylvia.

"I think I'd better, too," Sylvia agreed.

Sylvia slowly followed Judy's orange Volkswagen bug back to her nearby townhouse in Seabrook on Galveston Bay.[10] Judy had furnished it minimally, as befit an engineer. No pictures adorned the walls or knickknacks cluttered the shelves, but piles of technical papers lined her stairs and filled her study and living room. Judy prepared for a flight by relentlessly drilling herself on the mission after work each night and memorizing her checklists before bed.

"That was her way of making sure everything was okay," Sylvia said.[11] *If anything goes wrong with the flight, it won't be because of me,* Judy resolved. "I'm not scared," Judy once told the press.[12] "I think something is only dangerous if you are not prepared for it."[13]

Sylvia chatted as Judy swiveled in her favorite chair, a chocolate-brown Eames she had bought in Dallas with her sister Linda's interior designer discount.[14]

In the middle of the room was a black Steinway piano. It was pristine, like a museum piece. Though Judy never played publicly, she had finally gotten to a place where she enjoyed playing for herself.

"It soothed her. Like some of us go exercise, she played the piano," Sylvia said.[15]

Sylvia eyed the gleaming piano across the room. She was just tipsy enough to ask: "Will you play the piano for me tonight?"

"No!" Judy said.

"Come on—why do you even have it?" Sylvia asked. "I am begging you. Just once!"

"Nah, no," Judy said again. "Playing for me is very, very personal."

Judy did not talk about her childhood often, but Sylvia knew enough to understand the hurt behind her words. Still, tonight would be different.

"I'm not leaving here until you play at least *something*."

"Good, you can sleep on the couch," Judy said.

Sylvia did not budge. Her eyes said, *you can trust me with this.* Judy softened.

Ah geez. Okay. I see you are not going to let this go.

Judy sat at the bench, straightened her back, and stretched out her long, elegant fingers. Sylvia watched her friend transform as she played. Judy's hands spun down the keys in swift, deliberate movements—Rachmaninoff by heart. She had played it a thousand times. Her head bobbed from side to side as she swayed with the melancholy phrases. Her performance was grand, full, muscular; she was a virtuoso.

"Just watching I got goosebumps. I knew it was a special moment," Sylvia said.[16] It was an act of trust from a woman who kept the intimate details of her life closely wrapped around her. Sylvia would cherish that night for the rest of her days.

Clear Lake High School, Houston. Mid-January 1986.

He came for the soccer ball. Janelle Onizuka looked up from team practice to see her dad jogging across Clear Lake High's freshly cut green field. He was supposed to be in quarantine.

El broke out into a smile when he saw his daughter—now seventeen and looking more like him than ever. In addition to being an astronaut and a lieutenant colonel in the Air Force, El coached the Lady Falcons—Clear Lake High School's girls' soccer team. He never missed a game, at least not while he was on Earth. "[When] he was supposed to be in quarantine, he would sneak out just to see . . . the game," his wife, Lorna, said. "None of us would know until we'd see him at the corner of the fence. When we'd look up, he'd be gone."[17]

The Lady Falcons adored Coach El and his goofy antics. He could not assign drills or lecture the girls without breaking into a broad grin or lobbing a dad joke. Coach El inspired more laughter and joy than fear, which was why, all week long, the team had been signing a practice soccer ball for El. Their beloved coach was going to take their ball to space.

As soon as they spotted him, the Lady Falcons rushed over to present him with the gift. El took it graciously, reading each of their names off and the inscription carefully written in thick black marker: GOOD LUCK, SHUTTLE CREW![18] Janelle said goodbye to her dad, and he gave her one more of his toothy smiles. Before returning to practice, she watched him jog back to his car across the field, acting like a child playing hooky, jumping over a little ditch and disappearing behind a hedge.

Unlike El's classified first flight, 51L was the highly publicized "Teacher in Space" flight. Since he could not gather his large extended family for his first mission launch, he invited them all to this one—over sixty Hawaiian relatives flew to the Cape for launch. The Onizuka clan would then head to Houston to celebrate his touchdown in traditional luau style: kālua pig, lomi salmon, and Parker Ranch beef.

Back in crew quarters, El packed his black duffle bag with keepsakes: a family picture with the governor of Hawaii, a lotus charm for his Buddhist faith, a football from his alma mater, the University of Colorado, a pennant from his high school, and the signed soccer ball from his daughter's team.[19]

The first six American women astronauts, Rhea Seddon, Anna Fisher, Judy Resnik, Shannon Lucid, Sally Ride, and Kathy Sullivan, pose at a press conference announcing their selection.

The first three African American astronauts, Guy Bluford, Ron McNair, and Fred Gregory, pose for an official photo in their Extravehicular Mobility Unit (EMU) suits.

Sheraton-Kings Inn

To: the Astronaut Selection Committee

The task of answering the question "Why do you want to be an astronaut?" without sounding trite is a difficult one. There are several intellectual reasons that I can offer but my basic underlying motivation in applying for a position as a mission specialist is that becoming an astronaut has been one of my lifetime dreams. Somehow the idea and challenge of space travel captured my imagination and now the possibility of becoming an astronaut seems like "a dream come true." I feel that space exploration is perhaps the ultimate destiny of mankind and that being an astronaut is a lifetime mission which I can pour my entire being as well as my intellectual resources and academic background into. The chance to participate in the early stages of that destiny would be quite a privilege. Furthermore, the type of "generalist" training that is required exactly meets my combined interests in both the mathematical-chemical-physical sciences and medicine. Until now I have essentially had to choose between the two disciplines. The idea of being trained for a mission which places both physical and mental demands on me meets a need somewhere in the depths of my being. I realize that there will be certain significant sacrifices which I must make in both my personal and professional lives in order to become a mission specialist astronaut but these are sacrifices which I thoughtfully and willingly will make if given the opportunity to fulfill a lifetime dream.

Anna Lee Sims, MD

"Why do you want to become an astronaut?" Anna Fisher answers the question for the Astronaut Selection Committee.

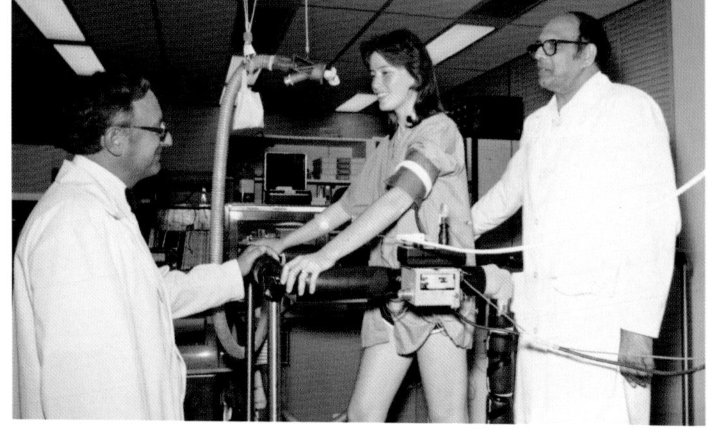

Astronaut candidate Anna Fisher undergoes cardiopulmonary testing at Johnson Space Center during interview week.

Nichelle Nichols, aka *Star Trek*'s Lieutenant Uhura, sits in Mission Control for a 1977 NASA recruitment advertisement aimed at encouraging women and minority applicants to the space shuttle program.

Nichols joins NASA officials and the rest of the cast of the *Star Trek* television series at the 1976 *Enterprise* rollout in Palmdale, California. With her are DeForest Kelley, George Takei, James Doohan, Leonard Nimoy, Walter Koenig, and series creator Gene Roddenberry.

The official 1978 class photo of Astronaut Group 8, or "The New Guys," taken in Johnson's Teague Auditorium.

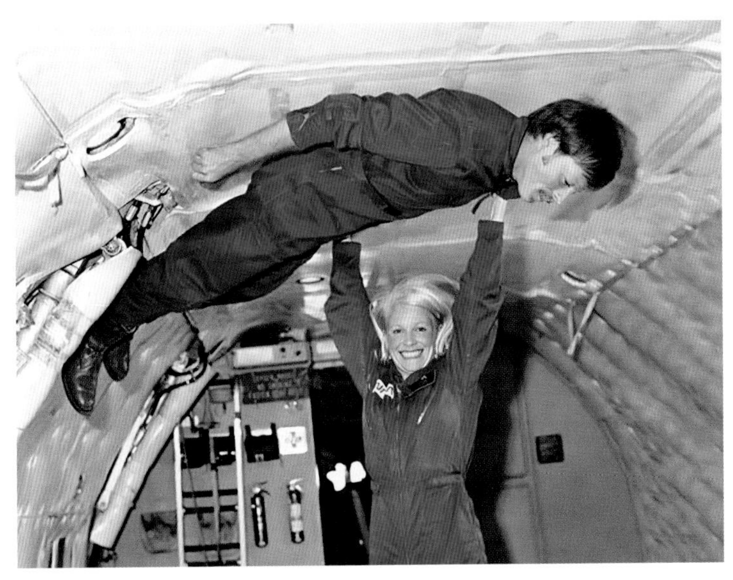

Rhea Seddon hoists classmate and future husband Hoot Gibson overhead while performing zero-gravity training in NASA's "Vomit Comet."

Sally Ride gets ready to jump into Biscayne Bay for water survival training at Florida's Homestead Air Force Base.

Astronaut candidates during water survival training at Florida's Homestead Air Force Base. Pictured from left to right: Shannon Lucid, Steve Hawley, Jeffrey Hoffman, Ron McNair, and Rhea Seddon.

Rhea Seddon (left) prepares to fly back seat in a T-38 with Hoot Gibson (right).

Director of flight operations George Abbey (center)—also known as the Dark Lord or UNO for "Unidentified NASA Official"—walks out with the crew of STS-5 in a rare photograph.

Sally Ride graces the January 1983 cover of *Ms. Magazine* ahead of her historic first flight.

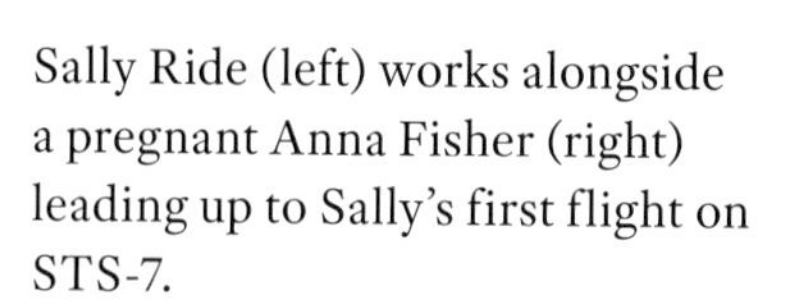

Sally Ride (left) works alongside a pregnant Anna Fisher (right) leading up to Sally's first flight on STS-7.

Guy Bluford, the first African American to fly to space, speaks with Vanessa Williams, the first African American Miss America, and Coretta Scott King at a press conference following his historic flight on STS-8.

NASA administrator James M. Beggs poses with an early model of the space station in 1984.

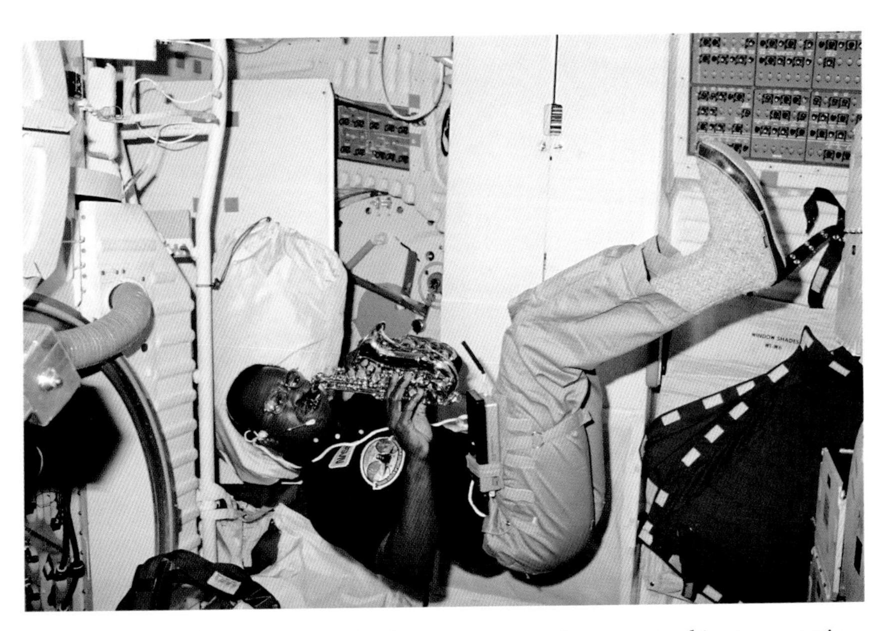

Ron McNair makes history as the first person to play a musical instrument in space on STS-41B.

Bruce McCandless II debuts the Manned Maneuvering Unit (MMU) on STS-41B, making him the first human satellite.

Judy Resnik, Steve Hawley, pilot Mike Coats, commander Hank Hartsfield, Mike Mullane, and Charles Walker of the STS-41D crew in orbit.

Kathy Sullivan leads the crew of STS-41G to the Astrovan for a ride to the launchpad. STS-41G, which included Sally Ride, was the first time two women traveled to space together.

Kathy Sullivan (right) makes history as the first American woman to perform a spacewalk with David Leetsma (left), as they perform a satellite refueling experiment.

Sally Ride kisses husband and classmate Steve Hawley after STS-41G lands at Cape Canaveral, Florida, on January 2, 1984.

El Onizuka, the first Asian American in space, wears a headband emblazoned with the rising sun symbol and the word "kamikaze" on STS-51C.

Dale Gardner holds up a FOR SALE sign after he and Joseph Allen miraculously retrieved two wayward satellites on STS-51A.

Couples Bill and Anna Fisher, and Hoot Gibson and Rhea Seddon hold their "Astro-tots," Kristin Fisher and Paul Gibson.

The *Challenger* crew attends a training session before the launch of STS-51L. Pictured from left to right: El Onizuka, Ron McNair, Gregory Jarvis, Teacher in Space winner Christa McAuliffe, Judy Resnik, and backup payload specialist Barbara Morgan, who was the runner-up for the Teacher in Space competition.

The crew of STS-51L, Mike Smith, El Onizuka, Judy Resnik, and Dick Scobee, runs through simulations in the Shuttle Mission Simulator.

Icicles form on the launchpad the night before *Challenger*'s doomed launch, due to record-low temperatures at Kennedy Space Center.

Fred Gregory (left) and Dick Covey (right), acting as capsule communicators (CapComs), struggle to understand what is happening in the immediate aftermath of the *Challenger* accident.

President Ronald Reagan and First Lady Nancy Reagan with the Scobee (left) and Smith (right) families at the memorial service for the fallen *Challenger* crew.

A diver is hoisted aboard the USS *Preserver* while recovering pieces of the space shuttle *Challenger* from the ocean.

The esteemed members of the Rogers Commission swear in on the first day of the hearings to investigate the cause of the *Challenger* accident. The members included Sally Ride, Neil Armstrong, and Nobel Prize–winning physicist Richard Feynman.

A recovered segment from *Challenger*'s right wing.

Kathy Sullivan prepares to don her EMU suit in the airlock of *Discovery* during the Hubble Space Telescope's deployment mission.

Shannon Lucid (left) jokes with the "Yuris," Usachov (center) and Onufriyenko (right), on *Mir* space station, where she spent a total of 188 days, setting a new space endurance record for most time in space by any American astronaut.

Guy Bluford, wearing an EMU suit, is lowered into Johnson's Weightless Environment Training Facility as part of his training for STS-39.

NASA deputy administrator Fred Gregory greets the coffins of *Columbia*'s fallen crew at Delaware's Dover Air Force Base on February 5, 2003.

Shannon Lucid serves as CapCom for STS-135, the final mission of the space shuttle program, on July 12, 2011.

The surviving first women astronauts, Shannon Lucid, Kathy Sullivan, Anna Fisher, and Rhea Seddon, reunite with their mentor Carolyn Huntoon for a Christmas party at Villa Capri in 2017.

Building 4, Johnson Space Center. 1985.

Jean-Michel Jarre strolled through the long hallways of the Astronaut Office, poking his head into each room as if searching for something in particular. He was clearly an outsider in this buttoned-up world of aerospace engineers, scientists, and test pilots. His flowing, shoulder-length black hair, styled like a 1980s rocker, whooshed against his black leather jacket. His whole outfit, in fact, was black: black jeans, black V-neck, black jacket. He looked as though he should be DJing at a dance club, not walking past bulletin boards laden with the latest interoffice memos and chili cook-off announcements.

Jarre, a multiplatinum, new age electronic musician from France, was known for creating sonic and visual spectacles, complete with blaring synthesizer music, laser beams, and enormous projections. He came to meet with NASA organizers to brainstorm ideas for a spectacular concert celebrating the 150th anniversary of the city of Houston and the 25th anniversary of Johnson Space Center.[20] Jarre was looking for inspiration, but he had not found it yet. Then he met Ron McNair.

As fellow musicians they were kindred spirits, humanists, and dreamers.[21] *In the total silence of space, what you hear the most is your own heartbeat,* Ron told him. Jarre latched on to the notion: He would compose a piece of music for Ron to play in space on his next mission and beam it down to Earth during a live concert in Houston. Ron's own heartbeat would serve as the percussive back beat to the piece.[22]

Ron's space performance would be projected on to a skyscraper during the sprawling party. A crowd of millions, celebrating at outdoor venues all over downtown Houston, would see and hear Ron's performance, a majestic interlude linking space and Earth in a new way—through the sound of a saxophone and Ron's very own heartbeat.[23]

Not only would Ron reclaim victory from the failure of his earlier inflight jazz concert, but this saxophone solo would also be his swan song at NASA. After his second flight on 51L, Ron planned to resign from the space agency and join the University of South Carolina's (USC) physics department.[24] He had not told anyone in the Astronaut Office yet. Ron and Cheryl had often discussed the idea of returning to his home state. He wanted to be closer to his family now that they had two rambunctious toddlers to juggle: Reggie was almost four and Joy was eighteen months. Ron yearned to give back to his community.[25] According to USC's dean, Ron wanted to prove that "Blacks don't grow up and prosper and move away. They go away and prosper and come back."[26]

Ron threw his heart and soul into preparing for the seven-day mission and its two satellite deployments. The first was the Tracking and Data Relay Satellite (TDRS), the second of three communications satellites that would enable Mission Control to remain in constant contact with the shuttle crew. When completed, the TDRS System would replace most of NASA's ground stations and would provide full orbital coverage of spacecraft-to-ground communications.[27] The second was a free-flying spacecraft named SPARTAN—the same deployable and retrievable carrier module that Shannon deployed on STS-51G—that would be released and retrieved by Judy and Ron using the remote arm. SPARTAN had an array of instruments to observe Halley's comet when it was closest to the sun, which only happened once every seventy-five years.[28] Although ancient civilizations as far back as the Babylonians had observed Halley's comet, its 1986 emergence would be the first time humans could send a vehicle to capture images.[29] The Soviets, Japanese, and Europeans were also sending unmanned probes to observe the comet, but America wanted to be first.[30,31]

In case something went wrong with these deployments, two crew members would train for spacewalks. Ron volunteered, despite the many underwater drills the job required. On his first flight, Ron was inspired watching Bruce McCandless make history with the Manned Maneuvering Unit. He wanted in. Soon he was requesting more time in the Johnson pool, where he worked on payload mockups twenty-five feet below the surface.[32] El joined him as the second spacewalker. Engineers monitored Ron and El's performance while an Intravehicular Activity officer guided the spacewalkers, the way a belayer might offer pointers to a rock climber.[33] Trainers were impressed with Ron's agility. Some attributed his abilities to his early karate training, which likely honed his psychomotor skills.[34] Watching Ron in the pool, no one would have guessed that he arrived at NASA terrified of the water.

Between preparing for his solo and prepping for an EVA, Ron's packed schedule did not leave much time for his new baby, Joy. *After this,* he told himself, *there will be time.* Training for a final mission proved bittersweet, but Ron committed to enjoy every moment. Friends and family were flying in to see him launch, and Cheryl planned to bring the kids, too. "We were just going to enjoy that last flight," she said. "We were looking forward to it."[35]

Kennedy Space Center. January 26, 1986.

On Superbowl Sunday, the crew awoke to a gorgeous day—a surprise since their launch had been scrubbed due to forecasts of bad weather.[36] Christa McAuliffe and Greg Jarvis took advantage of the unexpected free time, riding their bicycles around Kennedy. A television news reporter pointed a camera in their direction. After several additional scrubs and launch delays in the previous days, Christa offered up a weary smile. "It must be a slow news day," she quipped.[37]

The next day, they would try again.

January 27 was the anniversary of NASA's biggest tragedy—the Apollo 1 fire. The event had been seared into the New Guys' memories when they listened to the chilling audio of the incident as AsCans. Like Grissom, White, and Chaffee nearly two decades earlier, they were now at the mercy of a great mechanical beast.[38]

By 7:56 AM, the crew sat strapped in their seats, ready to fly.

The crowd in the grandstands, filled with anticipation after the previous day's scrub, waited for the countdown to begin.

The closeout crew engineer only needed to seal the hatch door before *Challenger* blasted off. As the world watched, a technician wearing white coveralls tried to remove the latch handle, but his wrench spun around in his hand.[39] *The goddamn bolt was stripped.* If they could not get the bolt off, they could not remove the latch handle, and the handle had to come off before launch.[40]

One engineer called for a drill, which took forty-five agonizing minutes to arrive. Not that it mattered—the drill would not power on. The batteries, having been out in the cold all night, had drained. With only ninety minutes left in the launch window, the team radioed for permission to improvise. *Let's cut the damn latch off with a hacksaw.*[41]

In the Oval Office, President Reagan, glued to a live feed of the pad, shrank in embarrassment as the closeout crew fumbled with an ordinary hacksaw, looking more like the Three Stooges, said news outlets, than space engineers.[42,43] By the time they got permission to saw off the metal handle, two and a half hours had passed.[44] Thirty mile per hour winds whistled around the orbiter, above the crosswind limit for launch. *Another scrub.*[45] The crew emerged from the hatch after a five-hour hold, tired and frustrated.

Challenger's postponement headlined the evening news. "Yet another costly, red-faces-all-around, space shuttle launch delay," reported Dan Rather on *CBS Evening News*. "Once again a flawless liftoff proved to be too much of a challenge for the *Challenger*," ABC's John Quinones reveled in wordplay on *World News Tonight*. The press portrayed the white room technicians as a bunch of bumblers in a "high-tech low comedy."[46] *Your tax dollars at work.* Some complained the snafu was Beggs's Shuttle Sweepstakes come to bear: Cost-cutting measures removed decision makers from the Cape, decreasing responsiveness to time-sensitive technical decisions—*like that damned latch.*[47]

The moment humiliated NASA and the Reagan administration, which had widely promoted the school-teacher flight. The president's annual State of the Union address was scheduled for the following evening, January 28. Reagan planned to point to Christa's trip as evidence of his commitment to the teachers' unions. "Tonight, while I am speaking to you," a draft of his speech read, "a young schoolteacher from Concord, New Hampshire, is taking us all on the ultimate field trip as she orbits the Earth as the first citizen passenger on the space shuttle." Reagan would also recommit to NASA, which, given the Draconian federal budget cuts that his administration touted to balance the budget, would be a lifeline for the agency. "Christa McAuliffe's journey is a prelude to the journeys of other Americans and our friends around the world who will be living and working together in a permanently manned space station."[48]

Not only did NASA managers want to keep Reagan happy, but the pressure was on to trump the Russians in the race to capture data on Halley's comet. Perhaps most importantly, they needed *Challenger* to fly and return in time to send their very expensive deep space probes, Galileo and Ulysses, to Jupiter and the sun—the trial run of the military's new Centaur technology.[49] If the agency could not launch *Challenger* by the following day, the domino effects would be punishing.[50]

However urgent the situation, NASA could not control the weather.

The howling winds that stalled that morning's launch were a harbinger of much worse to come. A dry arctic ridge crept south from Canada, bringing with it freezing temperatures.[51] NASA had never faced these kinds of conditions. Arnie Aldrich, the director of the National Space Transportation System, reached out to his NASA subordinates and the shuttle contractors to get a prognosis on how the cold might affect shuttle systems.[52]

The chain of command to get a go for launch ran from the contractors to the various centers that managed them (Marshall, Kennedy, and Johnson), then back through a series of Flight Readiness Reviews (joint meetings between the centers) and on to the Mission Management Team, which consisted of the launch director, the chief engineer, and the weather director, among others. The collective assessments then landed in the lap of the shuttle project office manager Stanley Reinartz, who reported to Aldrich. In turn, Aldrich would make his recommendations to Jesse Moore, the new head of Johnson Space Center who, at the time, was still operating as the associate administrator for spaceflight. As the associate administrator, Moore would give the final "go" for launch.

Larry Mulloy, head of Marshall's booster rocket program, began contacting his support staff and his contractors to understand how the potential cold might affect the shuttle's rocketry. Managers at Morton Thiokol, the firm responsible for the shuttle's boosters, rallied their forces to research any possible launch constraints the cold might trigger. Within Thiokol, a handful of engineers had been vocal about the effect of cold weather on the field joints' O-rings for the past year, but Thiokol managers continued to drag their feet on the issue. After thirteen years of Thiokol's monopoly on the boosters, NASA had yielded to Congressional pressure to open booster production to competitive bidding. NASA was scheduled to meet with Thiokol about the future of their relationship the very next day—January 28.[53] With other aerospace contractors circling, Thiokol dreaded any technical issue that might affect their reputation.

Allan McDonald, who directed the booster rocket project for Morton Thiokol, traveled to the Cape for the launch. When he learned of the coming cold front—meteorologists predicted the overnight temperatures would drop into the teens—he quickly arranged a teleconference between Morton Thiokol in Ogden, Utah; Marshall in Huntsville, Alabama; and the Thiokol and Marshall teams stationed at the Cape. Together, they would discuss if the O-rings would properly seal in these record-low temperatures.[54] In the hours leading up to the teleconference, McDonald grew anxious. "It was the first time in my two-year career in the Shuttle program that we were experiencing a last-minute crisis of this magnitude just before a launch," McDonald said.[55]

The telecon was set for 8:15 PM.[56] As meeting members streamed in, Thiokol's engineers in Utah were still faxing over their hastily handwritten charts.[57] Thiokol engineers and managers, as well as Marshall's top dogs, listened to an hour-long presentation from Boisjoly and other engineers from the O-ring Task Force. They explained that the primary O-rings had shown signs of erosion on prior launches, and that the secondary O-rings may not fully seal in freezing temperatures. Robert Lund, vice president of engineering at Thiokol, concluded that Thiokol did not recommend launching below 53°F, the lowest temperature at which the O-rings were known to function.[58] Weather reports predicted that the temperature at launch time would be a mere 26°F.[59]

Mulloy, to whom Thiokol reported, was outraged. He questioned Boisjoly's logic, arguing that the engineers' findings were inconclusive. *Yes, Thiokol had observed blowby during cold launches, but hadn't blowby been observed during a warm launch, too?* Mulloy was referring to Fred Gregory's 51B launch, for which the temperature had been in the 70s. The flawed argument, of course, implied that the O-ring design could be dangerous under any temperature.

"The eve of a launch is a helluva time to be generating new launch commit criteria!" Mulloy barked.[60] Mulloy's concern went beyond the launch of 51L. The temperature constraint would make dependable launches from the DoD's $3 billion site at Vandenberg nearly impossible.[61] Vandenberg, which sat on a small peninsula off the central California coast, frequently saw temperatures under 53°F—especially in the early hours when launches took place.[62] A new temperature constraint would likely make Vandenberg a nonstarter and shutter the NASA/DoD partnership.[63] *Was that why,* Allan McDonald wondered, *Mulloy had asked him to take out the references to temperature on his presentation to Headquarters a few months ago?*

"My God, Thiokol, when do you want me to launch . . . next April?" Mulloy demanded.[64,65] The last thing Mulloy wanted was to have to go back to Marshall Center director William Lucas, his demanding boss, and tell him the shuttle was not launching because of Marshall. In spite of Mulloy's frustration, Thiokol's engineers stood firm. It was 10:30 PM, and the group had reached an impasse.[66] Joe Kilminster, Thiokol's vice president for space booster programs, asked for five minutes to caucus offline with his group in Utah.[67] Back at Kennedy, Allan McDonald and Larry Mulloy argued. McDonald was firmly against the launch, calling Mulloy's argument "asinine."[68]

"If we're wrong and something goes wrong on this flight, I wouldn't want to have to be the person to stand up in front of a board of inquiry," McDonald said.[69]

Meanwhile, in Utah, Thiokol's five-minute caucus stretched to thirty minutes. Realizing Thiokol's managers were ignoring the engineers in the room, Boisjoly desperately bounded over to the managers' table with his stack of photographs. He threw down image after image of soot residue between the primary and secondary O-rings on El's flight, STS-51C. The soot from the blowby was pitch black, much worse than the splotches of gray soot seen on warmer launches. To Boisjoly, the cold temperatures put the O-rings and the crew in a whole new world of danger.[70] The managers ignored him. Finally, Boisjoly stopped arguing. Nobody was listening.[71]

"Take off your engineering hat and put on your management hat," Jerry Mason, the senior vice president of Thiokol's Utah plant, told Lund.[72] To Mason, this was not about engineering arcana, it was about business.

At 11:00 PM, Thiokol resumed the conference call with Marshall and reversed its position. Mulloy demanded that Thiokol fax its launch recommendation to him. Allan McDonald refused to sign the recommendation, so Kilminster, who was McDonald's direct superior, signed it.[73] The O-ring issue was decided.

The members of the Astronaut Office, and the flight crew themselves, were unaware of these late-night meetings.[74] Given the weather forecasts, the crew did not believe they would launch the next morning.[75] After dinner, Ron bid Cheryl goodbye and smiled. "See you tomorrow," he said.[76]

Judy spoke to Sylvia by phone. They had not caught up in weeks. Sylvia made Judy promise that they would have "a talk" about Frank when she returned from her mission.[77] Judy promised.

"I love you. You're like my sister," Sylvia said.

"I know," Judy answered. "I love you, too."

Overnight, the arctic ridge caught up with the cold front over the Florida Panhandle, pushing its way toward the Cape. Across the state, citrus growers, one of Florida's key industries, lost sleep. A year ago, a similar freeze had destroyed eighty thousand acres of crops and bankrupted many. This cold front would be worse.[78] That night, outside the space complex, farmers set out heavy oil heaters, piled soil up around fragile saplings, and flooded their fields with warm groundwater to save their crops.[79]

The Kennedy Space Center implemented its own freeze protection plan. To keep pipes from bursting, workers ran a consistent stream of water through the onsite showers and fire suppression systems. Astronaut Jim Bagian, who had been rejected during the TFNG class selection and later became part of the 1980 class, had been assigned as the Cape Crusader in charge of guarding the set configurations on the shuttle overnight. While on his shift, he heard the water running for hours. "It was like being in a monsoon," he said. The ice team poured fourteen hundred gallons of antifreeze into the overpressure water troughs in each of the solid rocket booster's exhaust holes.[80] These troughs held a total of 6,580 gallons of water, used to relieve pressure generated by the roar of the boosters as they ignited.[81] The antifreeze would, hopefully, keep the water in the troughs from solidifying.[82] Through the night, the ice team would clear ice that formed on the orbiter or launchpad to keep it from damaging the shuttle on liftoff.

As midnight approached, the moon rose high over a raging black sea.

Aboard the *Liberty Star* off the coast of the Cape, NASA Retrieval Operations Manager Jim Devlin wedged himself into a corner of the boat and grabbed a rail to steady his feet. His boats were headed east toward their booster recovery station, but conditions were making that journey impossible. Gale-force winds, blowing from fifty to seventy-two knots, howled across the sea from the west.[83] Thirty-foot swells pounded the stern of his shallow-draft boat, threatening to capsize it.[84] In his thirty-three years of sailing, including all twenty-four trips to recover the shuttle's solid rocket boosters, Devlin had never seen conditions like this, "outside of a hurricane or a winter gale on the North Atlantic."[85] For the safety of his men, he would have to head back to shore even if that meant turning into the wind, and facing the crashing, thirty-foot waves head on.[86]

"We've never been off station for a launch yet," he shouted to Thiokol's man beside him. "But I am afraid we're going to have to be this time."[87]

Just after midnight, Larry Mulloy called shuttle program director Arnie Aldrich to warn him that Devlin would not be on station to retrieve the booster rockets given the conditions at sea.[88] They could only buoy the boosters for retrieval later. Sitting nearby, Allan McDonald listened in on their phone conversation, but to his surprise Mulloy never mentioned Thiokol's concerns about the O-ring issue. As far as Aldrich was concerned, the issue of how shuttle systems would perform in the cold had been discussed, analyzed, and closed. Now he could get a few hours of much-needed sleep before launch.[89]

Hugh Harris, the public affairs officer who would narrate *Challenger*'s countdown for the world, headed into Kennedy at 2:00 AM on the morning of January 28. Usually, he would have to battle traffic on the way into the Cape, but this morning, the roads were empty. Given the record cold, most believed the launch would be scrubbed. "The few who had come were huddled inside their vehicles," he said. *Everyone was trying to keep warm.*[90]

The coldest temperatures came before dawn. *Challenger*, married to its rocket stack, sat on launchpad 39B like a lonely lighthouse on a cliff in a storm. Gaining northwesterly winds whipped the unshielded orbiter as temperatures dropped to 22°F.[91]

When Kennedy's ice team returned to check on conditions at the pad, they encountered an unbelievable spectacle: a frozen waterfall cascaded down the north face of the 250-foot-high launch tower. Eighteen-inch stalactites hung from walkways and girders, only sixty feet from *Challenger*'s left wingtip. Sheets of ice covered the emergency exits.[92] The scene was like "something out of *Dr. Zhivago*," one worker said.[93] Running the fire hoses and emergency showers had prevented the pipes from bursting but caused another calamity.

The ice team began to examine the launch vehicle from every angle to make sure no ice was forming on the various connections and the surface of the external tank. They broke off as many icicles they could reach to safeguard the orbiter during liftoff. When the ice team reported its findings to launch director Gene Thomas, he sloughed off their concerns, insisting that most of the ice could be cleared away or would melt once the sun rose.[94] They would push the launch an hour to give the sun more time to warm the pad.[95]

The antifreeze that the ice team poured into the booster troughs fell short of the task.[96] The water in the troughs turned to slush, and as the night got colder, froze completely. The ice team began manually breaking apart the floe, then fished out the large chunks of ice with fifty-foot long "shrimp nets."[97]

When Cape Crusader Bagian exited the crew cabin in the morning, he saw the catwalk iced over. "If somebody slips," he thought, "they're going right off the side. That'll be the end of that. Secondly, there's ice hanging all over that could damage the tiles."[98]

Bagian reported his concerns to George Abbey.

"Don't you worry about that," he told Bagian. "There are other people worrying about that," George said. "You['ve] got to be up in a couple of hours and take them out to the pad."[99] Bagian rolled his eyes, then headed to crew quarters. George may have been projecting confidence to his subordinates, but internally he had grave concerns about the weather. Like Bagian, he saw the icy conditions and hoped the launch would scrub.

The astronauts had a 6:00 AM wake-up call, but most rose by 5:00 AM. They called their families to let them know the launch was proceeding.[100] Shocked, Lorna told El that the milk she had accidentally left outside had frozen overnight—no cereal for the girls.[101] *Are you sure you're launching?*

At breakfast, the crew posed with two dozen red and white roses arranged with American flags, and a vanilla-frosted sheet cake decorated with the mission patch: Halley's comet streaking behind the orbiter, plus a bright red apple for the teacher. After a final medical check and suit-up, they collected their personal effects. El grabbed his black duffle bag with the signed soccer ball from his daughter's team.[102] Judy wore a charm around her neck, the sign-language symbol for "I love you" in honor of her nephew, who was deaf. In her flight suit, she tucked away a locket for her niece and her ex-boyfriend Len's cigarette lighter.[103] Ron grabbed his sax.

At 7:48 AM, George followed his astronauts out of crew quarters past the press to the Astrovan. Their breath steamed in the cold air. Reporters shouted questions. There were not any takers. George rode out as far as the Launch Control Center and then said goodbye to the crew.[104]

The floor of Launch Control was arranged like an indoor amphitheater. The top-tier operational managers sat at the highest level in the back of the room, in front of huge, thousand-square foot windows.[105] The engineers that executed their commands sat in the lower rows in front, on the vast firing room floor behind banks of greenish-blue consoles that dated from the Apollo Era. Managers, like George, sat cordoned off in observation booths that looked out over the entire floor. The managers' chatter would not interfere with the launch procedures, but they were on hand for consultation.[106]

From his seat, George glanced at the launch control desk, as Arnie Aldrich made a move across the room. As a program manager, Aldrich was responsible for polling his team at Kennedy and giving the final launch recommendation to the associate administrator. Aldrich was on his way out with William Lucas, the head of Marshall, and Jesse Moore, the head of Johnson and acting associate administrator for spaceflight. George deduced they were going to meet with Rockwell about the ice issue. As the orbiter's manufacturer, Rockwell engineers had to assess how ice on the pad might impact the orbiter and give their recommendation for launch.[107] George tried to read Aldrich's body language as he walked to the ops center down the hall.

At 9:00 AM, Aldrich and his Mission Management Team crowded around a T-shaped conference table with two-dozen of NASA's top management and its contractors. The ice team presented their findings. Ice was still hanging off the launch tower and forming in the boosters' sound-suppression troughs.[108]

"Rockwell cannot assure that it is safe to fly," the company's vice president Robert Glaysher said. They had not seen or modeled out conditions quite like this before and so "they would not give an unqualified go for launch."[109] Aldrich interpreted that to mean that Rockwell did not have sufficient data to recommend *against* launch. Aldrich did request that the ice team do an additional inspection as close to launch as possible to assess the danger. Barring any change in circumstances at the pad, Aldrich recommended to Moore that they proceed.[110]

Crew families arrived in buses at the Launch Control Center and were escorted to the launch director's office. Cheryl pushed baby Joy in her stroller while Reggie followed sleepily behind. Steve McAuliffe, Christa's husband, carried his daughter Caroline, six, and guided his son Scott, nine, ahead of them. The other children dutifully took their seats around a long conference table.

"My God," Mike Smith's wife, Jane, exclaimed when she saw the deflated balloons from yesterday's scrubbed launch. "I hope that's not an omen."[111]

The crew's families had already been through a stressful series of delays, and many were in their own thoughts—nervous, tired, excited, frayed. Lorna, known for jovial banter, tried to keep things light. The small children sat on the floor playing with toy trucks, stuffed animals, and coloring books that had been put out for them. Adults drank coffee and ate the fresh fruit and pastries NASA provided. Others stared out the large windows that offered views to *Challenger* three miles away.[112]

Sylvia joined about twenty members of the Astronaut Office, including Hoot Gibson and Charlie Bolden, in a conference room at Johnson Space Center to watch the launch.[113] Having returned from his mission ten days earlier, Hoot knew the uneasiness the crew must be feeling around the icy conditions. Some of the astronauts left the room, believing the launch would be scrubbed.[114] Hoot and Charlie decided to stay. So did Sylvia, who blinked back tears. She cried at every shuttle launch, but at this one especially because Judy was onboard.[115]

Fred Gregory arrived early at Mission Control before the crew loaded. He too was sure, given the weather, that they would see another scrub.[116] He joined Dick Covey at the CapCom console and watched the icy scene at Kennedy unfold on their newly installed, live-feed TVs. They got the "all clear" from Kennedy and shrugged. Time for launch.[117]

At 8:25 AM, the crew walked out of the launch tower elevator, and made their way across the causeway, 195 feet above the ground. The closeout crew shouted out a warning. *It's slippery—you're walking on a sheet of ice.*[118]

"What a great day for flying!" Scobee said, relentlessly cheerful.[119]

Judy and Christa put on their flight gear and danced around to keep warm. "The next time I see you, we'll be in space," Judy said, before heading into the upper flight deck, where she and El would sit with the pilots.[120] Christa would be on the middeck with Ron and Greg Jarvis.[121] The closeout crew suited Christa up and then presented her with a big red apple.

"Save it for me," she said, before entering the orbiter.[122]

By 8:36 AM, the crew had strapped in. Launch managers advised that they were going to wait until it "warmed up" to clear for launch.[123]

As the debate about launch time ping-ponged between Launch Control and the ice teams, the crew tried to keep their spirits high. A heating tube, controlled by the closeout crew, shot a blast of hot air through the cabin to warm everyone up. Judy broke out laughing.

"It was right up my you-know-what," she said. "Ellison thought it was great."[124]

By 10:30 AM, the crew was already lamenting the effects of the wait.

"I feel like I'm at the four-hour point of yesterday," Smith said.

"I feel like I'm past it. My butt is dead already," Judy joked. Frost blew off the external tank. "Is that snow?" she asked.

"Yep, that's snow," Scobee replied.

"You're kidding. You see snow on the window?" Ron asked from the middeck.

"We should've slept an extra hour this morning," Scobee said.

"They're probably making a fortune selling coffee and doughnuts at the viewing areas," Jarvis said.

"How about that? We should have gotten some," Scobee said.

"A few hot toddies . . . My bun is dying," Judy said.

"Ellison and I'll massage it for you . . ." Jarvis offered.

"Hah!" Scobee said.

"Ellison's not even interested," Judy said.

"I'll bet you could wake him up for that," Jarvis said.

"Hah!" Judy said.

"There goes a seagull," Scobee said.

"Better get out of the way," Judy said.

"He'll be long gone from here by the time we launch," Scobee said.

"He's built a nest by now . . ." Judy said.[125]

Back in Launch Control, George saw Aldrich return from his meeting and bolted up to him.

"What did you decide?" George asked.

"We're still a go," Aldrich said.

"Did Rockwell say they were a go?" George was suspicious.

"Yes," Aldrich replied.

How could Rockwell have said yes? George wondered. *Weren't they seeing what he was seeing?*[126] However, George did not press the issue further with his superiors. Yes, George was the Godfather to his astronauts, but he respected and followed the chain of command. His genius was not as a revolutionary, but as a master bureaucrat.

The new launch time was set: 11:38 AM.

Jim Beggs was watching the proceedings on closed-circuit NASA TV from his seventh-floor executive suite at NASA Headquarters in Washington, DC. With conditions as they were, he marveled that the countdown was going forward.[127] *I would not launch*, Beggs thought, but it was no longer his decision to make.

Beggs's political rival William Graham, the acting administrator, was not at the Cape or in Mission Control. For the first time in shuttle history, neither the NASA administrator nor the deputy were on site for launch. Instead, Graham was meeting in DC with the Republican leader of NASA's oversight committee, Representative Manuel Lujan Jr. of New Mexico, to solicit Lujan's support in his battle with Beggs.[128]

At T minus nine minutes, launch controllers traditionally held the countdown so administrators could make their final recommendation for launch. Just before the nine-minute mark, Aldrich thanked his ice team, still clearing ice from the pad, and told them to evacuate. The temperature had risen to 36°F.[129] High above the pad, John Young flew NASA's Gulfstream II jet doing a final weather check. He reported evidence of high-altitude wind shear, but nothing that skirted the bounds of launch criteria.

Jay Greene, the mission's ascent flight director in Houston, polled his eleven-member team for a final status—everyone in Mission Control was a go. The decision was back to Aldrich and his managers for a final poll. Their position had not changed. "Go," they said, and kicked the final decision over to Jesse Moore. Moore said *go*.

Armed security officers locked and guarded the firing room doors.

In the launch director's office, the families gathered around a large television to listen to the voice of public affairs, Hugh Harris.

"We're at T minus nine minutes and counting," he reported. "The ground launch sequence has been initiated."[130]

Crew family members looked at each other with mixed emotions. *How had NASA found a way to overcome the cold?* Adults put on their coats and hats and zipped up their children's jackets. The astronaut family escorts—astronauts themselves who provided emotional and logistical support during launches—shepherded the crew families to the roof. Frank Culbertson, Judy's boyfriend, led the way. Cheryl McNair pushed Joy's stroller down the long corridors of the fourth floor followed by the group of twenty. It would be cold on the roof, but they wanted to watch the launch in the open air, with naked eyes.[131]

The families huddled together for warmth. Cheryl still expected the launch to scrub. "It was just too cold," she said.[132] June Scobee stood with Steve McAuliffe and his children, posing for quick photos.

"The postcard view took my breath away," June said. "The shuttle was beautiful against the clean blue sky . . . it glistened white and sparkling in the light."[133]

George looked out the floor-to-ceiling windows of Launch Control at *Challenger* on the pad. *All that ice.* He did not like what he saw, but now, like citizens and schoolchildren all over the United States, all he could do was watch.[134]

The crew lowered and locked their helmet visors.

"Welcome to space, guys," Scobee shouted.

"Okay, there goes the LOX," Smith said. The LOX, or liquid oxygen arm, supplied the external tank with liquid oxygen fuel.

"Goes the beanie cap," Scobee added.[135] The "beanie cap" covered a vent at the top of the tank during fueling and carried away oxygen vapor.

"Doesn't it go the other way?" El said. They all chuckled.

"God, I hope not, Ellison . . ." Smith said.

"One minute, downstairs," Scobee radioed to Ron, Christa, and Jarvis on the middeck. A deep roar filled the cabin and swelled like an approaching subway car.

"Cabin pressure is probably going to give us an alarm," Judy said.

"Alarm looks good," Smith said.

"Thirty seconds . . . Fifteen," Scobee said. The main engines began to ignite. "There they go, guys," Scobee said.[136]

Hugh Harris's voice boomed over the loudspeakers to all the spectators, braced against the cold wind.

"T minus ten," Harris chanted. "Nine, eight, seven, six . . . we have main engine start."

At T minus six seconds, *Challenger*'s main engines ignited.[137] The first blast rolled over the grandstands. White clouds formed under the orbiter.[138]

"Three at a hundred," said Scobee as all three main engines fired at 100 percent.[139]

". . . four, three, two, one . . ." Harris continued. *Challenger*'s solid rocket boosters lit, forcing the shuttle to break free from its mooring bolts. For a split second, *Challenger* hovered over the pad, then made its first push toward the sky.

Go!! Go!! Go!! screamed the spectators.

"And liftoff! Liftoff of the twenty-fifth space shuttle mission and it has cleared the tower," Harris reported.[140]

"All right!" Judy yelled.[141]

As Rockwell technicians had feared, the reverberations from main engine ignition shook loose a shower of ice from the Fixed Service Structure. Sheets as large as thirty-six square inches cascaded around the shuttle. Smaller ice cube–sized pieces struck the left-hand booster rocket. Luckily, none of the ice damaged the orbiter.[142]

At 0.678 seconds after liftoff, as *Challenger* hovered over the pad, a puff of ashen smoke eked out of the lowest field joint on the right-side booster.[143] The smoke came and went in an instant, obscured by the wafts of white clouds billowing from the rockets.

Even though the ambient temperature registered at 36°F, the temperature of shuttle hardware was colder, having had little time to warm up from the cold night before. As the booster segments ballooned with the stress of ignition and the joints rotated against one another, the primary O-ring within that field joint was too cold and rigid to fall into place. The gasses roiling at 5,000°F inside the booster blasted past the primary seal. In previous flights, like Fred's and El's, the secondary seal slipped into place and sealed the joint. Now the secondary O-ring was also too cold and too rigid to fill the gap. No barrier existed to prevent the hot gasses from leaking through.[144] Over the next two seconds, as *Challenger* began its ascent, eight more puffs of smoke—each blacker than the last—leaked from the field joint. The dark color indicated that hot gasses were vaporizing the grease, joint insulation, and the O-rings themselves.

At just under three seconds after liftoff, the black smoke stopped, before anyone could have noticed it with the naked eye.[145] Glass aluminum oxides created by the burning propellant temporarily sealed the joint, preventing flames from seeping through.[146] A carefully calibrated symphony of explosions continued to propel *Challenger* upward. At full thrust, the shuttle shook, rattling the crew's bones.[147,148]

"Aaall right!" Judy exclaimed.

"Here we go!" Smith screamed.[149]

"Houston, *Challenger*, roll program," Scobee said. At seven seconds, the shuttle began its roll maneuver, coiling backward so the orbiter flew horizontal to ground. Sunlight streamed through El's flight deck window.[150]

"Roger roll, *Challenger*," Dick Covey answered from Mission Control.[151]

"Go, you mother!" Smith yelled.

Ron sat on the middeck, behind Christa and Jarvis. He did not have an active role in takeoff, so he got to be a spectator. Since his first flight, Ron had played a tape of his own liftoff over and over again. When he visited his brother Carl in Atlanta, he would sit in front of Carl's expensive stereo and let the noise of his launch wash over him, so loud that he blew out the speakers.[152] Now he was back for the real thing.

"Shit hot!" said Judy. Only fifteen seconds into their climb, and they were already going over two hundred miles an hour.[153]

"Ooo-kaay," Scobee said, reacting to his friend's choice of words.[154]

"Looks like we got a lot of wind up here today," Smith said.[155]

At nineteen seconds, the shuttle hit a change in wind speed—the turbulence that John Young had detected earlier.[156] At thirty-five seconds, *Challenger* completed the roll maneuver. Onboard computers throttled the main engines down to 65 percent to reduce pressure on the vehicle as the orbiter approached max q, the moment the shuttle would be under the greatest mechanical stress during its ascent.

"There's Mach 1," Smith said as they ripped through the sound barrier.[157]

At fifty-one seconds, the orbiter successfully passed through the max q threshold and Scobee began to throttle up the main engines again. For the next ten seconds, *Challenger* passed through heavy wind shear.[158] The orbiter shook. The shuttle's internal computers stabilized the vehicle, but the vibrations were strong enough to shatter the temporary aluminum oxide seal in the damaged booster joint.[159]

A second and a half later, a small, nearly imperceptible flame seeped through the aft joint and licked the side of the right booster, where the puffs of smoke had been.[160] Within another second, the flame grew, becoming visible to the naked eye. The fiery plume deflected downward, scorching the external tank's lower strut—one of three that connected the aft end of the booster to the tank. The chamber pressure in the right booster registered lower than the left booster, a sign that there might be a leak. At sixty-two seconds, an automatic control system gimbaled the left booster to counter the yaw of the right. As far as the astronauts and mission controllers knew, the shuttle's computers were doing their job.

"Feel that mother go!" Smith exclaimed. "Woooohoooo!"

"*Challenger*, go at throttle up," CapCom Dick Covey said in Mission Control.

"Roger," Scobee said. "Go at throttle up."[161]

At seventy seconds, the main engines once again roared at full force. As the shuttle soared higher, the flame leaking from the right booster changed both shape and color. A bright, sustained glow emanated from the external tank, indicating that the flame had breached the tank and was now mixing with liquid hydrogen.

At seventy-two seconds, the flame ruptured the hydrogen tank that sat inside the base of the external tank and below the intertank where oxygen and hydrogen mixed to power the main engines. The aft dome of the hydrogen tank dropped away, and liquid hydrogen cascaded out, creating a sudden forward thrust of two to three million pounds. That thrust pushed the hydrogen tank into the intertank structure, and even farther up into the oxygen tank, which sat at the top of the external tank.[162]

"Uh-oh," Smith said, perhaps responding to that unexpected jolt.[163,164]

In Mission Control, Fred monitored the data on his computer, when something caught his eye. He looked up to check the internal TV feed that sat above the rows of their computer banks and saw a flash at the base of the right solid rocket booster. *Wow, we've really improved our visuals.*

Those must be the solid rocket boosters separating. Then Fred thought again. *No! It's too early.*[165] Fred jostled Covey. "Look."[166]

At seventy-three seconds, the 150-foot-tall right booster broke free of its scorched lower strut and rotated around its still-attached upper strut. The nose of the booster pierced the delicate, quarter-inch-thick aluminum skin of the external tank, rupturing the intertank. A sudden brilliant flash came and went between the shuttle and the external tank. The external tank began releasing the remaining million plus pounds of propellant. Oxygen and hydrogen met in midair, igniting instantaneously. Hypergolic fireballs rolled through the sky. Then the external tank crashed into the solid rocket booster where they had been attached, setting off a massive explosion.[167]

In Mission Control, Covey's jaw dropped.

"Data was still coming in normally on our monitors, which was what most of us were focused on and then the system caught up—the data stopped coming," Fred said. "Then I looked up, and all around the room. All our monitors had gone static."[168] No one had any idea what they were looking at. *Are we in a contingency abort? What should we tell the crew?* A single crackling noise gurgled over the air-to-ground radio as ground transmitters searched for the shuttle's frequency range for a signal.

"We have negative contact. Loss of downlink," the ground communications officer reported.[169]

At Mach 2, at an altitude of forty-six thousand feet, a bright flash of light appeared in the vicinity of the orbiter and engulfed the ship in a brilliant orange flame. At the Cape, spectators and newscasters saw an expanding fire cloud of red and burnt orange. For the many first-time viewers in the audience, especially the kids who had come to see the first teacher in space, the optics were confusing. *Was this part of a normal launch?*

Nearby, Scott McAuliffe's third grade classmates cheered at the wild pyrotechnics taking place overhead.[170] Seconds ticked by, but *Challenger* did not emerge from the blooming cloud of orange smoke above them.[171]

"Where is it?" a child asked.[172] Christa McAuliffe's parents, the Corrigans, continued to stare skyward, waiting for the shuttle to reappear.[173]

At Mission Control, Steve Nesbitt, the public announcer who picked up after Hugh Harris's countdown, continued narrating in his trademark rapid-fire, flat tone.

"One-minute-fifteen-seconds. Velocity-twenty-nine-hundred-feet-per-second. Altitude nine nautical miles," he said looking at the data still coming in on his screen. His matter-of-fact delivery was so incongruous with the spectacle, it confused everyone.[174]

Nearby, he heard a Navy flight surgeon, who was watching the television monitors, blurt out, "What was that?"[175] Nesbitt paused his narration and finally looked at the huge television monitor above Mission Control, seeing the two solid rocket boosters "veering off on wild separate rides."[176]

"Flight controllers here are looking very carefully at the situation," he said, stunned. "Obviously, a major malfunction." Then Nesbitt turned off his mic.[177]

Jim Devlin stood on the bridge of the *Liberty Star* off the coast of the Cape. The sea swelled around him. Out of the fire, the two boosters emerged making a V-shape and then changed direction, crisscrossing each other to form a loose knot of smoke in the sky. The right booster, now to the left of its counterpart, carved a path back to the shore. *They better blow that thing up before it reaches land!*

At the Cape, the range safety officer armed the boosters' explosives, waited ten seconds, and then sent the FIRE command. He hoped and prayed that the orbiter, if it still existed, was nowhere nearby. The boosters detonated in midair, showering the Atlantic with thousands more pieces of smoking debris.[178]

In DC, William Graham, mid-meeting with Senator Lujan, looked over his shoulder at the television. He understood something was very wrong but could do nothing but watch wordlessly.[179] Phil Culbertson, the senior ranking administrator at Kennedy, tried for over an hour to get in touch with Graham, eventually calling Beggs instead. Beggs headed downstairs to Graham's office at NASA Headquarters.[180]

"Where the hell is he?" Beggs asked Graham's assistant.

"He's on the Hill," the secretary answered. Beggs fumed. *He should have been at the Cape.* Beggs demanded Graham's "gals" call him as soon as Graham returned.

"You should get on an airplane and get your ass down to Kennedy immediately," Beggs barked when Graham returned to headquarters a half hour later. "When you get there, set up an internal investigation board now, before data from the launch is lost. Here's who should be on it," Beggs said, shoving a piece of paper at him.[181]

"Those were literally the first words he had spoken to me since he got indicted [two months earlier]," Graham said.[182] Graham took the list and flew to Kennedy as Beggs suggested.

President Reagan was in the Oval Office when his director of communications, Pat Buchanan, walked in. "The space shuttle just blew up."

"Isn't that the one with the teacher on it?" Reagan asked.[183] Buchanan was befuddled. Was he asking in earnest, or was he in shock?

On the roof of Launch Control, Cheryl McNair, Lorna Onizuka, Marvin Resnik, June Scobee, and the other crew families watched, comprehension escaping them. Out of the explosion, hundreds of capillaries of white soot wove smokey threads down toward the ocean. They searched the sky for the orbiter—but *Challenger* was nowhere to be seen.

"Oh, God! It can't be," June whispered under her breath.[184]

"Is he going to be okay?" Lorna asked. Then she asked again, and again.

Richard Nygren, assistant director of flight crew operations, was unable to give her a straight answer. "We'll have to see," he said.[185]

"If there's any way my dad can get everybody out of it," Scobee's son Rich said confidently, "he will."[186]

Cheryl believed Ron would survive whatever happened.[187]

In Mission Control, Fred put his hand on his mic and stared at the stalled data on the screen in front of him. He felt a hollowness creep in. *Had they been lost? What could Mission Control have done? What could he have done?* At the very least he should have let them know they were not alone. *Goddamn it, why didn't I say it?* One word, a blessing on a new journey, an archaic way of saying goodbye, a benediction. He could have given them that. And so he whispered it now to himself.

Godspeed, Challenger.[188]

CHAPTER 19

SPEEDBRAKE

Johnson Space Center/Kennedy Space Center. January 28, 1986. 8:39 AM CDT/11:39 AM EDT.

In Mission Control, NASA commentator Steve Nesbitt, still reeling from shock, had new information to convey. Reluctantly, he turned on his microphone.

"We have the report from the flight dynamics officer that the vehicle has exploded," he said.[1]

"Lock the doors," flight director Jay Greene barked. "Everybody, no communications out. We're staying here until we've determined what happened and if what happened could have been prevented by Mission Control."[2]

Fred Gregory and Dick Covey started parsing their data—as did everyone around them.[3]

A thousand miles away, George stood frozen in Launch Control. *They're gone.* As devastated as he was, he knew he had to step into action quickly. *The families. I've got to get to the families before the press reaches them.* He seized the nearest telephone and called for the buses to take the families back to crew quarters. Then he raced up the four flights of stairs to the roof, where crew quarters manager Nancy Gunter was helping to position guards in a line to shield the families from other NASA guests.[4,5]

Cheryl McNair broke down sobbing, as she clung to her young children. Mother Pearl embraced the trio. El's wife, Lorna, could not believe what she was seeing. In her heart, she felt the crew could escape. Christa McAuliffe's husband, Steven, held on to their son, Scott, and daughter, Caroline.[6] Judy's boyfriend, Frank Culbertson, reached out to support Judy's father, Marvin. Devastated as anyone there, Culbertson was nevertheless called to act as an astronaut escort, caretaking the families.[7]

"I want my father!" Mike Smith's youngest daughter Erin sobbed. "He told us it was safe!"[8] George scooped the eight-year-old up in his arms and led the rest of the group downstairs to the buses waiting to take them back to crew quarters.[9]

June Scobee, no longer in command of her body, stumbled. Her son guided her and his sisters down the stairs. June silently asked God for strength. "They'll be rescued," June mumbled. "They'll be all right."[10]

In the grandstands, a woman's woeful "Oh my God" punctured the silence. As if giving permission for others to cry out, her wail triggered a wave of screaming and sobbing. A young girl pointed at the sky, crying, "The teacher! The teacher is up there!" Spectators wept and held each other. Others shook their heads no, as if in an argument. *No, this wasn't supposed to happen. No, this wasn't fair.* Many stared up at the sky in disbelief, unable to look away.[11,12]

A NASA official climbed over rows in the VIP grandstands to reach Christa McAuliffe's stunned parents, who had opted to watch the launch with their grandson's classmates rather than join the other families on the roof. "The craft has exploded," the official said to Grace Corrigan. Dazed, she parroted back his words, "The craft exploded?"[13] Then she repeated the message to her husband, Edward, still blinking up at the sky, a picture of a smiling Christa pinned to his lapel. Grace rested her head on Ed's shoulder, clutching his and her daughter Lisa's hands. Photographers in the press box zoomed in on the Corrigans, snapping photos that gave the world a close-up view of their anguish.[14]

Within minutes, all three national news networks were covering the story, even though they had not telecast the launch. CNN, already broadcasting, initiated full-time coverage.[15] The print reporters on site raced to the payphones to dictate their stories.[16]

In Houston, the air had been sucked out of the Astronaut Office conference room.[17] Sylvia looked across the table at Hoot Gibson and Charlie Bolden and knew instantly from their expressions that her best friend, and the rest of the crew, had been lost. Unlike many at NASA with military backgrounds, she had no training in modulating emotions. She burst into tears and had to be helped from the room.[18] Others left without a word.[19]

Anna had taken a break from a robotics training session for her upcoming flight to join the group watching the launch in the conference room as the accident unfolded.[20] As she walked from the conference room, she saw staff in the Astronaut Office weeping openly, walking up and down the hallways, shell-shocked.[21]

Rhea was visiting an offsite contractor with the crew of her next flight when she heard. She drove to Johnson, crying over her lost friends. Her thoughts turned to her husband, Hoot, whose flight had landed only ten days earlier. *It could have been Hoot,* she thought.[22] *It could have been any of us.* When she arrived at Building 4, Hoot came running out the doors. She flew into his arms.

The astronauts still in Houston gathered to listen to George around the speakerphone in his office. George told some to get on an airplane and come to the Cape. He wanted them for the nascent investigation. To others, he said, *We need you tonight at Ellington.*[23] George planned to fly the crew families home to the Houston Air Base that evening, and he wanted astronauts on hand to escort families home safely, away from the press.[24] The astronauts hung on to George's words like a life raft. "We were like those school kids watching at the Cape—lost," TFNG Brewster Shaw reflected.[25]

Kathy Sullivan was traveling home from California where she had been training for her upcoming mission that would deploy the Hubble telescope. On a layover in Dallas, Kathy pressed a payphone to her ear, unable to believe what her secretary was telling her. "Unfathomable," Kathy said. "The world seemed to stop completely. I could not make any sense of her words."[26]

Kathy's flight from Dallas to Houston was packed with reporters "racing to the scene of the great tragedy for the story of their lives."[27] Reeling from the incomprehensible news, Kathy squeezed into a middle seat, surrounded by press on all sides. "The reporters had no idea they were sitting next to an astronaut," Kathy said. "I am picking up bits and pieces and forming a picture of what's happened by listening to them. I'm getting madder and madder about how ghoulish they're being. And I'm just thinking, seven people are dead. Four of my classmates are dead and all you guys can think about is how fast can I get there to get the juicy story."[28]

As soon as she landed, Kathy drove straight to Johnson. The world around her became surreal. "I felt as though someone had paused the movie I was in, but I was somehow still moving through the frozen, silent scenery."[29] She did not allow herself to cry until she finally reached her classmates on the third floor.[30]

At the Cape, George loaded the families on to buses and rode with them back to crew quarters, weaving in and out of traffic. Cars had stopped everywhere—in the middle of streets, along curbs, and even on sidewalks. Some people left their vehicles and embraced one another.[31] Others stood by the roadside, pointing at the white plumes still visible in the eastern sky.[32]

At crew quarters, George gathered the family members—adults and children, nearly thirty in all—in the lounge. The room proved too small for the large group; there were not enough chairs. Not having the presence of mind to seek a more comfortable spot, they crammed in next to each other, crying, shaking, waiting for news.[33]

Astronauts John Young, Jim Bagian, and Sonny Carter met George at crew quarters. George handed Erin Smith back to her mother and listened to their latest reports: There was no daring last-minute rescue, no boat bringing the crew back, no miraculous reversal of fortune.[34] Shoulders slumped, eyes bloodshot, George returned to the lounge and delivered the news, plainly and unequivocally.[35] "All the crew members are dead," he told them. "They could not have survived."[36]

"What am I going to do?" Cheryl McNair hugged her mother-in-law, Pearl. Reggie and Joy, still babies, were now without a father.[37] Lorna Onizuka sobbed next to her.[38]

"Are you all right, honey?" Marvin approached Lorna.

"Yes," she said, "I'm okay," but she was not. Lorna slid down the wall, knocking a few light switches. A flight surgeon rushed to her.[39]

Marvin turned in grief to his stepdaughter, Linda Reppert, who had talked to Judy over the phone the day prior. *Judy had been excited to fly*, Linda reminded Marvin. *She died doing what she loved.*[40]

June Scobee withdrew to her husband's crew quarters bedroom and sobbed into his clothing, breathing in his scent. Even though Valentine's Day was not for a couple of weeks, she found a card he had written her, and a message on a scrap of paper quoting space author Ben Bova.[41]

> We have whole planets to explore. We have new worlds to build . . . And if only a tiny fraction of the human race reaches out toward space, the work they do there will totally change the lives of all the billions of humans who remain on Earth.[42]

The note was not the miracle June had prayed for, but it gave her comfort to know her husband had died for a cause in which he believed.[43]

Jim Bagian stepped away from the grieving families and headed to the kitchen. There, he found John Young wobbling from one foot to the other, mumbling to himself. George sat, staring out the window without a word.[44] The younger astronauts laughed nervously: They felt like children watching their parents have a nervous breakdown.[45]

"Hey, they're going to send [a] helicopter out," astronaut Jim Wetherbee announced, breaking the tension.

"I'll go," Bagian said, jumping out of his seat to volunteer. "I don't want to be here."[46]

Flying over the Atlantic Ocean, Bagian quickly spotted debris from *Challenger*'s external tank. "We're seeing rocket fuel boiling up from the ocean," Bagian said. "We see the nose cap of [the external tank] floating, inverted. Floating on water."[47] Bagian hurriedly began sketching a map of the area in a notepad, hoping it would help the investigation that would follow.

While the recovery operation began, Vice President George H. W. Bush, accompanied by senators Jake Garn and John Glenn, flew to the Cape to offer their condolences. George escorted the VIPs to the fourth-floor conference room where the families waited.[48] The vice president, visibly shaken and near tears, communicated that the families were in the president's prayers. Senator Glenn shared that he, too, had lost a loved one to the program—his best friend, Gus Grissom—during the Apollo 1 fire. "I know that the seven brave heroes were carrying our dreams and hopes with them," Glenn said. "And I know that we will continue to carry their memories with us."[49]

Before the politicians left, June Scobee, still clutching her husband's note in her hand, made an impassioned plea on behalf of the *Challenger* crew. "Keep space exploration alive," she begged. "Continue the space program."[50] It was a sentiment shared by all the families even in the first blush of tragedy.

At the Banana River Turn Basin, three miles from the launch site, journalists flooded into the geodesic dome that served as the Cape's press office. In the 350-seat grandstand, flanked by telephone hookups and wire services, the journalists hollered out rapid-fire questions to the small public relations staff on hand: *Were there survivors? What caused the accident? Was* Challenger*'s destruction an act of terrorism?* Public affairs had no answers. The volume of inquiries crashed the phone system, not just at the space agency, but throughout the surrounding area as far away as Orlando.[51]

Five hours after the accident, Johnson director Jesse Moore, looking mournful and drawn, finally spoke to the growing crowd of reporters from a table set up outside, in front of a larger grandstand.[52] The delay broke NASA protocol, frustrating reporters accustomed to a more immediate response. The NASA contingency plan prescribed that within an hour of an emergency, a senior NASA official would address the press.[53] Moore stated that NASA was forming an interim investigation board and was impounding all relevant information. "We will not speculate as to the specific cause of the explosion based on that footage," he added.[54] Moore was referencing CNN's live news coverage of the launch, which the network had been replaying in slow motion: "You could see, right before the explosion, at the external tank where it attaches to the shuttle, a flame appeared to break out between the external tank and the shuttle," observed CNN's anchor.[55,56] "It will take all the data, careful review of that data, before we can draw any conclusions on this national tragedy," Moore cautioned the reporters.[57]

At 5:00 PM Eastern time, people all over the country gathered around their televisions to watch the president's address from the Oval Office. Instead of delivering his scheduled State of the Union, Reagan chose to speak to the American people about the accident, especially the millions of children who had witnessed it live in person or in their classrooms.

"For the families of the seven," the president began, "we cannot bear, as you do, the full impact of this tragedy. But we feel the loss . . . Your loved ones were daring and brave, and they had that special grace, that special spirit that says, 'Give me a challenge and I'll meet it with joy.'"[58] He assured the families that the crew's sacrifice was not in vain. "The future doesn't belong to the fainthearted," he intoned. "It belongs to the brave." He promised the nation, as June Scobee and the other families had petitioned, that the space shuttle program would continue.

To speak to the emotion of the moment, Peggy Noonan, the thirty-year-old wunderkind speechwriter who had drafted Reagan's remarks, suggested that the president quote "High Flight," a poem by John Gillespie Magee, a Canadian Air Force pilot who had died in a midair collision in World War II.[59] *Did the president know that poem?* Noonan asked. *He did.* "We will never forget them, nor the last time we saw them, this morning," Reagan continued. "As they prepared for their journey and waved goodbye and 'slipped the surly bonds of Earth' to 'touch the face of God.'"[60]

With that four-minute speech, Reagan touched the soul of the nation. The address, which he delivered with presidential poise, pathos, and a glint of cowboy grit, is today considered one of the best of the twentieth century, and the reference to the poem became its most iconic and identifiable passage.[61] His words struck the opening note of a long and difficult process of healing. With the echo of the president's remarks in mind, the crew families flew home into the western sunset, the cabin lights off. "The only sounds were soft sobs and the moan of airplane engines, as though they too were grieving," June said.[62] Most slept through the flight, depleted from the day's emotions.[63]

Bill Bailey, a constable whose jurisdiction covered Johnson Space Center, prepared for the families' return to Ellington. Bailey sent his deputies to the astronauts' homes to keep the press at bay. He positioned journalists far from tarmac and turned down the lights at the hangar to obscure the media's view of the arriving families.[64]

Fred Gregory and Dick Covey wound down their long shift at Mission Control after turning in all their data. So far it looked like mission controllers could have done nothing to prevent the disaster.[65] Fred and Covey joined others in the office—Anna, Rhea, Hoot, Carolyn Huntoon—and headed to Ellington.[66] Together, they consoled the families, exhausted and broken, and guided them back to their homes, shielding them from the unrelenting press.[67]

"They had the television remote facilities already set up outside of the Scobees' house," said Fred, who lived next door. "I went over and invited those [reporters] over to my house, and I talked about absolutely nothing to get them away, so that when June Scobee and the kids got back to the house, they wouldn't have to go through this gauntlet."[68]

"The national press was just god-awful," astronaut Pinky Nelson said of the scene at the Onizukas', where his kids used to play with El's kids. "We had to unplug the phones."[69]

After the families left Ellington, the NASA jet was towed back to the hangar. Alone on the darkened tarmac, Bailey was about to leave when he felt a presence behind him. He looked back to find George. "He looked as gutshot as any human being I'd ever seen," Bailey said. "He had lost everyone." Not sure what else to do, Bailey hugged George and walked him to his car.[70]

"It was a killer for him just as it was for the rest of us," said Fred, "but . . . I think he blamed himself. He wouldn't point and blame other people. [He] had assumed the role of protector."[71]

After Bailey left, George turned the key in his car's ignition and the engine idled. George had given everything to his work. His wife had long since left him. His three oldest children were grown and out of the house, leaving two teenage boys behind. They would need to go to bed soon. Questions haunted him. *What went wrong? Could I have stopped this?* He had seen the icy conditions on the pad, but he had faith in the system. *Challenger* was gone, along with the astronauts he had come to care for as if they were his own children. He had chosen them, readied them in ways they knew and did not know, and he had sent them on this fatal mission. Tonight, George was a man adrift. He did not know where to go. So he headed back to work.

Johnson Space Center. January 31, 1986.

Rhea drove past hundreds of flower arrangements lined up in front of the large entrance sign to the Johnson Space Center. A blue-and-white cross sat amidst dozens of red and white wreaths. American flags surrounded a shuttle-shaped flower display. A heart sign read VIRGINIA CARES. Another simply said KAWAII in honor of El. The world was reaching out to them.

Her eyes stinging from tears, Rhea pulled into the parking lot and headed into Building 4. Like her, the other astronauts and the staff were shell-shocked. *Just going through the motions.* Rhea had a grim job ahead of her: to invite all the retired astronauts to a memorial service scheduled for that Friday, January 31, where President Reagan would eulogize the fallen crew.[72]

Outside her office window, workers transformed their campus for the president's arrival. Reagan's advance team chose a stretch of lawn in front of Building 16 as the setting for the presidential address. They constructed a raised platform, setting the stage at an angle that would be perfect for the light of his midmorning speech, and arranged hundreds of chairs in a semi-circle around the dais. When the advance men determined the trees behind the presidential podium needed to be cleared to create a better visual, gardeners arrived with chainsaws to hack down the oaks. Rhea watched forlorn as the trees fell to the ground: *More life thoughtlessly destroyed.*[73]

That Friday, ten thousand people gathered under bright, sunny skies to honor the dead. Mourners comforted each other while an Air Force band played a Naval hymn, "Eternal Father, Strong to Save," a blessing for seafarers who were lost on a perilous ocean.[74] The President and First Lady led the crew families to their seats in the VIP section. Rhea held Hoot's hand tightly and prayed. Nearby, Sally and Steve, Anna and Bill, Guy, Shannon, Kathy and Fred, and their other classmates grieved for their friends.

"We are gathered here today as a family," Steve Hawley's father, the Reverend Bernard Hawley, began. "A family that shared a common dream, a common goal, a common commitment—and now a common tragic and devastating loss."[75]

After Reverend Hawley concluded his remarks, President Reagan ascended the podium. After honoring each of the crew members by name, he paid a tribute to their diversity: "They were so different, yet in their mission, their quest, they held so much in common."[76] When the President finished speaking, the band played "America the Beautiful." Some sang along, others could not muster the words. Four T-38s sped overhead in the "Missing Man" formation. One lone pilot broke away from the group, streaking up toward the heavens and leaving a symbolic absence in its wake.[77]

Even as Reagan was memorializing the fallen astronauts, his administration had begun censuring NASA. To avoid the public appearance of bias, President Reagan appointed an external board to review the disaster rather than rely on NASA's internal Mishap Investigation Team. On February 3, three days after the memorial, Reagan formally announced the thirteen-member panel would be led by William Rogers, the former secretary of state under President Richard Nixon.[78]

Meanwhile, George stood ready to fill the power vacuum left by NASA's fractured leadership. "George found himself in a fairly unique position after *Challenger*," TFNG Mike Coats said. "In essence, George was the senior individual as the director of flight crew operations . . . we didn't have a center director and effectively didn't really have a NASA administrator."[79] Moving quickly, George installed his astronauts as support staff for the Rogers Commission. He was not only offering the astronaut perspective on the investigation, but also keeping himself central to the flow of information.

George assigned Steve Hawley the task of reviewing the orbiter's prelaunch processing.[80] Mike Coats and Brewster Shaw joined the photo study team, which would examine launch visuals to see if they provided any clues as to the cause of the accident.[81] George designated Bob Crippen as the official astronaut representative at the Cape, responsible for overseeing *Challenger*'s search and salvage effort. These four astronauts, and a dozen more already at the Cape, would report directly to the Rogers Commission.[82]

Within minutes of the accident, the *Liberty Star* and *Freedom Star*, Thiokol's solid rocket booster recovery boats, received the coordinates of *Challenger*'s impact zone in the Atlantic.[83] NASA assumed that either the main engines, the external tank, or solid rockets were responsible for the accident, so initial recovery efforts focused on collecting pieces of rocketry before they were lost in the tumultuous winter sea.[84]

When debris stopped falling at half past noon, the range safety officer gave the all-clear for aircraft to search the area. The effort swelled as nearby search and rescue vessels were pressed into duty. By 7:00 PM, twelve aircraft and eight ships were participating in the search from the Florida coast all the way up to South Carolina.[85]

The next morning, Mike Coats and Brewster Shaw began studying the footage of *Challenger*'s breakup. "As soon as we ran the first video," Coats said, "we knew what had happened."[86] The black smoke puffing out of the right booster field joint at liftoff seemed to be a literal smoking gun. By the end of the day, recovering the right booster debris became a top priority.[87,88]

Although NASA had a leading theory as to the cause of the accident, the agency was not sharing that news with the press.[89] Two days after the accident, NASA's accident review board decided to embargo any information about the 51L crew.[90] NASA public affairs officers stonewalled reporters, forcing them to go through a lengthy Freedom of Information process to have their substantive questions addressed, a process that NASA usually reserved for complicated legal or policy queries.[91] In the coming weeks, the *New York Times* would sue NASA when the agency stalled on giving the press a tape recording of communications between ground control and *Challenger*'s crew before and during the flight.[92]

NASA was not providing answers, so veteran NBC reporter Jay Barbree leaned on his deep connections within the space community for a lead. He convinced an old friend, a recently retired NASA engineer at the Cape, to don his old badge and listen in on chatter about the accident at Kennedy headquarters. The friend reported back to Barbree: NASA indeed had a leading theory as to the cause of the accident. At 6:30 PM, February 2, Barbree broke the story on *NBC Nightly News* with Tom Brokaw: A malfunction in the solid rocket booster had likely destroyed *Challenger*.[93]

The black smoke video seemed to indicate that hot gasses had breached the right booster's joint. George knew that Thiokol manufactured these boosters, but the ultimate responsibility lay with Marshall.[94] His anger swelled at the thought that Marshall managers may have known about the booster defects but stayed silent.[95] If George had been a workaholic before, now he toiled like a man possessed. His family grew concerned about his inordinately long hours.[96] "He never came home," his daughter Joyce said.[97] After leaving the office, George would visit the *Challenger* families to lend support. He became a surrogate father for the grieving children, even if it meant sometimes neglecting his own.[98]

Off the Coast of Cape Canaveral, USS *Preserver*. February 1986.

Astronaut Jim Bagian, who had gone out with the helicopters hours after the disaster, became a de facto part of the investigation at the Cape. He never went back home to Houston, except to attend the memorial for the *Challenger* crew at the Johnson Space Center, scrounge together a bag of clothes, and kiss his wife goodbye. Assigned by George to be a part of the investigation, Bagian and Sonny Carter, a former flight surgeon, fighter pilot, and astronaut from the 1984 class, lived in crew quarters, where the *Challenger* Seven had spent their last nights.[99,100] Bagian and Carter worked alongside Navy and contractor divers on the USS *Preserver*, a massive World War II–era salvage ship that ran two hundred feet long, weighed one thousand tons, and carried a crew of eighty-five.[101]

The breakup of *Challenger* scattered debris over 420 square miles of the Atlantic Ocean. The Navy divided the search area into a grid, with each section one mile long by two hundred yards wide.[102] Now divers and submersibles were combing the sea floor at depths that ranged from ten to over twelve hundred feet for those disparate parts.[103] Sixteen watercraft assisted in the recovery of the debris, including boats, submarines, and underwater robotic vehicles from NASA, the Navy, the Air Force, the Coast Guard, as well as several commercial operators. The Navy enlisted its high-tech NR1 nuclear submarine, which specialized in underwater search and recovery.[104] The hunt for *Challenger* became the largest salvage operation since World War II.

The search was made more difficult because the shuttle debris—three million pounds of it—landed in an area off the coast of Florida that was cluttered with trash. Captain Charles Bartholomew, the Navy's supervisor of salvage, said the searchers ". . . chased a lot of junk," including "a Pershing missile, half a torpedo, a refrigerator, a file cabinet . . . and a toilet . . ." Not to mention "eight shipwrecks . . . a DC3, World War II vintage [plane] . . . [and] a floating duffle containing 25 kilograms of cocaine."[105] It was painstaking work. "You'd look out at a salvage boat, and it moved a couple of hundred yards every day—which isn't much," said Bagian. "Each day we checked off a small piece of the map until the puzzle had been put together."[106]

Through February and March, crews found the Tracking and Data Relay Satellite (TDRS) that the *Challenger* crew would have deployed on orbit, as well as parts of all three main engines, the rear fuselage, and large segments of both solid rocket boosters, some of which had to be hauled up from twelve-hundred-foot depths.[107] Pay dirt did not come till late April when the right hand booster joint, the very joint from which the black smoke had leaked, was discovered outside the initial search area and hauled to the surface. Seeing the badly burned joint, NASA now had definitive proof as to the source of the accident.[108]

Johnson Space Center. Mid-February 1986.

Rhea peered closely at the grainy, unfocused image of *Challenger*'s midair detonation and this time she saw something new. A visiting expert from the National Transportation Safety Board was presenting Rhea and her colleagues with photographs of the breakup captured by a camera recording the launch. The dominant belief was that the orbiter vaporized in the fireball that spectators had seen that day. The crew had not suffered; they died immediately.

Now, looking at the photos, doubt crept into Rhea's mind. Pitching downward and away from the explosion was a part of the shuttle that resembled the front of the orbiter, where the crew compartment was located. "Here is the nose," another astronaut pointed out. "It's facing us now . . . and here are the forward windows, rimmed in black."[109] Whatever she was looking at had not been consumed by the fireball, but instead had carved a path out of the explosion. *Could the crew compartment have escaped intact?* Rhea wondered. *Could my friends have survived the initial rupture?*

The question sparked heated debate. Some insisted the photo was too blurry—nothing could be seen from that angle. Others were sure they were seeing the capsule. To find out, Rhea and her colleagues went back and examined the radar tracks of all the shuttle's major components—identifying the boosters, the external tank, and several large pieces of debris that might be from the orbiter. Sure enough, they found something that was roughly the size of the crew compartment making a ballistic trajectory out of the breakup.[110]

Off the Coast of Cape Canaveral, USS *Preserver*. February 1986.

On the first days out at sea, Bagian spotted a familiar mosaic of silica tiles floating amidst white caps. When hauled up, Bagian identified the object as part of the forward shell of the orbiter. Given the prevailing theory that the orbiter had been vaporized, Bagian was moved by their discovery. At crew quarters that night, he debriefed Crip, who was staying there with him and Sonny Carter.

"I think we're going to find the cabin," Jim Bagian told Crip. Crip got angry.

"I don't want to hear anybody say that again," he replied firmly.[111] Bagian was taken aback by his reaction. Bagian knew that Crip had a close friendship with Judy and sensed that Crip's overwhelming grief about her loss was clouding his judgment about the investigation. *He didn't want to find the crew cabin; he didn't even want to believe for a second that she had suffered.*

"We are not going to talk about it anymore," Crip told Bagian.

"What do you mean we're not going to talk about it? We are pulling it up," Bagian said, frustrated. If they had found one piece of the orbiter, Bagian thought, they would find more.[112]

Jetty Park at Port Canaveral, Florida. Late February 1986.

The divers and salvage crews called them "jetty rats," but Bagian thought they were more like vultures.[113] The four hundred journalists who had reported on *Challenger*'s launch had quadrupled in number and, like birds of prey, migrated south to Jetty Park. A man-made channel that connects the Banana River to the Atlantic Ocean separates Jetty Park from Kennedy's launch complex. Cruise ships, military vessels, and weekend boaters find safe harbor inside the waterway, protected by the piled rocks of the jetty. The pier juts out like a black teardrop into a vast turquoise pool and shields a long, sandy beach from harsh waves and erosion. Nearby, Jetty Park Campground reigns popular with tourists who want to park their RVs oceanside, below the warm Florida sun and near a popular local fishing pier. In the evenings, boats parade past the wharf, returning to dock after another day at sea.

In February and March 1986, this little piece of Florida heaven became ground zero for those working to uncover the truth about *Challenger.* Frustrated with NASA's stonewalling, journalists from newspapers, magazines, television stations, networks, and wire services set up posts along the rocks at Jetty Park.[114] From these perches, reporters watched the salvage ships head out to sea at dawn. Then they waited, training their binoculars and telephoto lenses on the horizon and listening for radio reports between the ships and shore for any information they could glean about the search. Come nightfall, rock-and-roll music floated through the camp while reporters scarfed down slices of local pizza. When the ships returned to port under the glare of the press's powerful spotlights, reporters shot footage and still photographs of debris-laden ships unloading at the docks across the channel. Stitched together, the images of the search vessels might help journalists uncover the story of what had happened to *Challenger* and its crew. Reporters kept vigil around the clock. RVs outfitted with huge satellite dishes stood by for live feeds in case of a juicy discovery. Some reporters spent the night in vans at Jetty Park before resuming their espionage campaign the next day.[115]

News organizations, normally in competition, now collaborated to gather information. William Harwood of United Press International, who worked the story seven days a week, ten hours a day, came up with the idea of intercepting and taping ships' radio communications using a high-frequency receiver.[116] Other journalists clamored to use his antenna.[117] The CBS News crew, with a larger budget, set up a permanent camp at Jetty Park and outfitted themselves with night vision cameras used by the Israeli military in order to view the debris being hauled to shore.[118]

Bagian, like the others in the search effort, grew disgusted with the press's relentless hunt for wreckage imagery. To throw off news organizations, he, along with colleagues in the Navy and the Coast Guard, came up with "cracker-jack" codes, and false reports that sent journalists chasing hopelessly bogus leads. Several weeks into the search, Bagian's crew radioed a vague and ominous message to the Coast Guard on shore: "We have a bird of prey on board and need to transport it," Bagian said. Their "bird of prey" was a living hawk raptor nursed back to health by the ship's cook after it crashed onto their boat. Bagian and the crew knew the press would mistake the message as code and begin to speculate. That night, *CBS Evening News* anchor Dan Rather reported that the salvage effort may have found the "final remains." The crew watched in the wardroom and roared with laughter. *Gallows humor.* The next day, the divers gave the CBS crew a picture of their very real hawk raptor—and told them to take it to Rather. These cat-and-mouse games went on for weeks, then months.

Despite the antics, the search teams worried about the moment they might find the orbiter. "There was a lot of talk about what was going to happen if we found the crew compartment," said Bagian. Both he and Sonny Carter, as physicians, divers, and astronauts, could recognize the various parts of the vehicle and even handle the remains of the crew, if it came to that. To ensure privacy from the press, the team agreed on a code for when they found the module. The team had already located *Challenger*'s speedbrake, a mechanical flap on the shuttle's tail that acted like a rudder and slowed the vehicle down during landings. There was not a second one, but only the insiders knew that. A simple message—"we found the speedbrake"—radioed out would indicate they had identified the target.[119]

Hetzel Shoal Buoy, Cape Canaveral. Friday, March 7.

The "Lucy"—a weathered but capable Air Force landing craft unit (LCU)—bobbed over the waves of a turbulent Atlantic Ocean, on the hunt for wreckage. Stormy weather the previous week had created dangerous conditions, thwarting the recovery effort. Today, the seas were calmer, but they were still expecting four-to-six-foot swells by sundown.[120]

Cutting a line fifteen miles northeast of Cape Canaveral, the Lucy arrived at the shallows near Hetzel Shoal buoy, where Florida fishermen often moored for the night, and dropped its anchor a hundred feet down to the ocean floor. Mike McAllister, a local diver hired by Air Force contractor Petcham Inc., sat on the deck wishing he were back at port, for another Cocoa Beach happy hour.[121] McAllister was a cheerful young man in his twenties with a deep tan and a dimpled smile. He often took contract jobs with NASA to make ends meet, having worked at Kennedy Space Center as a shuttle hardware inspector and as a technician testing the orbiter's thermal tiles.[122]

McAllister peered across the vessel's fourteen-hundred-square-foot deck, his curiosity piqued by a fracas near the stern. First mate Louis Brinn bubbled up from the ocean floor. Brinn, who was battling a sinus infection, had sprung a messy nosebleed while under the surface. Someone else had to take over for him and complete the last dive of the day. Although McAllister had already completed two dives earlier that day, he resigned himself to the job, and grabbed his gear.

Terry Bailey, his fellow diver, was already waiting. McAllister suited up in a weighted, waterproof canvas suit, a metal diving helmet, and shoes to counteract buoyancy. He and his fellow divers performed "hard-hat" dives. Unlike regular scuba divers, hard-hat divers were tethered to their boat by an oxygen-supply hose and limited by the length of that hose. Hard-hat diving used for deep underwater work allows divers to stay under for longer than the air-supply from a scuba tank would permit. Still, the divers only worked in forty-minute shifts, to avoid having to make long decompression stops on their way back to the surface.

McCallister climbed aboard a small metal platform and was lowered a hundred feet to the ocean floor. The descent usually took ten to fifteen minutes, leaving only twenty minutes to search. Today, they would have to fight against strong currents, leaving even less time on the seafloor.

As McAllister and Bailey walked across the ocean floor, small sea creatures, blooms of phytoplankton, and silt churned up by the spring storms clouded their field of vision. They strained to see even an arm's length ahead. They were not finding much and only had a few minutes left before they needed to start their ascent. Then out of the darkness, a mound of rubble appeared. At first, neither diver knew what he was seeing. They inched closer, and the pile—the size of a large truck—grew more distinct. Large pieces of metal tangled together with wire were jutting out everywhere.

McAllister could hear his heart beating above the exhaust of the scuba regulator.[123] Something flashed white and caught his eye. He and Bailey moved toward it, then suddenly stopped short. McAllister's whole body stiffened. Two white human legs stuck out from the debris, swaying slowly in the water, like a ghost. Had McAllister and Bailey found *Challenger*'s astronauts? Bailey lurched back in terror; he was not sticking around to find out.[124] He tugged on the divers' platform for the crew to pull him up. When he emerged through the surface, he was gasping for air.

"He was pretty freaked out," said Walt Hardman, the Lucy's skipper. "He looked like a Pekingese dog. His eyes were bugged out."

"I'm not diving in a graveyard," Bailey sputtered as he climbed aboard the deck.[125]

Even though his time was up, McAllister was not leaving. He gathered his courage and approached the apparition. His heart beat out of his chest. His breath quickened. Finally, he reached out, grabbing at the foot . . . and it blew away from him. It was not a man at all. It was not a ghost. It was an empty spacesuit, blown up by the current. McAllister knew spacesuits were kept in the crew compartment. If this spacesuit was here, he must be close to the capsule, and perhaps the crew. That mystery would have to wait—McAllister was out of time.[126]

By the time McAlister surfaced, Captain Hardman knew he had explosive information. Hardman laid a buoy at the dive spot and prepped a small speedboat to head to the *Preserver* to relay the full story in person. Then he radioed the crew of the *Preserver*: "The speedbrake has been found."[127]

CHAPTER 20

ALL WE KNOW OF HEAVEN

Off the Coast of Cape Canaveral, USS *Preserver.* March 8, 1986.

At dawn, Bagian suited up in scuba gear with the *Preserver*'s Captain John Devlin. The *Preserver,* the larger boat, would take over the investigation from the Lucy. Bagian and Devlin would do a reconnaissance mission before deploying the hard-hat divers. As the sole NASA representative onboard (Sonny Carter was on leave), only Bagian could properly identify the orbiter.

Overnight, a strong current had pushed the *Preserver* away from the diving spot where Bailey and McAllister discovered the spacesuit. "We only had forty minutes from the time your head hit the water. So that clock was already ticking," Bagian said. Bagian and Captain Devlin swam against the drift to find the location. "We dive down. And the first thing I see is a transitional hand controller. I was pretty sure it was from the robotic arm." Bagian saw the joystick-shaped controller lodged in the silty ocean bed and motioned for Devlin to follow him. "This is pay dirt. Let's keep heading this direction because we're starting to see more stuff."[1]

The next thing the Air Force physician saw made his blood run cold: a leather flight boot. "It was broken right off at the boot top," said Bagian. The name of the crew member was on the side of the boot: Ron McNair. Unprepared and lacking the proper equipment to haul debris back to the surface, Bagian unzipped the top of his wetsuit and tucked the boot close to his naked chest.

"The visibility was horrible, maybe thirty feet at this point," said Bagian. "So by the time you saw something you were pretty damn close. I started seeing this ghostly apparition. It's maybe a forty-by-sixty-foot pile mound."[2]

Unlike McCallister, he knew right away that he was seeing the crew module. "It looked like you blew it up with a bomb and then shoved it back together with a bulldozer into one pile," Bagian explained. "Which means it hit intact. It didn't blow up."[3] The metal pieces he saw were "granular as if they had smashed apart, not shiny, like when you rip a tin can. When metal hits the water that hard, it doesn't rip, it shatters." He surmised that the shuttle had hit the water at two hundred miles per hour. Bagian saw the full instrument panel, the windscreen structure, the command console, and the control levers. Though everything had imploded on impact, it was all still held together by the network of wires that ran throughout the shuttle. Everything was jumbled, but recognizable, like a Cubist reconstruction of the crew compartment.[4]

What Bagian saw next broke his heart—the bodies of the crew members, sitting strapped into their seats. "We saw three or four definitely, different individuals, still in their flight suits, some with nameplates still on them." An arm extended from one, with a class ring on the finger. "You see his ring and you just knew who he was. And their flight suits are intact and there it was. There's no question."[5]

When Bagian returned to the *Preserver*, he struggled to find the words to describe the horror he had witnessed. He pulled the boot from inside his wetsuit to show the search team what he and the captain had discovered. *That said everything.*

With Bagian instructing them, the Navy crew conducted the next search. The divers only had their small pipe platform, not more than four feet by four feet, to bring up what they found. Bagian waited expectantly on the boat for the first diver to return, watching the bubbles from his regulator surface. As the bubbles came faster, he saw the diver rise from the sea headfirst, standing wide-legged on the platform. A single body lay at his feet. *Judy.*

"There she was in her powder blue flight suit." Bagian reached forward, kneeling to grab her from the diver and bring her onto the boat. But the body, having been in warm salt water for weeks, took on a jellylike texture, scavenged by shrimps and crabs. "Everything started coming apart," Bagian said. He tried to save the remains as they fell through his arms back into the water.[6] "This was day one. Day one," Bagian lamented. "We were not prepared."

Divers on the *Preserver* worked for hours to surface the crew remains, along with as much of the orbiter as they could. Computers, tape-recorders, anything that would help them understand what had happened to the astronauts, were high-priority items.

"We had large plastic buckets on the fantail, and depending on what item was brought up, it would go in one of them," Devlin said. One container held iced salt water, another iced fresh water, and another ambient fresh water. The divers worked fast, nonstop, for hours trying to recover as much as they could and keep the remnants' exposure to the air at a minimum. "The whole end of the fantail of the boat was full of wreckage," Bagian said. "We worked until we couldn't put anything more onboard."[7]

Well after dusk, the *Preserver* motored back to shore, its stern filled with wreckage. Even though Captain Devlin had not radioed the shore about their discovery, the jetty rats noticed the *Preserver* had not returned to port at its usual time, sundown. Just before nine o'clock that night, word spread across the rocks that the *Preserver* was finally headed home. News crews and photographers dashed into action, setting up their shots and directing a giant spotlight on the channel. Shining their own floodlights, two Coast Guard boats motored back and forth between the jetty and the approaching *Preserver* to obscure the press's view. The *Preserver* passed by in an instant and docked with its fantail toward the landing to further screen the wreckage from photographers.[8] Bagian jumped off the boat and headed straight to the nearest payphone to call George.

"We found the crew module—there's no question," Bagian said, relaying the details of his gruesome day. *We have the crew remains.* George had trouble accepting the news at first. The crew had not vanished into the ether after *Challenger*'s explosion. The bodies had been found on the ocean's floor. He had little time to take it in, but he also understood that the media would likely break the story soon and that he needed to relay the information to the families before the press did.[9]

Bagian and Bob Overmyer—an MOL astronaut who had joined NASA with Crip and the senior NASA official on site—needed to know where to take the bodies. It was not a simple question.[10] Kennedy Space Center was not an exclusive jurisdiction. Any remains returned to the Cape legally should have gone to the Brevard County medical examiner, but NASA was not having a local perform its astronauts' autopsies. NASA feared that information—or even worse, photographs of the bodies—might leak to the press, violating the privacy of the deceased astronauts and their families. NASA officials told Bagian and Overmyer to take the bodies to Patrick Air Force Base, where they would be under federal jurisdiction.[11] Crip ordered that the transportation of the remains be as inconspicuous as possible, so they traveled by truck, which would be less noticeable than by helicopter or larger military transport.[12] The divers loaded the remains into the back of a US Navy truck, and Bagian headed off for the hour-long trip.

It was Saturday night, after midnight. As they made their way south along A1A, Florida's beachfront highway, the bars and nightclubs on Cocoa Beach were open. None of the patrons could have guessed that, passing them in that dark truck, lay the answer to an agonizing mystery that had haunted the nation for weeks.[13]

The journalists on the jetty ran to develop their photographic film to see if they had captured any revealing images and the TV guys ferried their videotapes to their stations. Other news crews fruitlessly chased ambulances they saw heading to Patrick Air Force Base Hospital, to no avail.[14] Despite NASA's misdirection, by Monday, March 10, the press had pieced together some of the story, blanketing the covers of newspapers nationwide. "The wreckage of the cabin was found Friday in a hundred feet of water and identified by Navy divers," William Harwood's UPI story read. "The *Preserver* returned to port Saturday night under the cover of darkness and with no running lights. Divers from the *Preserver* were able to provide positive identification of the *Challenger* crew compartment debris and the existence of crew remains."[15]

George had prepped the immediate family for the announcement. Some were relieved that the search was over.[16] For others, the news opened wounds that had started to heal.[17] "In our hearts, we had already buried our loved ones," June Scobee said. "With this discovery, the anguish returned along with the media at my door and on my telephone."[18]

For the next six weeks, Jim Bagian, Sonny Carter, Bill Shepherd—an astronaut from the class of 1984 who joined the effort as a diver—and the *Preserver* crew worked every waking moment to surface the rest of the remains and the crew module.[19] Now that they knew what they were up against, they prepared with the proper equipment. Divers placed the remains in long aluminum boxes, the same kind used to retrieve the bodies of MIA soldiers, and draped the boxes with American flags. Astronauts themselves ferried the remains back to port by boat and then unloaded the aluminum containers from behind a screen that offered shelter from the press's probing floodlights. Unmarked vehicles transported the remains the rest of the way to a KSC hangar.[20,21]

Crip forbade Bagian and the divers from taking photos of the orbiter wreckage as they pulled it from the water. The debris was sent back to the Cape under tarps, hauled to an Air Force warehouse, and moved into a cavernous hangar near the Vehicle Assembly Building, where the painstaking process of reconstructing the shuttle began.[22] *No press allowed.*[23]

Kennedy Space Center. Early April 1986.

On a warm spring morning, Rhea rounded the corner to the open hangar door and was struck by the dockside odor of dead fish. Watching the technicians reassemble *Challenger* piece by piece, Rhea felt unmoored from reality. She had no idea so much of the orbiter had been recovered. The sight of the wrecked vehicle, held together only by braces and scaffolding, put a lump in her throat. Parts of the shuttle still appeared pristine; others were unrecognizable. The insulation blankets covering the fuselage and tail had become a new home for barnacles and other sea creatures. The right wing and tail were scorched. The fuel pumps of the main engines—the parts that kept blowing up prior to STS-1—were fully intact. The external tank had crumpled like a tin can and the aft end of the shuttle's right wing and right side of the tail were badly burned, indicating that the right solid booster was indeed the locus of the accident.[24]

As the astronaut in charge of crew equipment, Rhea had come to identify and tag items from the crew compartment. Everything was set up on long metal shelves, much like it would be on the orbiter. In the wreckage she counted five computers, the navigation equipment, the spacesuits, food packages, and checklists.

Among the personal mementos found were Judy's "I love you" necklace, El's soccer ball, and Ron's saxophone.[25] Rhea thought of her friends—of Ron and those days spent practicing laps in the community pool, when they first started AsCan training. Ron's children were the same age as hers. *Was this how his story ended? It was so unfair.*

A few days after she arrived at the Cape, Rhea joined Dr. Joe Kerwin, Johnson's director of Space Life Sciences, who was overseeing the medical team performing the autopsies.[26] As Rhea entered Hangar L, which served as a temporary morgue, the odor jolted her.[27] Even though stored in a cold container, the remains had been soused in the ocean for over a month. She tried to detach herself from the job ahead. "Their spirits [are] gone from here," she told herself. "These [are] parts of the mystery left behind for us to discover the answers."[28]

In an era before DNA sequencing, the pathologists had to use what physiological clues they had to identify the bodies. Women's pelvises were a different shape than men's. If someone had broken a bone, the scars might still be visible. "So total was NASA's denial that they didn't have fingerprints or footprints from us," Rhea explained, so they had to pull from the crew's dental records and health histories.[29]

The toughest part for Rhea came after the medical work. Pilot Mike Smith's widow, Jane, had requested that her husband be buried in the NASA flight suit he proudly wore for launch. Rhea personally went to retrieve the suit from the morgue to make sure it was cleaned before the funeral. Mike's suit was tattered and dingy from its long stay in the sea and his body's deterioration. Rhea could not give the suit to Jane, not like this. She located an industrial sink inside the morgue, grabbed a scrub brush and detergent, and went to work.

"The more I scrubbed, the sadder I got," Rhea said. "Why did this have to be happening? Why to a wonderful man like Mike? Why to a picture-perfect family like the Smiths?" The security guard sitting next to her tried to make small talk. Rhea answered in monosyllables. "Soon tears began pouring down my face, splashing into the wash water."[30] The flight suit got cleaner and cleaner, but somehow Rhea could not make it look like she wanted, so she cried even harder. "The guard noticed he wasn't receiving responses from me anymore. At this point, he glanced at me and must have realized that Mike Smith had been more than a distant acquaintance, that he had been my friend who is now gone. He left the room and let me scrub in silence until my knuckles were raw."[31]

Kennedy Space Center. April 29, 1986.

Nearly three months after the disaster, NASA released a statement officially confirming that the remains of all of the *Challenger* Seven and the crew compartment had been recovered.[32] Ten days later, NASA employees at Kennedy Space Center lined the roads, some bowing their heads, others weeping, as a convoy of seven black hearses drove by them en route to the same landing strip where, thirteen weeks earlier, space shuttle *Challenger* should have successfully completed its mission.[33,34] George Abbey, John Young, and Dick Truly stood watch on the runway while seven flag-draped coffins were transferred from hearses to an Air Force C-141. The men accompanied the *Challenger* Seven on a somber two hour and fifty-minute flight to Delaware's Dover Air Force base.[35,36] As silver hearses lined up to meet the coffins, there were no speeches, no music, no sound but the snaps of camera lenses. The bodies now belonged to the families, who could finally lay their loved ones to rest.[37]

Although the *Challenger* crew had been committed to their final resting places, NASA had yet to produce an accident report. To avoid more bad press, Dick Truly, who had recently been named NASA's associate administrator for space flight, wanted the report to be brief. "Just two sentences," Bagian said sardonically of Truly's request. "We found the crew. They were dead."[38] Bagian would not sign the report. "We just did the most exhaustive, the most high-profile accident investigation ever and we're going to put out a two-sentence report?"[39] He confronted Truly. "It might not be a lie, but it's deceitful," he said.[40] "People feel terrible about what happened. If anybody's on the fence that we [NASA] are a bunch of incompetents or lying SOBs, then this will remove all doubt."[41]

"You write what you need," Truly capitulated. "I don't want anybody having this wrenching in their gut."[42]

Contemporaneously, in Washington, DC, the presidentially appointed Rogers Commission was excoriating NASA's management for the way it managed safety risks. Including details of the crew's last moments in the accident report would raise questions about the shuttle's escape contingencies, of which there were none. Due in part to the shuttle's "operational" label, ejection seats were never installed in *Challenger.* The decision by NASA managers had been made long ago to save on cost and time, appeasing Congress's budgetary demands. Now the nearsightedness of that decision was laid bare. "Ain't none of them ole boys ever died because a desk crashed," John Young fumed.[43]

Those leading the investigation were driven to uncover the truth about the crew's final moments not out of a macabre curiosity. Understanding how the crew died could help improve the safety of spaceflight. *If astronauts can survive an accident like* Challenger, *shouldn't NASA have had an escape plan in place?* "If we have them die and we don't learn anything from it, that's the travesty," Bagian said. "Talk about dishonoring their sacrifice."[44]

Piecing together the final moments of the *Challenger* Seven became the final task of the pathologists. They already knew the crew compartment had been largely intact when it hit the water, but had the crew been conscious for their fall to the ocean, or did the cabin depressurize such that they passed out? If the compartment had split open, the pathologists might find hydrazine fuel from the nearby jet tanks in the human tissues.[45] The forensic specialists found no signs of decompression, and no hydrazine in the tissues.

In the orbiter's wreckage, investigators discovered the crew's

Personal Egress Air Packs (PEAPS). These emergency air packs provided the astronauts with six minutes of breathable air in the case of an emergency. Bagian discovered that three of the crew's packs had been turned on. More surprising, approximately forty-three seconds' worth of oxygen had been breathed down. The recovery team sent the air packs to Corpus Christi Army Depot, where those initial findings were confirmed.[46] The astronauts had been alive for their fall to the water and may have even been conscious for at least part of their descent. "We know that there wasn't an explosive decompression," Bagian said, because the crew would've passed out in less than two seconds, leaving no time to switch on their air packs.[47] How long were they conscious? "Was it ten seconds? Was it twenty seconds?" Bagian wondered. "We don't know."[48]

To the crew families on the roof of the Launch Control Center that day, the accident had looked like an explosion, which enveloped the orbiter. What they were actually witnessing was an aerodynamic breakup. That distinction might seem like a matter of semantics, but the difference was essential to understanding the forensics of the accident.[49]

A fire plume had snaked through the right booster joint, destabilizing the pressure in the right booster and piercing the hydrogen cask of the external tank. Flammable hydrogen, then oxygen, billowed out of the tank.

"Uh-oh," Mike Smith had said, the first indication to the ground that anything was wrong. Then the right booster wobbled and broke off its lower strut, rotating around the upper strut and crashing into the external tank. The external tank released its entire load of propellant, which instantly ignited in the electrically charged atmosphere, generating that massive fireball in the sky. To those on the ground, *Challenger* seemed as if it had vaporized, but in fact the orbiter had been obscured by burning liquid fuel.

The shuttle, amidst midair collisions and under severe aerodynamic pressures, split into several large pieces. The right-hand booster swung around again and, like a blowtorch, cut through the orbiter's right wing.[50] Another impact, perhaps with the external tank, sheared off the left wing. *Challenger*'s appendages cartwheeled end-over-end in either direction, while the main engine compartment held together, falling in one large piece toward the ocean, engines still smoldering.[51] *Challenger*'s fuselage split open like a tube of toothpaste with its top sliced off. Still flying at twice the speed of sound, the resulting rush of air filled the payload bay and overpressurized the structure, tearing it apart from the inside out. The TDRS satellite in *Challenger*'s cargo bay was blown free into the open air, as was the Spartan-Halley spacecraft.[52]

The orbiter's nose, with the crew compartment, severed from the payload bay. A mass of electrical cables from the cargo hold fluttered behind the crew cabin like an umbilical cord as the module shot through the air, climbing ever higher.[53] This wingless capsule, with great velocity, continued upward in the sky reaching an altitude of twelve miles, then pitched downward in a long arc to the wintery sea.[54]

The moment was the fruition of the fear that Judy, Ron, El, and Scobee had all lived with since joining NASA. They tried to compartmentalize the danger, joke it off, drive it out of their minds in order to live normal lives, in order to say goodbye to their children and their lovers and not fall apart every time. All the while, they knew that one day, one act of providence, one risk unassessed, one mistake overlooked could lead to this. Ever the possibility, that destiny now rose to meet them.

Despite that awareness—maybe even because of it—they worked toward perfection. Expedition, inquiry, exploration, these were the worlds in which they lived, and for that privilege, this was the price the fates might ask.

Those who have visited that undiscovered country and had the good fortune to return report that one's past unfolds like an old home movie reel. Life flashes before one's eyes—loves and losses, triumphs and travails, and the quiet moments that meant everything. The ocean mist caressing one's face during a morning run on the beach. The taste of red wine and rare steak on a Friday night. The smell of freshly cut grass before a soccer match. The slow rush of a total eclipse, witnessed at nine-tenths the speed of sound. Some say that lost loved ones appear, welcoming the dead to the hereafter. Others describe an overwhelming brilliant light that basks the spirit in calm and euphoria. Some believe these visions are God welcoming the departed to heaven. Others scoff, attributing the experience to a complex mix of chemicals that flood the brain and ease one to the end—a strange benevolence bestowed on humans by evolution.

El was a Buddhist and believed in rebirth. Ron was a Baptist and knew Jesus would be on the other side. Judy did not believe in God; the beauty of a well-ordered universe was what moved her. That, and maybe a challenging concerto. Rachmaninoff, played alone, in the dark. Dick Scobee, though a Christian, was not ready to meet his maker. He loved his life, his wife, his kids, too much. "Scob fought for any and every edge to survive," said his friend Bob Overmyer. "He flew that ship without wings all the way down."[55]

Judy or El, who sat behind the pilot and commander, knew they were in trouble. One of them pulled the switch on the pilot's personal air pack, releasing a flow of oxygen needed to survive such an extreme depressurization.[56] They hoped to give the pilot time, a chance to save them. It was a final act of survival from someone who believed, even in the darkest of moments, when hope was seemingly lost, that giving up was not an option.

CHAPTER 21

NATURE CANNOT BE FOOLED

**Hartsfield-Jackson Airport,
Atlanta, Georgia. January 28, 1986.**

On the morning of January 28, 1986, Sally Ride, for the first time in her career, was missing a shuttle launch.[1] Instead she was driving to Atlanta's Hartsfield-Jackson airport, after spending a long weekend celebrating Tam's thirty-fourth birthday.

The two women had been seeing each other for over a year and Sally's priorities were shifting. From the moment she joined NASA, Sally strove for unqualified excellence—mastering the robotic arm, keeping her body lean and strong, delivering on her mission. She had grinned through post-work happy hours when she would rather have been at home and helped the astronaut team win its share of softball games. She had been the ultimate team player, sublimating her interests to the job. Now she wanted something else. Now she was skipping non-mandatory NASA events to visit Tam on the weekends.

NASA no longer ranked as the most important thing in her life. After one last shuttle flight in July, Sally planned to resign. A quiet life in academia, far from the public eye, beckoned. So, too, did a divorce from her husband, Steve. Fate, however, was about to put those plans on hold.

Sally, hoping to enjoy a normal, civilian flight with the other passengers onboard, took her seat and buried her nose in a book. On approach to Houston, Sally heard the news break over the plane's PA system.

The space shuttle Challenger *has had a major accident.*

Before the pilot finished his announcement, Sally was on her feet, charging toward the front of the plane. She flashed her NASA identification badge, and a flight attendant ushered her into the cockpit. Surprised to see the first American woman in space, the pilot immediately accommodated her, handing her an extra pair of headphones to listen to the radio transmissions from the ground controllers. The reports were grim: "No crew members survived. All are lost."[2] With only about thirty minutes left in the flight, Sally continued to monitor the chatter about the accident, her heart sinking with each update.[3]

As soon as she landed in Houston, Sally sped forty miles south to Johnson. Four of the seven souls onboard—Judy, El, Ron, and Scobee—were her classmates, her friends.[4] In their 1978 press conference photo, Sally sat in the front row between Judy and Scobee, all three of them with bright smiles and hopeful expressions. They had known each other for nearly a decade and ridden the highs and lows of the program together. Now they were gone.

Sally did not have long to grieve before she was called back to duty. A few days after the accident, the phone rang during dinner with Steve and his father, Reverend Hawley, who had come to lead the prayer at the Johnson memorial. Sally answered. NASA administrator William Graham was calling to tell her President Reagan was announcing an independent commission to investigate the accident. Would Sally be willing to serve?

She would be the only active astronaut—the only current NASA employee, in fact—named to the commission. They wanted her expertise:

She had flown on *Challenger* twice and had a firsthand understanding of the shuttle program. Left unspoken was Sally's unimpeachable public standing. The nation could trust America's first woman in space to find out what went wrong, and to make it right.

Sally stood with the phone to her ear, staring at the floor. The grave responsibility of the job would have a scrambling effect on her life plans. *I was going to resign from NASA. I was going to leave Steve. I was planning a life with Tam.* Now all that had to wait.

"Every one of the astronauts wanted to be doing something, contributing in some way to getting NASA back on its feet and investigating what went wrong," Sally said.[5] Even though she promised Tam a new life, she knew she could not walk away from her duty. "No matter how painful it was," she said. "If nothing else, I needed closure. I needed to know what the answer was, and that NASA was going to do something to fix it."[6]

Sally hung up the phone. "I need to do this," she said quietly.[7]

National Academy of Sciences, Washington, DC. February 6, 1986.

On a cold, clear morning, a limousine with tinted windows pulled up outside the National Academy of Sciences on Constitution Avenue. Sally stepped out, wearing a violet blouse and a dark plaid skirt and carrying a tan spiral notebook, decorated with the mission patch of *Challenger*'s fallen crew.[8,9] Scores of press swarmed outside the formidable building's green lawns and marble walkways. She cut her way through the crowd, squinting in the bright lights of the cameras.[10] She was on her way to her first day on the Rogers Commission.

Three days prior, President Reagan had formally announced his blue-ribbon commission.[11] "It's time now to assemble a group of distinguished Americans to take a hard look at the accident, to make a calm and deliberate assessment of the facts and find ways to avoid repetition," said the president.[12] The Rogers Commission had 120 days to find answers and make recommendations for change.

The head of the commission was William P. Rogers, Nixon's former secretary of state. "This isn't going to be another Warren Commission,"

Rogers told his group, referencing the infamous investigation into John F. Kennedy's assassination. "We want to find the answer, the true answer, and we need to do it the right way."[13]

"Keep independence from NASA," Sally interpreted in her notes.[14]

Although the commissioners certainly were an impressive group, some in Congress and the press were concerned about that very independence.[15] Sally was still an astronaut with NASA and Neil Armstrong, the commission's vice chairman, was a NASA hero. Air Force General Donald Kutyna was the former manager of the DoD's space shuttle program. Chairman Rogers was a career politician and a Reagan supporter.

Other commissioners included heavyweights like Albert "Bud" Wheelon, who helped develop the CIA's network of spy satellites; solar physicist and Stanford professor Art Walker, who was also Sally's former mentor and the only person of color on the commission; legendary test pilot and retired Air Force general Chuck Yeager; and Nobel Prize–winning physicist Richard Feynman, who had worked on the Manhattan Project. Feynman was known to be a notoriously independent thinker and a bit of a wild card.

The only commissioner not present was Chuck Yeager, who was off preparing to break a speed record for turboprop aircraft.[16] Yeager would rarely show up for proceedings, so much so that Sally thought he should have resigned from the commission altogether.[17] Sally and the other commissioners were sworn in and led into a large auditorium, where they were guided to wide, leather chairs on a raised dais, like a courtroom jury. Sally sat in the middle of the front row, at Chairman Rogers's left, facing a bank of TV cameras. Her former Stanford professor Dr. Arthur Walker sat beside her.[18]

The first task of the Rogers Commission was to uncover the cause of the accident that cold January morning. As the commission convened, NASA believed that the catalyst had been that "unusual plume" seen escaping from the lower part of the right-side booster.[19] However, what had caused the flame was still unknown.

"Was the potential impact of low temperatures on the SRBs discussed with you?" Bud Wheelon pointedly asked Johnson Center director Jesse Moore, who was first to testify.[20]

"It was not presented to me as a matter of concern," Moore said.[21] The recently appointed Moore tried to keep up with the commissioner's queries, but he was in an unenviable position of being so new to the job that he had little technical expertise to offer the conversation. Instead, he referred most questions back to his subordinates, who would testify later that day.[22] Shuttle manager Arnie Aldrich was up after Moore.[23]

"How does the ice team document what it sees?" Sally asked Aldrich. "Do they take cameras out with them?"

"They take cameras, I believe, Sally." Aldrich called her "Sally," not the more formal "Dr. Ride," even though he addressed Rogers as senator and Feynman as professor. Regarding the ice that was seen on the pad, Aldrich said, "The total recommendation from all parties concerned was that we did *not* see a credible threat to the orbiter." Aldrich then paused ever so slightly. "Except for the Rockwell International Orbiter contractor, who in that meeting expressed some concern that there might be a slightly higher risk for the orbiter, because this was a condition we had no experience with before. That is lifting off with ice on the launchpad."[24]

Sally's ears perked. *Did Rockwell have a problem with flying?* At the time, no one on the commission pursued the question. However, later the recollection of the conversation between Aldrich's team and Rockwell would be heavily debated—both as to the facts of what was said and to the interpretation of those facts. Rockwell managers felt they had voiced their concerns loud and clear when they said, "Rockwell cannot assure that it is safe to fly."[25] Aldrich had not heard it that way: In his opinion, Rockwell could not avow that it was safe to launch, but they also could not guarantee that it was *unsafe* to launch.[26]

When asked if he had any concerns about the effect of the cold weather on the solid rocket boosters, Aldrich deferred to Dr. Judson Lovingood, the deputy manager of the shuttle projects office at Marshall and Larry Mulloy's boss.[27]

Lovingood confirmed that the effect of cold temperatures on the O-rings was discussed during the evening teleconference with Morton Thiokol on January 27, but that Thiokol recommended proceeding with launch.[28] When General Kutyna asked if Thiokol had seen damage to the

O-ring joints in previous flights, Lovingood admitted that they had, but insisted the secondary seal had never been breached.

"Was that any cause for concern?" asked Kutyna, referring to the blowby and erosion on the primary O-ring.

"Oh, yes, that is an anomaly, and that was thoroughly worked," Lovingood said.[29]

Whether Lovingood knew it at the time, the secondary seal had in fact eroded before, on STS-51B, Fred's first flight.[30] That very mission led to Morton Thiokol's O-ring Task Force, which presented its findings to Marshall, where Dr. Lovingood was a deputy manager.[31]

Chairman Rogers left the questioning there. Given that the solid rocket boosters were the prime suspect as to the cause of the disaster, some onlookers, like Thiokol's Allan McDonald, thought it odd that Rogers did not press the issue.[32] Unbeknownst to the other commissioners, Rogers had received a request from President Reagan himself: *Whatever you do, don't embarrass NASA. They're national heroes. We're going to need them. They're going to have to launch again.*[33] Reagan's Star Wars initiative relied on the shuttle, and he could not have the agency dragged through the mud.

Unlike most of the commissioners, Sally had a direct line to NASA: her friends in the Astronaut Office and all the engineers and support staff that she had come to know over her career. Sally personally sorted through the data and interviewed her own sources, often working eighteen-hour days, seven days a week.[34]

She spoke on the phone with Crip, who was leading NASA's investigation at the Cape, and Steve—still referred to in her notes as "A^2." Steve combed through the processing data for that last launch, hunting for anything that could have possibly contributed to the accident.[35] "She, through me, was able to get a sense for what the astronauts felt," Steve said. With Steve's insight, Sally could also assess whether the data on the ground matched what witnesses were saying on the stand.[36]

Steve would not be Sally's only source of insider intelligence. One anonymous individual possessed information that would alter the entire course of the investigation. Their identity has never been revealed, but they studied Sally, watched her, and they came to trust her with the crucial information they had to divulge. Soon after the commission began, this person contacted Sally and slipped her a report. They did not want to be known. Indeed they must have believed the leak would end their career. Nevertheless, they wanted the commission to know the truth. The document listed two columns, one with temperatures and the other with measures of the O-ring resiliency.[37] A trend emerged from the data: O-rings hardened when exposed to low temperatures. Sally connected the dots: In the subfreezing temperatures of that January morning, the O-rings failed to seal the gap in the field joint, allowing hot gasses to seep through to the external tank, causing the accident.

The document also confirmed Sally's worst suspicions about NASA: People at the agency knew the risk they were taking that fateful morning. *They knew and they launched anyway,* Sally realized. Had Judy been aware of the danger? Or Ron? Or El? The fact that the whistleblower did not want their identity known, and that the document had not surfaced through the regular channels of communication, suggested a cover-up.[38]

Sally had to find a way to surface this evidence, without implicating her source or herself. If the document came from her, NASA management or the press might be able to trace the report back to the whistleblower. If she were the one to break the case, it would be a nightmare for NASA and a field day for the press. NASA's "golden child," the first American woman in space, could not be the person to bring down the agency. No, the information needed to come from someone else.

Sally had become "really good buddies" with four-star Air Force General Kutyna; she trusted him instinctively.[39] At fifty-two, Kutyna styled his gray hair in a military-approved regulation cut. He had a dimpled chin and enigmatic blue eyes that saw far more than he let on. To Sally, he seemed "a boyish, avid worker" who took the metro rather than a private car, and always had chocolate in his briefcase.[40] *My kind of guy,* she thought.

On Friday, February 7, Sally plugged through a long, closed-door session with the FBI, who had not found any evidence of terrorist activities at the launch. After the hearing, Sally followed Kutyna down a basement corridor of the Old Executive Office Building.[41,42] Checking to make sure they were alone, she walked alongside him, not making eye contact. “She opened up her notebook and with her left hand, still looking straight ahead, gave me a piece of paper,” Kutyna recalled. “She didn’t say a single word.”[43] Before the general could respond, Sally disappeared around a corner.

Kutyna, who had a master’s in aeronautics and astronautics from MIT, understood her intentions once he examined the document.[44] She was trusting him to get this information out there, *without* incriminating her or her source. Kutyna had developed a rapport with fellow commissioner Richard Feynman, who had the benefit of being both a Nobel laureate and a person unconnected to NASA. That weekend, Kutyna invited Feynman over for dinner.

When Feynman arrived, Kutyna took him to his garage, where he kept a 1973 Opel GT, a vintage autobahn cruiser. He showed Feynman a carburetor that he was cleaning. “Professor, these carburetors have O-rings in them,” he said slyly. “And when it gets cold, they leak. Do you suppose that has anything to do with our situation?”[45]

Feynman did not say a word, but the wheels in his brain were turning. The O-rings had already been on his mind. When Feynman first arrived in Washington, he had been eager to hit the ground running, but the slow pace of the commission left him frustrated. Taking matters into his own hands, as was his nature, he called NASA administrator William Graham, who arranged for NASA engineers to answer Feynman’s questions about the finer mechanics of the shuttle.[46] In these meetings Feynman learned how critical the resilience of the rubber was to the design of the O-rings. Since then, he had wondered what might have compromised the O-rings when *Challenger* launched on January 28.

Kutyna’s question led Feynman to the answer.

On February 9, before Feynman had a chance to present his new thinking to the commission, an exposé in the *New York Times* rocked the investigation: “NASA Had Warning of a Disaster Risk Posed by Booster.”[47] According to two intrepid whistleblowers, engineers at NASA headquarters and at Marshall had known about the O-ring issue as early as 1982, when the seals were added to a "critical items list."[48] The article also quoted a memo from NASA budget analyst Richard Cook, written to management in July 1985: "Failure during the launch would certainly be catastrophic."[49] The article did not mention that Richard Cook was not just the memo's author, he was one of the whistleblowers.[50]

The piece further revealed that twelve prior instances of O-ring erosion had been recorded by Thiokol engineers. The reason that reform had been stymied boiled down to time and money. A late 1985 briefing for top-level NASA officials listed charring of the rings as one of the top "budget threats" to the solid-fuel booster program.[51]

In the memos presented to the *New York Times,* however, the reason for the O-ring failure was not attributed to the cold temperatures. Instead, the memos focused on "erosion and heat effects" observed on the seals after previous flights, and one memo mentioned, almost as an afterthought, that environmental effects could be a factor. Sally's smoking gun memo linking the cold to the brittle O-rings still needed to be revealed and Feynman was armed and ready.

The article blindsided the commission, which infuriated Chairman Rogers. *How was he supposed to protect NASA when the agency was withholding information?* He called for a closed-door session with the NASA officials and contractors who ran the solid rocket booster program. Perhaps they would come clean in an environment away from the prying eyes of the press.[52]

Old Executive Office Building, Washington, DC. February 10, 1986.

Larry Mulloy led the docket of the closed-door meeting. A small group of engineers including Allan McDonald, from Morton Thiokol accompanied him, sitting in the audience. Mulloy advised McDonald to keep mum unless Mulloy needed help answering a technical question. With Mulloy holding the floor, McDonald took a seat on the back bench against the wall.[53] Sally filed into the room with the other commissioners.

"The *New York Times* and other newspaper articles have created an unpleasant and unfortunate situation," Chairman Rogers began. "We would hope that NASA and NASA officials will volunteer any information in a frank and forthright manner. This is not an adversarial procedure."[54] NASA spokesmen ignored this olive branch, opening their remarks by casting doubt on the credentials of Richard Cook.[55] They claimed Cook's memos were "not accurate" and that Cook, as a budgetary analyst, had "no technical training."[56] Then they surrendered the floor to Mulloy.

To the bewilderment of everyone in the room, Mulloy described the current design of the booster's field joint as "an acceptable situation" and claimed he had "no data today to change that."[57]

"Not even today?!" Rogers asked incredulously.

"No sir, not even today," Mulloy said.[58]

During a break in testimony, Sally took a call from a *Washington Times* reporter. *Was the commission aware,* the reporter began, *that the night before the tragic launch, a shuttle contractor told NASA it had misgivings about the cold temperatures?* Sally was floored by the accusation. She did not have an answer for the reporter, but she was going to get one.[59]

Before Mulloy could resume his presentation, Sally interrupted.

"Well, before you start sharing," she said, "I answered a few of my phone calls. One was from a reporter here in Washington who said that they heard some rumors that one of the contractors may have recommended not even launching. Is that really true?"[60]

"Thiokol presented to us the fact that the lowest temperature that we had flown an O-ring or a case joint was 53°F, and they wanted to point out that we would be outside of that experience base," he admitted.[61] Mulloy was referring to El's 51C flight, in January 1985.[62]

Sally probed further. "We read in the *New York Times* about NASA internal memos where people within NASA were suggesting problems with erosion before, and I am wondering whether similar memos exist relating to problems of launching with the O-rings at low temperatures."

"I'm not aware of any such documents at Marshall," Mulloy answered, "but that isn't to say there aren't any. I need to go research my files to make sure."[63]

From his seat on the back bench, McDonald was stunned. "Even though what Mulloy said was literally true, his comments were misleading at best and, at worst, a flat-out lie," McDonald said.[64] Mulloy had been in the room during the teleconference on the night of January 27—

he knew perfectly well that Thiokol had initially recommended *not* to launch because of the cold temperatures. Any investigator would be able to find that original recommendation in NASA's files. In a hearing designed to encourage honesty, Mulloy was obfuscating the truth. "Listening to Mulloy," McDonald said, "I came to a resolute personal decision."[65] McDonald raised a shaky hand. At first, nobody noticed McDonald, who was sitting in the back of the room with the rest of the nonparticipants.[66] McDonald watched the back-and-forth, looking for his moment to break in. Finally, he got up from his seat and waved his hand frantically. At last, Mulloy acknowledged him.

"Mr. Chairman," Mulloy said, "Allan McDonald from Morton Thiokol wants to make a point."[67]

"Mr. Chairman," McDonald said in a wavering voice. "May I say something?"[68]

"Of course," Rogers said.

"The recommendation . . . from Thiokol was *not* to launch below 53° F because that was our lowest acceptable experience base . . ." McDonald said."[69]

With those words, the room went still.[70]

"Who in the hell are you?" Rogers asked.

"I'm the director of the solid rocket motor project at Morton Thiokol," McDonald answered.

"Would you please come down here and repeat what you've just said?" Rogers asked him. "Because if I just heard what I think I heard, then this may be in litigation for years to come."[71] McDonald walked from the back bench down to the floor, past Mulloy, who glowered at him. Sally noticed that McDonald was trembling. He spoke courageously, nonetheless.

Over the next half hour, McDonald explained to the commission, and an increasingly exasperated Rogers, that Thiokol had initially recommended *not to launch* in temperatures below 53°F, that they had seen worse blowby at lower temperatures, specifically at El's launch a year ago, and that they were generally concerned about how O-rings would seal at lower temperatures. The ambient temperature at the Cape was 36°F when *Challenger* launched.[72] Chairman Rogers interrupted McDonald.

"Am I hearing you say that you recommended against launch, and you never changed your mind?" Rogers asked.

"We did change our mind afterwards," McDonald said.

"What brought you to that decision?"[73]

McDonald told the commission about the evening teleconference between Marshall and Thiokol and the half-hour offline caucus at Thiokol headquarters in Utah, a discussion he was not privy to since he was at the Cape. When Thiokol managers came back online, they had changed their recommendation. Yes, they were concerned about the cold, but the data, by Thiokol's own admission, was inconclusive. Blowby had also been seen on some of the shuttle's warmest launches—Fred's STS-51B mission and Guy's second flight, STS-61A, a Spacelab mission from 1985 that launched at 75°F.[74]

"They'd reassessed all the data," McDonald said, "and had come to the conclusion that the temperature influence, based on the data they had available to them, was inconclusive; they, therefore, recommended a launch."[75] Marshall, McDonald reported, then requested the new recommendation be put in writing.[76]

"You recommended against a flight on one night," Rogers repeated, slowly, parsing the absurdity of what he had just heard, "and then have meetings with NASA people and they seem anxious to go ahead, or at least they were asking questions about it . . . and you checked back to your home office and you got word back from home office to go-ahead because the evidence is inconclusive!"[77]

McDonald confirmed that Rogers had the story right. Stunned, Rogers asked for a recess.

During the break, Sally noticed McDonald crying and shaking. All the adrenaline and stress of the moment had crashed down on him at once. Despite their informed reservations, McDonald and the Morton Thiokol engineers had been overridden by their management. By reversing its initial recommendation, Morton Thiokol had helped seal the fates of the *Challenger* Seven. Stoic demeanor be damned, Sally walked over and hugged the man. The uncharacteristically emotional moment surprised McDonald and even Sally herself.

"God, I'm glad somebody finally told us what really happened," she told him. "That took a lot of guts."[78]

"Her appreciation for what I had done put blood back into my face, air into my lungs, and steel back into my backbone. I knew I could not become a quitter," McDonald resolved.[79]

As shocked as Rogers was by McDonald's testimony, he still chose to give NASA executives some breathing room at the following day's public session.[80]

"Let's not talk about the weather at this next hearing," he said, still trying to spare NASA any further embarrassment. He cleared his throat. "Let's give NASA a chance to explain themselves."[81]

McDonald's confession felt like a crack in the case from which there was no turning back.[82] Sally made a note to look for internal memos that would show temperature's effect on O-rings, and to ask Crip what he knew about the previous examples of O-ring erosion. She scribbled to herself a reminder to ask about "'budget threats' vs. safety concerns . . . interplay of safety and budget."[83] Sally had joined the commission to find out what really happened, and to give herself a sense of closure.[84] Now the truth was becoming clear to her: Even after years of knowing that the seals were defective, NASA had prioritized its budget over the lives of its astronauts. Sally and the other astronauts had been kept in the dark about these dangerous engineering problems. Yes, every astronaut understood that spaceflight was a perilous business, but not sharing the known risks with the people who bore them was unconscionable.[85]

"It was a very, very difficult time for me," Sally later said.[86] She had to take to task the NASA titans that had put her into space and made her famous. Her crisis of faith was soothed only by watching her classmates pick up the mantle: "Just seeing how the accident hurt them, how much effort they put into the investigation, and how important it was to them to understand what had happened . . . restored my confidence."[87]

Harry S. Truman Building, Washington, DC. February 11, 1986.
The next day a few inches of snow dusted the nation's capital. Gray clouds obscured the sun as Sally walked into Tuesday's public hearing.[88]

Chairman Rogers got right to the point: "This morning we will start the meeting . . . with the matter of seals on the booster rockets. And I would like as much as possible to limit our discussions today to that one subject matter."[89]

Foremost among Rogers's concerns—other than the booster seals—was how the press, many members of which were present in the audience, would portray the commission given McDonald's recent confession. "I hope that we don't develop any friction between the media and NASA and this commission, because we are all working to the same end," he added to his remarks. "I hope we can cooperate with them in all ways to deal with this very difficult and tragic accident, which is of such importance to the nation."[90]

After a brief preamble from William Graham and Jesse Moore, during which Graham promised NASA's cooperation with the commission, Larry Mulloy, the man of the hour, rose to speak.[91] Clutching a sheaf of papers, dark circles under his eyes, and his eyebrows furrowed, Mulloy raised his right hand for the swearing in.[92] Mulloy walked the commissioners through a diagram of the booster rockets, displayed on a television in the front of the room. His confident Cajun accent took on a staccato quality and he constantly licked his lips as if his mouth had gone dry.[93] Despite an obvious case of nerves, Mulloy knew his stuff. He rattled off the technical specs of the O-rings, picked up a section of the booster's joint that he had brought in to show to the commissioners, and explained how it all fit together. Over a foot long, the segment contained two parts—the tang and clevis—that were meant to be sealed by the O-rings.

Sally asked for a closer look and the commissioners passed the segment around to the group. When Feynman got a hold of the hardware, he asked Mulloy, "This rubber thing that is put in, the so-called O-ring, is supposed to expand to make contact with the metal underneath so that it makes a seal, is that the idea?"

"Yes, sir. In the static condition . . . it should be in direct contact with the tang and the clevis of the joint, and be squeezed twenty thousandths of an inch," said Mulloy.

"And if it weren't there, if it weren't in contact at all and there was no seal at all, that would be a leak. Why don't we take the O-rings out?" said Feynman.

"Because you would have hot gas expanding through the joint," said Mulloy.[94]

"[And] if this material weren't resilient for say a second or two, that would be enough to be a very dangerous situation?" Feynman questioned.

"Yes, sir," said Mulloy.[95] From her position in the front row, Sally's ears pricked up. Feynman was leading Mulloy into a trap.

Feynman fell silent, allowing Mulloy to resume his testimony. Meanwhile, Feynman flagged down someone to bring him a glass of ice water.[96] When the joint made its way to where Feynman sat, in the back row, instead of merely glancing at the segment and passing it along, as every other commissioner had done, Feynman extracted the rubber O-ring and squeezed it in a C-clamp. Then he plunged the clamped O-ring into his cup of cold water.[97] As Mulloy continued his presentation, Feynman agitated in his seat, barely able to contain his excitement. After a brief recess, Feynman pushed the button on his mic.[98]

"I took this stuff that I got out of your seal, and I put it in ice water," Feynman told Mulloy. "And I discovered that when you put some pressure on it for a while and then undo it, it doesn't stretch back." Feynman took the now icy-cold O-ring out of the C-clamp and showed the other commissioners that the O-ring did not bounce back to its original form. "It stays the same dimension. In other words, for a few seconds at least and more seconds than that, there is no resilience in this particular material when it is at a temperature of 32°F."[99] Feynman paused in the gravity of the moment. "I believe," he said, settling back into his seat, "that has some significance for our problem."[100]

"That," Chairman Rogers cut in, "is a matter we will consider, of course, at length in the session that we will hold on the weather."[101] Rogers did not appreciate Feynman's theatrics. Two rows behind Rogers, Kutyna and Feynman grinned in the afterglow of a hand well played. Later, Kutyna remarked that only a Nobel laureate could have gotten away with the churlish ice water experiment. Feynman's show-and-tell blanketed the news that night and brought the public's attention to the cold weather conditions for the first time.

As Rogers silently cursed the day he had ever agreed to let Feynman on his commission, Sally sat quietly and inconspicuously beside him, the mastermind behind it all.

Kennedy Space Center. February 14, 1986.

"It just concerned me terribly," Roger Boisjoly told the commissioners, ". . . with how temperature, low temperature, affects the timing function and the ability of the seal to seal. Low temperature—and I stated this for over a year—is away from the direction of goodness."[102]

Three days after Feynman's flashy demonstration, the Rogers Commission—minus Chuck Yeager—flew to Cape Canaveral to talk with the ice team about the temperature readings on the morning of January 28 and the decision to launch. Now they were hearing directly from Roger Boisjoly, Thiokol's Cassandra—who called attention to the black soot between the O-rings after El's 51C flight, and who had worked on the O-ring Task Force.[103] Boisjoly had been the one in that Utah conference room, the night before the disaster, pleading with the Thiokol managers not to launch. Having felt sidelined for years, Boisjoly finally had his chance to tell the commission what happened during the thirty-minute offline caucus at Morton Thiokol, on the night before the doomed launch. His statements disturbed the commissioners.[104]

"And the O-ring was your concern?" Dr. Walker asked.

"Yes," Boisjoly answered. "I felt we were very successful up until early evening, because it culminated in the recommendation not to fly, and that was the initial conclusion."

"What was the motivation driving those who were trying to overturn your opposition?" General Kutyna probed.

"They felt that I had not conclusively demonstrated that there was a tie-in between temperature and blowby," Boisjoly responded. "My main concern was if the timing function changed and that seal took longer to get there, then you might not have any seal left because it might be eroded . . . I just don't know how to quantify that. But I felt that the observations made were telling us . . . that temperature was a discriminator, and I couldn't get that point across."[105]

Boisjoly told the commissioners how he had shown the managers pictures of the jet-black soot from the booster joints on El's flight, revealing worse blowby than on warmer flights. He echoed Feynman's concern that the seconds it took for the brittle O-rings to regain their resiliency was enough time for hot gas to leak past the primary and secondary O-ring seals. This escaping gas became visible during the launch—as black puffs of smoke seeping from the booster's joints. "I presented those feelings very strong—I get very emotional about these things—and I was quite strong over the [teleconference] about it," Boisjoly said.[106] Despite Boisjoly's warnings, the managers labeled his data "inconclusive."

"I basically had no direct input into the final recommendation to launch and I was not polled," Boisjoly said.[107] His damning testimony directly contradicted what Mulloy and Lovingood had reported to the commission.

"Very concerned. Recommended not to fly," Sally wrote in her notes.[108] That weekend, the Rogers Commission, in a seismic move, removed Mulloy, Aldrich, and Moore—and anyone who had been involved in the decision to launch *Challenger*—from the command and reporting structure of the investigation. The commission felt that these individuals could no longer be trusted to report the facts.[109] As one NASA official said, "It looks like they're no longer investigators. They're investigatees."[110]

Harry S. Truman Building, Washington, DC. February 26, 1986.
A week and a half later, Larry Mulloy returned to the witness stand, prepared to defend his decision to ignore Thiokol's warnings around *Challenger*'s launch. He took a seat at a long table in the front of the auditorium to the left of the commissioners.[111] Sally's eyes narrowed as Mulloy sat down at the long table at the front of the auditorium.[112] Her demeanor had changed since the first hearing. Gone was any innocence. Her brunette hair was cut into a sleek bob and she wore a dark turtleneck. She was prepared—eager, in fact—to close in on an essential remaining question: Had the managers at the Marshall Space Center been negligent?

Chairman Rogers led the questioning, "Wasn't [the situation before *Challenger*'s launch] a pretty bad case? With the weather and all of the alarms that you had, and the recommendations from the engineers at Thiokol? Wasn't this what seemed like a pretty dangerous situation?"

"It did not seem that way to me then, sir," Mulloy replied.[113] A few seconds later, Mulloy casually mentioned how "Mr. [Allan] McDonald testified yesterday," adding, "I cannot assert to the factualness of what he stated."[114] Mulloy's feeble attempt to cast doubt on McDonald's testimony must have made Sally's blood boil. She had watched McDonald cry and tremble: She knew how much bravery it took for the engineer to speak the truth, regardless of what his bosses might say.

"What are the solid rocket motors qualified to, what temperature?" Sally entered the fray.[115]

Mulloy explained that the booster rockets were qualified to operate at up to two hundred thousand feet, in temperatures between 40°F and 90°F. However, he also cited the appendix of a Johnson document saying that the boosters had to "be capable of operating at an ambient temperature down to 31°F."[116] Even this obscure citation did not help Mulloy's case since Mulloy had supported launching in 26°F during the evening teleconference with Thiokol.[117]

"So what you're saying is there was a spec that NASA imposed saying that the SRM should be qualified to launch at 31°F. Now, was that taken into account in the qualification test program for the SRM?" Sally pounced.[118]

"Dr. Ride, we did not come into this discussion today with all those specifics," Mulloy said.[119] Sally pressed ahead.

"[The engineers] . . . felt that their temperature data was inconclusive," Sally said. "Wasn't that a major basis for their recommendation not to launch in the first place, that they simply felt they just didn't have the proof it was safe? Did you think *you* had the proof?"[120]

Between nervous sips of water, Mulloy fumbled through an answer, returning to his story that the data was "not conclusive." Sally then zeroed in on her argument for the commission.

"The primary O-ring is what is defined as a Criticality 1 item," she pointed out. If Criticality 1 items failed, by definition, the orbiter would be lost. "All Criticality 1 items are reviewed and signed off all the way up the NASA chain, always up to Level I, and have to be signed off and understood at a very high level," she said. "It would concern me if I thought that on the day before launch . . . that engineers were allowed to decide."[121]

In other words, NASA protocol dictated that managers like Mulloy report any changes to Criticality 1 items. Mulloy should have raised the issue with Arnie Aldrich and received his approval for a waiver to launch. He did not.

"They [the engineers] may be right," Sally continued. "But . . . that decision hasn't been signed off at the levels that the original decision was signed off at, and it would concern me to think that Criticality 1s could be handled that way by our system."[122]

Right on, Sally Ride! thought Allan McDonald. She had framed the issue clearly: In not informing the NASA higher-ups, there could be no doubt that Marshall had been negligent.[123]

On the morning of March 6, Sally sat down with journalist Lynn Sherr to give an unvarnished interview about the commission for ABC News.

"I think that we may have been misleading people into thinking that this is a routine operation, that it's just like getting on an airliner and going across the country and that it's safe. And it's not," Sally said. "Now, maybe they did understand the risks, maybe they didn't, and I think that that's NASA's responsibility and I'm not sure that NASA had carried out that responsibility."[124]

"You've flown on the shuttle twice," Sherr said. "Knowing what you know now, would you fly again?"

"I am not ready to fly again now," Sally replied evenly. "I think there are very few astronauts who are ready to fly again now."[125]

Two days after Sally's soul-baring interview, the recovery team found the crew module and the bodies of *Challenger*'s crew.

Testimony wrapped on May 2, 1986. On May 8, 1986, NASA announced that Larry Mulloy would be reassigned within Marshall.[126] Two months later, Mulloy retired.[127] Marshall's director, William Lucas, announced his own early retirement in June.[128] Former administrator James Beggs officially resigned a month after the *Challenger* accident, while the acting administrator, William Graham, left his post in October 1986, becoming the director of the White House Office of Science and Technology Policy. In early November 1986, Arnie Aldrich was moved from head of the shuttle program office at Johnson to NASA headquarters.[129] Jesse Moore resigned as the head of Johnson Space Center on October 12, 1986, and left NASA for private industry in February 1987.[130]

Initially, both Allan McDonald and Roger Boisjoly were reprimanded by Morton Thiokol for coming forward with their testimonies.[131] McDonald was demoted and Boisjoly placed on leave, but when the US Representative from Massachusetts Edward Markey learned about this, he threatened to bar Thiokol from all future government contracts.[132] McDonald was then promoted to a VP position and put in charge of the booster rocket redesign project for which he had advocated since 1985. Boisjoly decided not to return to Thiokol, but instead became a consultant on corporate ethics.[133]

On June 9, 1986, William Rogers officially delivered the commission's report to President Reagan. The message from the commission was clear: Thiokol's flawed booster design, Marshall's lack of transparency, the shuttle's unsustainable flight rate, and NASA's unreliable system of safety reporting were all to blame. "The decision to launch the *Challenger* was flawed," the commissioners determined.[134] "[T]estimony reveals failures in communication that resulted in a decision to launch 51L based on incomplete and sometimes misleading information, a conflict between engineering data and management judgments, and a NASA management structure that permitted internal flight safety problems to bypass key Shuttle managers."[135] The report acknowledged that mistakes made were symptomatic of a larger cultural issue at the agency. The *Challenger* disaster was, as the report read, "an accident rooted in history."[136]

Dr. Feynman wrote an addendum to the report. He was pointed in his excoriation of NASA's administration, saying that NASA had minimized the risk of space flight "in order to ensure the supply of funds."[137] This, he said, "has very unfortunate consequences, the most serious of which is to encourage ordinary citizens to fly in such a dangerous machine, as if it had attained the safety of an ordinary airliner."[138] Dr. Feynman concluded: "For a successful technology, reality must take precedence over public relations, *for nature cannot be fooled.*"[139]

In what seemed like an attempt to soften the blow, Chairman Rogers accompanied the report with his own letter to the president: The commission was not criticizing the shuttle program as a whole, and fully recognized "that the risk associated with space flight cannot be totally eliminated."[140]

As for Sally, she knew she would not take that risk again. After participating in twelve weeks of hearings, listening to hundreds of hours of testimony, reviewing over six thousand documents, and traveling across the country to NASA's many outposts and then back to DC again, Sally said she "looked tired and just kind of gray in the face."[141] Sally was adrift, disillusioned with the agency that had launched her into space and celebrity. With the report completed, she revisited her conversation with Tam. A plan to retire from NASA and decamp to academia had long been forming in her mind, but the revelations of the investigation crystallized her resolve. When she told George she was resigning from the astronaut corps, he understood. He could not blame her.[142]

George testified to the Rogers Commission only once, alongside Bob Crippen, and John Young.[143] Sally did not utter a word during that hearing.[144] John Young conceded that the planned launch schedule for 1986 might have been overly ambitious.[145] Crip, and later Young, acknowledged that they were aware of the blowby on El's 51C flight—although both said they had not understood the seriousness of the issue.[146,147] Like Young and Crip, George knew there had been issues with the booster joints, among probably hundreds of similarly weighted engineering issues with the shuttle, but he had never been presented with information showing how catastrophic the O-ring issue could be.[148]

Sally listened silently as the men who had shaped so much of her career were questioned. She must have hoped that George, Young, and Crip would have spoken up to ensure astronaut safety and prevent *Challenger* from taking off that day. After all, they represented the crew at the Flight Readiness Reviews and could weigh in on whether NASA was a go for launch. To some, the commissioners went too easy on the leaders of the Astronaut Office. TFNG Bob Overmyer, who assisted Jim Bagian in recovering the bodies, argued the commission should have probed into the role George played in "preventing the flight crews' memos and long-standing concerns from getting properly aired before NASA management."[149] In Overmyer's opinion, George had been sheltered from more aggressive questioning because of his relationship with Tom Tate, a long-time friend of George and the administrative assistant to Don Fuqua, the chairman of the House Science and Technology Committee.[150] However, as Crip pointed out, the Astronaut Office's decisions could only be as good as the data they were presented with. George blamed the Marshall managers for not adequately framing the O-ring issue for the Johnson and Kennedy folks.[151] In that respect, Sally could not fault her former bosses.

After leaving Johnson, Sally planned on staying in DC for several more months to write what would become the Ride Report, her visionary recommendations for the future of NASA. The other New Guys gradually learned that Sally was leaving Houston. "Normally, Shannon, Rhea, Anna, and I would take her out for a beer," Kathy said. "To see off your colleague, however super-friendly or competitive you've been. A gracious goodbye from this unique thing you've done together."[152]

Sally was not interested in saying goodbye.

"Shannon and I had to trick her into meeting us at a small Mexican restaurant for a bite of lunch," Kathy said. "She was clearly incredibly uncomfortable about it."[153] Even though Sally had been through so much with these women, she was not the kind of person who liked big emotional moments. Despite hoping to slip out of town with as little fuss as possible, she endured the farewell lunch with her classmates.

Before Sally finalized her move, she needed to have one more life-altering conversation.

In winter 1986, Sally was visiting Houston for the weekend, on leave from DC. On Sunday, about ten minutes before she had to go to the airport, she turned to Steve.

"You're probably going to kill me for this," she said, "but I don't want to be married anymore."[154]

Steve had not exactly seen this coming. Yes, there was an undeniable friction in the Hawley-Ride marriage, which had been building since before *Challenger.* In the first weeks after the accident, Steve believed he and Sally would table their personal issues and focus on the accident investigation. He followed the hearings on TV and was proud of Sally's work on the commission.[155] The two had often compared notes with each other over the phone.[156]

Now that the investigation was over, Steve believed Sally might be leaving NASA—and asking him how they would navigate a long-distance relationship. He had not expected a divorce. There was not much of a discussion after Sally's bombshell announcement. She had a flight to catch. "Pretty much she left," Steve said, but not before promising to take care of all the "legal issues."[157]

Sally did not say a word about her affair with Tam. At that point, Steve only knew Tam as Sally's friend, "somebody she'd go visit when she was in California."[158] He did understand that he was no longer a priority in Sally's life.

"Honestly," Steve said, "for quite a while, it was clear to me that I wasn't really that important to her."[159]

More than a year after her trip to Atlanta for Tam's thirty-fourth birthday, Sally revisited their earlier conversation. Even with the press scrutiny during Sally's tenure on the Rogers Commission, Tam and Sally had been seeing each other on the weekends, with Tam flying to DC and Sally flying to Atlanta. They walked around the National Mall and visited the masterpieces in the Hirshhorn Museum. They went for runs and chased pizza with ice cream.[160] Somewhere between the commission and the divorce from Steve and the Ride Report and all these weekends with Tam, Sally was becoming comfortable in her own skin. She was in love with another woman, and it was (finally) time for them to start their lives together.

In September 1989, Sally's beloved father, Dale Ride, died of a heart attack during surgery.[161] Tam flew to California to be with Sally for the memorial. During that awful week, full of grief, Sally turned to Tam and said, "I can't take the long distance anymore."[162] Tam quit her PhD program and moved to California, where she got a job teaching biology at San Diego Mesa State. Sally took a nearby professorship at the University of California, San Diego. Just north of town, the women moved in together in a townhouse in La Jolla.

"So, tell me how you're thinking about us," Tam once asked the famously cool Sally. "Is this forever for you? What do you feel?"

"You know I can't think more than five years ahead," Sally replied.

"What? Excuse me?" Tam sputtered. "You're supposed to say, 'Yes, it's forever! You're the love of my life!' You know? 'This is it.'"

That was not Sally's way. They came up with a system. Every five years, they checked in. Tam would ask, "Okay, are you going to sign up again? Are we renewing?"[163]

Sally always said yes.

On May 6, 1987, Jane Smith, the widow of *Challenger*'s pilot Mike Smith, filed a $1.5 billion lawsuit against Morton Thiokol, appealing for reparations and punitive damages, and demanding the contractor be barred from any future work on the shuttle.[164] Thiokol settled for an undisclosed amount.[165] Four of the astronauts' families—the Scobees, the Onizukas, the Jarvises, and the McAuliffes—settled with the Justice Department for $7.7 million, divided in a split pot, of which Morton Thiokol was responsible for 60 percent.[166,167] The Resniks and the McNairs both sued and settled with the company separately. Those settlements took years.[168]

By the end of the salvage mission, 120 tons of *Challenger* wreckage had been raised off the floor of the Atlantic Ocean. The retrieval accounted for 30 percent of *Challenger*'s total structure, including about 75 percent of its crew cabin and surrounding fuselage. Other than the landing gear—which would be used for research and testing at NASA Langley—the entirety of the *Challenger* wreckage was buried deep underground, a decision by the man who had headed the investigation at the Cape—Bob Crippen.[169]

After the Rogers Commission Report was published, 102 wooden crates containing *Challenger*'s remnants were put into two abandoned underground Minuteman missile silos, silos 31B and 32B, at the Cape Canaveral Air Force Station.[170] *Challenger,* which had taken Crip to space three times, flown the first American woman and first African American to space, and served as the backbone of the fleet, would be hidden from the world for decades to come. *Perhaps,* thought the man who buried it, *some memories are better put to rest.*

With the *Challenger* tragedy, NASA's once stellar reputation had been tarnished. Sally Ride, one of the agency's biggest and brightest stars, would never again fly aboard the space shuttle. However, the shuttle program would not be buried like *Challenger.* As for the astronauts who remained—whose memories of the *Challenger* Seven were still fresh sources of grief—they were determined that their friends' deaths would not be in vain. NASA may have lost its way, but the New Guys were committed to restoring the agency to its proper place among the stars.

CHAPTER 22

GOD HELP YOU IF YOU SCREW THIS UP

League City, Texas. June 1987.

On a warm Saturday evening, Hoot Gibson took the stage in the big banquet room at Walter Hall Park. In his TFNG T-shirt, his hair slicked back and an electric guitar slung across his waist, he looked like a 1950s rockabilly.[1] Behind him, Jim Wetherbee kicked off the back beat on the snare drum and high hat. TFNG Brewster Shaw lit into a blues riff on rhythm guitar. Hoot tipped his head toward lead singer Pinky Nelson, who started to croon: "I'm going to Kansas City / Kansas City here I come."

The crowd squealed with delight, punctuating the roar of the rock-and-roll guitar with their best teenybopper impressions. The all-astronaut band, a tradition the New Guys would bequeath to future astronauts, was making its debut. They went on to play a medley of 1950s hits from their youth, Chuck Berry: "Maybellene," "Johnny B. Goode," and "Rock and Roll Music."

We are Max Q! the group yelled to the cheering crowd of secretaries, astronauts, and other NASA personnel. "A lot of noise, a lot of vibration, a lot of buffeting," Hoot joked.[2]

Even John Young—a private guy who usually did not attend work parties—attended the good old-fashioned sock hop, his sleeves rolled up with a cigarette pack and hair styled in a ducktail. Seeing the enthusiasm from the boss boosted morale. The Astronaut Office needed this.[3]

The last year and a half had been a gauntlet for everyone who remained in the Astronaut Office. They spent weeks arranging funerals and grueling cross-country trips to attend memorial services. At work, they endured painful reforms—gutting old, familiar systems and building new ones—all while wondering if they would ever fly again.

As the astronauts and staff swayed in the audience listening to Pinky sing, "Oh Maybellene, why can't you be true," they felt like things were getting back on track, like they were finally moving forward again.

A year earlier, in summer 1986, as *Space Camp* and *Top Gun* lit up the box office with exhilarating aviator adrenaline, the Rogers Commission handed down its sobering recommendations to NASA administrators. Aside from a mandate to fully redesign the booster joints, the commissioners called for a new set of safety standards, improved communications to and from the Marshall Space Flight Center, a reduced flight rate that did not overextend NASA's resources, and new maintenance safeguards that would keep a closer eye on critical items like the main engines and rocket boosters.[4] Far more daunting, the commissioners demanded systemic cultural change across the agency, starting at the top, and they pressed for leadership from the people who understood the risks and rewards of human spaceflight: NASA's astronauts.

In 1978, Fred Gregory, newly selected to Astronaut Group 8, had packed up the family's Dodge Ramcharger—affectionately nicknamed "The Beast" for its ability to withstand all abuse—and driven down to

Houston from DC. Now, in the heat of August 1986, Fred was returning to Washington, DC, to lead NASA's newly created Office of Safety, Reliability, and Quality Assurance at headquarters. George had given Fred the news personally. After *Challenger*, Fred openly criticized NASA's cult of secrecy, especially when he learned how close he himself had come to death on 51B. *Is this a punishment or a promotion?* he thought. Even though Fred preferred to fly again, he knew he needed to be a part of making spaceflight safe.[5]

As Safety Review Officer, Fred spearheaded an operational safety program to address the organizational and mechanical safety failures that led to the *Challenger* disaster. He instituted an anonymous reporting system through which employees could confidentially raise safety concerns, without fear of retaliation.[6] In the event of future accidents, he wrote a contingency plan to identify an internal investigation board, to avoid the haphazard post-*Challenger* scramble and the need for a presidential commission.

Not everyone wanted to stick around like Fred. A half-dozen astronauts exited the office. Some joined private industry, others returned to their military careers. Among them were three New Guys—Dale Gardner, Robert "Bob" Stewart, and James "Ox" van Hoften.[7] Sally exited a year later, in 1987.

Those that remained worked with NASA's contractors to rethink safety measures for the shuttle's return to flight. The *Challenger* accident made it clear that a crew might be able to survive a launch mishap, which meant the crew needed a way to bail out.[8] Astronaut Jim Bagian along with New Guys Pinky Nelson and Steve Nagel aided engineers in creating and testing a new shuttle escape system. The system they devised added a nine-foot-long telescoping pole that would extend from the middeck's outer hatch, allowing astronauts to slide past the shuttle's wing and parachute to safety.[9,10] Others helped redesign the shuttle's brakes and nose-wheel steering system, changing out the old beryllium brake pads for more durable carbon pads.[11,12] The astronauts' simple flight suits were upgraded to bright orange high-altitude pressure suits that, in case of emergency, would protect the astronauts' organs on launch and reentry and provide each astronaut with their own source of oxygen.[13]

Anna Fisher helped run flight simulations with mission controllers to keep them sharp. Guy Bluford worked on Spacelab and followed the development of the new boosters. Despite the often empty and spiritless office, he never considered leaving. "I owed it to my classmates to hang around and make sure we flew this thing again," he said.[14] The lovable Delta House—New Guys Hoot Gibson, Pinky Nelson, and Steve Hawley, plus Charlie Bolden and Franklin Chang-Díaz—participated in partial countdowns to keep the launch crews at Kennedy on their toes.[15]

The return to flight effort would likely take years. Without a launch to train for, the astronauts had time to focus on their lives outside NASA. Both Rhea and Anna had second children. Guy earned an MBA. Kathy embarked on a new adventure, joining the Navy Reserves and becoming a lieutenant commander.[16]

Crip moved to DC headquarters, and then to Kennedy Space Center, to lead the return to flight effort. As the shuttle's most experienced flier with four missions under his belt, Crip brought an astronaut's eye for safety and an insider's perspective to his new job. James Fletcher returned as NASA administrator, swearing into office in May 1986.[17] Fletcher had served as administrator from 1971 to 1977, when Congress first gave the green light for the shuttle program. The former University of Utah president and devout Mormon had supported Morton Thiokol's monopoly on the solid rocket boosters. Now he would oversee its redesign.

Carolyn Huntoon had risen to associate director of Johnson Space Center in 1984. She stayed in the position through *Challenger* and its aftermath, providing stability during those turbulent years. In April 1987, New Guy Dan Brandenstein replaced John Young as chief of the Astronaut Office. Young, who had been vocal about NASA's failings during the *Challenger* investigation, was reassigned as a special assistant to Johnson's director for engineering, operations, and safety. Brandenstein chose Steve Hawley as his deputy. The New Guys were now running the show.

What about George Abbey?

In April 1986, the *Washington Post* published a scathing article alleging George Abbey and John Young had cultivated a deeply flawed work environment in the Astronaut Office. The *Post* interviewed nine current and former astronauts, all off the record. One called George "an autocratic boss who brooks little disagreement." "There is certainly a lot of frustration and unhappiness" around crew assignments, said another. "It's not quantitative, it's not something you could see on paper. It's qualitative. The program wastes people: it doesn't develop their careers; it doesn't develop people well." One astronaut dressed down leadership for stifling safety concerns: "If you raised your voice about safety. You were told [that] if you were too big a coward to fly, they'd find somebody else."[18] In a later interview, astronaut Story Musgrave reflected on that time, "There was absolutely no functioning organizational structure within the astronaut office. Communication was generally handled one-on-one from George Abbey to separate individuals in the Astronaut Office. John Young wrote memos, but the information never got beyond George's office."[19]

The astronauts also complained about George's preferential treatment for Navy pilots. "Of the twenty-eight CDR and PLT seats available on the first fourteen missions," Fred observed later, "only six have been filled with Air Force pilots. Fifteen went to Navy pilots."[20] Some surmised the bias stemmed from George's loyalty to his alma mater, the Naval Academy, and his still-smoldering resentment of Air Force ace Chuck Yeager. Despite the negative press, George suffered almost no consequences for nearly two years.

Then in October 1987, the consummate administrator was felled by the bureaucracy he so loved. After more internal complaints about George's management style bubbled to the top, Johnson's new director Aaron Cohen, who had replaced Jesse Moore, removed George as the flight crew operations director and transferred him to his own staff.[21] George had little choice in the matter. Even Dick Truly, who had been one of George's bubbas and had been promoted to associate administrator for space flight in February 1986, backed Cohen's decision. As Fred echoed, "It was time for George to go."[22] A few months later, Truly transferred a reluctant George once again, this time out of Johnson and to DC Headquarters—the equivalent of getting put out to pasture.

The astronauts held a George Abbey Appreciation Night at Pe-Te's Cajun restaurant to bid farewell to their outgoing boss. Mark Lee, an astronaut from the 1984 class and master of ceremonies for the night, turned the evening into a roast. Fitted with a wig to imitate George's buzz cut,

Lee cracked jokes about the former astronaut boss while George watched stoically from a table nearby. Then, to the surprise of many, Lee got sentimental. "You can't have a boss for so many years and not be choked up by his departure," Lee told the crowd. "And for those of you who might be feeling a lump in the throat and getting all misty-eyed thinking about George leaving . . . just remember what an asshole he can be!"[23] The room erupted in laughter. George cracked a thin smile. Watching from across the room, Mike Mullane wondered what cruel revenge the Dark Lord was secretly plotting.[24]

In early 1988, George left Houston and headed to DC. At NASA headquarters, a morose George sat at a desk with a phone that never rang.[25] The man who had ruled manned spaceflight for nearly a decade had been toppled by his own.

Wasatch Mountains, Utah. August 1988.

All was quiet in the remote reaches of northern Utah's Wasatch Mountain Range. The dry August heat had thinned out the green vegetation that covered the rolling foothills in summer. Tall golden grasses swayed in a gentle breeze.

Allan McDonald huddled together with Morton Thiokol's management team, as well as NASA's managers, including Fred Gregory, in a control center overlooking the foothills. They had gathered at Thiokol's isolated testing facility, about sixty miles north of Salt Lake City, to witness the final test firing of the solid rocket booster's newly redesigned joint. To no one's surprise, a legion of reporters attended, questioning whether NASA would ever fly again.

After *Challenger*, Thiokol engineers, led by Allan McDonald, had spent two and a half years designing a new and better field joint—a breakneck pace in the world of rocket design. One year earlier, McDonald had personally volunteered to crawl through the nozzle of the giant redesigned rocket, twelve feet in diameter and surrounded by over one million pounds of the solid propellant that when lit burned at 6,000°F. After spending every waking hour on the issue, he wanted to inspect the newly designed and installed joints and O-rings himself to make sure they had come together correctly during assembly.[26]

The redesigned boosters used casings fitted with a capture system—an outer lip around each field joint—recommended by Allan McDonald's task force pre-*Challenger*. Each joint was fitted with a third O-ring: a primary, plus two backups to catch escaping hot gas.[27] Today's test would officially certify the new design.

As the technicians initiated the test, Fred Gregory and Allan McDonald peered at the live feeds of the rocket. McDonald focused one monitor on the most severely flawed field joint. Engineers had intentionally gouged and immolated the primary O-rings before installing them, to replicate the most significant damage possible in a real launch scenario. If the redesigned joint held here, it would hold during launch. McDonald kept his eyes peeled for hot gas escaping the joint, as it had in the *Challenger* accident.[28]

First, they saw light, then they heard the explosion and felt it reverberate in their bones. A massive red blaze erupted against the backdrop of the yellowed mountains. Over a million pounds of burning ammonium perchlorate, aluminum, and iron oxide billowed out behind the test mount. Acidic hydrogen chloride exhaust burned orange for a solid two minutes before finally coming to a sputtering, crackling finish. The cloud behind the rocket bloomed hundreds of feet in the sky until it became the largest thing on the landscape, dwarfing the mountain range below.[29]

In the control center, spectators kept a keen eye on the video feed: *No one saw smoke leak from the joint!* The room erupted in cheers. McDonald rushed from the control center to jump atop the still-smoldering booster and get a close-up look at the joint. The O-ring was intact![30] Of course they would need to verify that in the lab, but all the evidence so far indicated victory. At a press conference later, McDonald joyfully announced that the booster had passed its certification and that the shuttle was on track for launch next month. After more than two years of painstaking work, *it was time to fly*.

Kennedy Space Center. September 29, 1988.

At crew quarters, TFNG commander Rick Hauck and his crew donned their new, bright orange, high-altitude pressure suits and walked out to the Astrovan. For every previous shuttle launch, George Abbey had escorted the crew to the pad. Now Dan Brandenstein trailed them while George sat in an office seven hundred miles up the coast.

The world was watching. The press covered every minor hiccup that popped up in the months and weeks leading up to the launch, including "faulty welds" in a shuttle engine seal, a failure with a fuel pump, and a malfunctioning engine sensor.[31] The *Challenger* tragedy had cast a long shadow over NASA, obliterating the trust that the public once had in the agency. Headlines across the country, and across the world, asked the question: "Is the shuttle ready to fly?"[32]

Commander Hauck was confident: "The Mission Control team is ready, I know the launch control team here in Florida is ready, the bird is ready and we're ready."[33] Once the crew reached the launchpad, before they ascended the gantry, Brandenstein pulled all five astronauts aside for a short prayer. With Hauck, David Hilmers, John Lounge, and New Guys Dick Covey and Pinky Nelson gathered around him, Brandenstein bowed his head and said solemnly, "God help you if you screw this up."[34]

As the crew reached the scheduled hold at T minus nine minutes, the launch director polled his supporting engineers and administrators over the radio. One-by-one, launch controllers sounded off: "Go," "Go," "Go." Mission manager Bob Crippen had the last word: It was the first time that an astronaut would make the final decision to launch.[35] Since July 1987, as NASA's new deputy director of shuttle operations, Crip had been stationed at Kennedy, in wait for this moment.[36] Crip paused ever so briefly—high upper-atmosphere winds had threatened the launch earlier in the day, but they had died down. Now, full of emotion, Crip said, "Go."

As the countdown proceeded, nearly one million people crowded the beaches and highways surrounding Kennedy, and millions of others watched the launch on television.[37] At 11:37 AM on September 29, 1988, two years, eight months, twenty-three hours and fifty-nine minutes after *Challenger* was lost, *Discovery* leapt from the pad.[38]

For Crip, who had lost his friends and experienced the most painful moments of the *Challenger* recovery, giving the final "Go" for the shuttle's return to flight was a triumphant, emotion-filled moment that exceeded even the joy he felt on the shuttle's maiden voyage.[39] Over the last two years, he had sometimes lost faith, questioning if NASA would ever get back here. As *Discovery* soared past the gantry into the brilliant Floridian sky, he felt a huge weight lift from his shoulders.

Inside the orbiter, the crew still held their collective breath. Silently counting down the seconds, they sat in their rattling seats as *Discovery* approached throttle up and booster separation, the critical moment when all had gone wrong for *Challenger*.

Sixty thousand feet above the Atlantic Ocean, they heard the bang, saw the flash of the boosters separating, and felt the rockets break away. Eight and a half minutes after launching from the pad, whoops, hollers, and yahoos rang out from the flight deck. On the middeck, Pinky Nelson exhaled a sigh of relief—he had not needed to activate the escape hatch.[40] Reaching orbit was the mission's most important objective, proof that the last two and a half years had all been worth it. Back on the ground, the famously reserved flight controllers let out unbridled cheers.[41]

"Gooooooood morning, *Discovery*!" the unmistakable Robin Williams bellowed, in the vein of his role as a radio DJ in the movie *Good Morning, Vietnam*. His wake-up call cajoled the sleeping crew after their first night in orbit.[42] CapCom Kathy Sullivan had orchestrated the celebrity surprise. After the crew stopped laughing, Hauck keyed the mic on *Discovery* and let loose an enthusiastic "Gooooood morning, Houston." After having grieved the loss of her friends, Kathy could finally celebrate the shuttle's return as the crew floated among the stars.

On the crew's fourth and final day in orbit, they pointed *Discovery*'s camera toward Earth and remembered their fallen brethren as the shuttle passed over the continental US. "Today, up here where the blue sky turns to black," Hauck said, "We can say at long last to Dick, Mike, Judy, to Ron and El, and to Christa and Greg: dear friends, we have resumed the journey that we promised to continue for you . . . dear friends, your spirit and your dream are still alive in our hearts."[43]

Though the *Challenger* accident had given the nation pause and called many to question the future of the country's space program, young, hopeful astronaut candidates kept applying to NASA in droves and veteran astronauts lined up to fly again. Between 1990 and 1991, the shuttle flew twelve times, its busiest pace ever, save for its barn-burning 1985 schedule.

In 1987, shortly before George left Houston for headquarters, he recruited his fifth class of astronauts, which included Mae Jemison, the first African American woman to become an astronaut.[44] Two years later, Fred Gregory became the first Black man to command a space flight on STS-33. The 1990 Group 13 astronaut class was even more diverse than Jemison's class, bringing in former Air Force instructor and test pilot Eileen Collins, who would become the first woman to pilot and command a shuttle mission; a Chinese American chemical engineer named Leroy Chiao; Ellen Ochoa, a Mexican American engineer who would become the first Hispanic woman to go to space; and physician Bernard Harris, who would become the first African American to perform a spacewalk.[45] Four years later, the Group 15 class included Kalpana Chawla, an Indian-born, American mechanical engineer who would become the first Indian woman in space.[46]

With the shuttle returned to the skies, NASA needed to redefine its goals for the program and sketch a vision for the nation's future in space.

On August 15, 1986, after months of debate within his administration, President Reagan had announced a new launch strategy for NASA.[47] NASA would no longer ferry commercial communications satellites, which had made up the bulk of the shuttle's manifest.[48] Beggs's and Reagan's dream of a viable space business, of a delivery truck to the stars, had come to an end.

Recognizing the risky nature of the shuttle, Reagan also put an end to NASA's relationship with the Department of Defense. The Pentagon would shift its satellite launches back to expendable rockets. The Air Force would mothball its $3 billion Vandenberg shuttle launch facility before it ever went online and vacate its secure control room in Houston's

Mission Control.[49] Before the relationship ended, however, the shuttle would need to deliver the remaining defense payloads that had been specifically designed for its spacious cargo bay.

After the DoD deployments, important scientific satellites waited in the wings, including the Magellan planetary probe to Venus, the Chandra X-ray Observatory telescope, and the Hubble Space Telescope. Most importantly, NASA recommitted to building an orbiting space station. The Soviets had already launched the first module of their own space station *Mir* on February 20, 1986, less than one month after the explosion of *Challenger*. The USSR had beaten America in the space race once again. Now NASA stood ready to answer *Mir* with a space station of its own. The shuttle would play a critical role in constructing a new international space station, ferrying many of its modules to space in the shuttle's cargo bay and providing astronauts to connect those modules to each other once in orbit, like space-age construction workers.

While the shuttle program was rebuilding, engineers at Johnson and Marshall worked on a detailed, timesaving, cost-effective redesign of the proposed space station, scrapping the dual keel model for a "single truss" design that would be more practical given the shuttle's reduced flight rate. Before President Reagan left office, in July 1988, he christened the project: Space Station *Freedom*. The outgoing president painted a glowing picture of the country's future in space, dusting off Wernher von Braun's dream of a lunar base and setting sights on a Mars mission by the early twenty-first century.[50]

Many doubted the feasibility of these goals. "Mars is bull," said the plain-spoken Chris Kraft. "Trying to sell a Mars mission is the way to kill the space program. It would cost billions and billions and no damn fool in Congress is going to sign a check."[51] However, a space station seemed like an achievable first step toward the bigger, bolder goals of exploring deep space.

On September 29, 1988, the same day that *Discovery* leapt from the pad at Kennedy, NASA administrators signed an agreement with America's European allies, Canada, and Japan outlining the details of their international cooperation.[52] With the agreement in place, the international space station advanced from an idea to real-world hardware development.[53] The shuttle, now with a new raison d'être, would be vital in ferrying parts and astronauts to orbit to build mankind's newest home in space. As the original class of shuttle astronauts, the New Guys would be an essential part of turning this long-held dream into a reality.

CHAPTER 23

THROUGH A GLASS DARKLY

Johnson Space Center. April 1989.

Kathy watched the water in the pan at the center of the room start to bubble, then boil violently. Dressed in a Class One spacesuit, normally reserved for missions, she was locked inside a looming, three-story, white metal cylinder known as Chamber B, as technicians pumped out all the air to simulate the vacuum of space.[1] Kathy stared at the fast-bubbling liquid. In an instant, the water seemed to levitate, flash-freeze into solid ice, fall, and clatter into the metal pan. It looked like a magic trick, but it was pure science.

"That's what would happen to the fluids in your body if you weren't in that suit," the test conductor said dryly. Yep, Kathy got the message, loud and clear.[2]

A segment of the Hubble Telescope sat next to her. If anything went wrong on Hubble's deployment, Kathy would have to spacewalk out to fix the billion-dollar spyglass. Today, she would test the tools she had helped create for such an event.

The Hubble Telescope was the consummation of astronomers' fever dreams since Galileo—a telescope that would sit above Earth, unencumbered by the atmospheric phenomena that had clouded humanity's vision since we first looked up at the stars.[3] With Hubble, scientists would not only be able to determine the age of the universe, but also how the universe had evolved. They would get an unobstructed view of Earth's neighboring planets in the solar system and would be able to observe the most far-flung stars and galaxies. Astronomers were most intrigued by the questions no one had yet asked—by the unknown. If humanity had been trapped in Plato's cave, astronomically speaking, Hubble would light the way out.

Putting a school-bus-sized telescope that weighed over twelve tons and contained some four hundred thousand parts was pure science fiction until the shuttle came along. The shuttle provided a large, safe transport to usher the behemoth to space and to enable astronauts to maintain and upgrade it. Once deployed, Hubble would never return home. This was Hubble's genius—its ability to evolve and improve in orbit. The *Challenger* disaster in 1986 had postponed Hubble's scheduled deployment, but by late 1988, planners in Houston re-manifested Hubble for launch in December 1989, and Kathy began training again in earnest.[4]

For years, Kathy and her partner, Bruce McCandless, had spent countless hours learning every square inch of the complicated instrument. Hubble was now an old friend.[5] Today was the most important, and most dangerous, part of her preparation.[6] Until now, Kathy had trained in Marshall's Neutral Buoyancy Simulator, a massive, forty-foot-deep pool.[7] Chamber B felt like a death trap by comparison.

Situated in an enormous warehouse, the vertical cylinder stood forty-five feet tall and thirty-five feet in diameter, encompassing forty thousand cubic feet. Inside the cavernous space, a layer of liquid nitrogen shrouds cooled the chamber to 80 Kelvin (-315°F), to recreate the temperature of space. Inside each panel, helium cryogenic pumps depressurized the chamber to simulate the vacuum of space.[8] If Kathy's spacesuit malfunctioned here, she could not swim to safety as she could in the Neutral Buoyancy Simulator. No scuba divers watched her every move. She would be alone in Chamber B as the air was pumped from the room and the temperature dropped.[9]

Time to suit up.

The test engineers offered to swap jokes or play country music to help Kathy pass the time while she performed a mind-numbing, hours-long pre-breathe. During prep, her body acclimated to 100 percent pure oxygen, as opposed to the 80 percent nitrogen, 20 percent oxygen mixture humans breathe at sea level on Earth.[10] At such low pressures, her body's blood and tissues could not hold as much nitrogen as they do at sea level in Houston. As depressurization occurred, any excess nitrogen in her body would escape forming bubbles in her bloodstream, like carbon dioxide bubbles when one cracks open a can of Coke. She could die from a nasty case of "the bends." Kathy passed on the country music and opted to meditate in silence.[11]

Four hours later, her body pumping 100 percent oxygen, Kathy knuckled down to work. *Connect cord. Disconnect cord. Loosen bolt. Tighten the bolt. Just keep going,* she told herself. "I pity anybody who had to watch me work over the next six hours. There is nothing exciting about watching someone—even a woman in a spacesuit working in a room without air—repeatedly disconnect and reconnect electrical connectors (fake ones, at that)," Kathy said. "It wasn't even exciting, in an adrenaline-rush kind of way, for me, and I was the woman in the spacesuit. It was, however, all-absorbing."[12]

Five hours in, Kathy's hands and feet started to go numb. Without the warmth of the regular forty-five-minute sunrises she would experience on orbit, her body trembled. When she exited Chamber B another hour later, her body was rigid and her jaw was frozen shut. She was moments away from getting frostbite. Nevertheless, her tools had held up in the freezing vacuum, as had she.[13] She felt a swell of pride and relief.

Kathy first met Hubble four years earlier, in April 1985 at Hubble's birthplace: Lockheed's Sunnyvale plant, a four-hundred-acre campus of featureless buildings wedged between the southern edge of San Francisco Bay and the runways of the Navy's Moffett Field. Twenty thousand employees worked at the plant, laboring on classified Air Force and Navy aviation programs.[14] One of them, coincidentally, was Kathy's father, Donald, an engineer on the highly-restricted Trident Missile Program for the Department of Defense.[15] Kathy occasionally visited Donald and his new wife for dinner at their small condo near Lockheed's campus.[16]

A team of Lockheed engineers, who would help familiarize the astronauts with the space telescope, greeted Kathy and Bruce McCandless on their first day. After putting on a white bunny suit and getting blasted by a jet of air that blew off any contaminants, Kathy stepped into the vast, pure white, nine-story assembly chamber. "Hubble looked like a piece of precious sterling silver," Kathy observed. "It could be a gift from Tiffany's."[17]

Standing before Hubble, Kathy felt "like an ant on an iceberg."[18] The telescope, named for the twentieth-century astronomer Edwin Hubble, was a forty-three-foot-long cylinder, fourteen feet in diameter.[19] Twin solar arrays, each ten feet wide and twenty feet long, bookended both sides of the central barrel housing the telescope.[20] The high-gain antennas, which would send Hubble's images back to Earth, were four feet across.[21] Unseen was the nerve center of the whole thing, a power control unit with twenty-six thousand miles of entwined wires.[22] Kathy's eyes traced the bright yellow handrails lining the barrel of the telescope. She imagined pulling herself along them, hundreds of miles above Earth's surface.[23] She hoped she would get her chance to experience it for herself.

On March 18, 1990, Kathy arrived at Cape Canaveral for her impending launch. Hubble was glowing, as beautiful as when she first laid eyes on it. "It always sends chills down my spine to finally see the payload I had worked on for months, if not years, bolted into the cargo bay of the space shuttle, on the pad, and poised for launch," Kathy said.[24]

Kathy climbed along the launchpad support structure to get a better look at their precious cargo. Hubble had taken advantage of the shuttle's spacious payload bay: Only a fist's width was between it and the bay's wall.[25] Steve Hawley would deploy Hubble with the robotic arm, using extra care to extract the telescope without hitting the bay's walls—like a high stakes game of Hasbro's *Operation*. For Hubble to properly make its celestial observations, their orbiter *Discovery* needed to climb to an altitude of 340 nautical miles above Earth, twice as high as a normal shuttle flight, a record altitude for the shuttle.[26]

On April 25, 1990, twenty-four hours after their launch, Kathy and McCandless floated through the airlock setting up their spacesuits and repair equipment, then zoomed back to the flight deck for the deployment. Kathy, a self-described "camera fanatic," excitedly took pictures of the early stages of deployment, hoping to capture every aspect of the telescope for all the engineers on the ground.[27]

Pilot Charlie Bolden released the payload bay doors, then Steve grappled Hubble with the remote arm. They switched the telescope over to battery power and unplugged its electric umbilical cord, kicking off a critical countdown—six hours to deploy Hubble and release its new power source: solar arrays.[28] Any longer and Hubble would effectively become a very large paperweight.

Steve grappled Hubble to lift it out, but the robotic arm began swinging from left to right while poor Hubble remained crammed into the payload bay chockablock. An astronomer himself, Steve knew what Hubble meant to the broader astronomy community.[29] Spooked, he stopped the arm entirely rather than risk damaging the payload.[30] The arm's software had gone haywire. Now Steve would have to manually command each joint of the arm to get the job done.

Inch by inch, he pulled Hubble from the orbiter, taking slow, deliberate breaths as he went. He might have been holding a joystick, but this was not a video game. Over a billion dollars and thousands of hours of human effort lay at the other end of that controller. Finally, with one last flick of his wrist, Steve released Hubble into the vast expanse.[31]

The crew crowded around the aft deck windows and held their breath. *Time to activate the telescope.* Mission Control sent the first command and a curtain of shining Kapton, embedded with solar cells, unfurled. With one wing extended, twenty-three-by-eight-and-a-half feet, ground controllers now sent the second command. Silver tubes along the spacecraft began to telescope out, unfolding the golden array. This time, though, the tubes stopped short with only a foot of the solar panel deployed. Kathy waited with bated breath, all too aware of Hubble's rapidly draining battery.

The ground again commanded the array to open, and again, it caught. Now Kathy did not hesitate. She and McCandless zipped onto the middeck to suit up, eager for their chance to rescue Hubble. After their pre-breathe in the airlock, Bolden closed the hatch separating Kathy and McCandless from the shuttle cabin and opened the valve that allowed the oxygen inside the airlock to escape. Kathy listened to Steve Hawley and commander Loren Shriver discuss next steps with Mission Control, waiting for that crucial command: *GO for EVA.*[32]

Racing against time, the engineers in Mission Control finally identified a faulty sensor as the issue. They disabled the sensor, then re-sent the command. The second solar array unfurled fully. A cheer went up in Mission Control. Moments later, still inside the airlock, Kathy felt the bump of *Discovery*'s engines firing, indicating they were leaving Hubble safely in position. Her crewmates bantered joyously, snapping pictures of Hubble, the newest star in the sky.

"I only worked for five years on this thing, and I'm not even getting to watch it being deployed!" Kathy lamented.[33] She knew the mission came first, and she tried to set her disappointment aside.[34] As Hubble disappeared into the distance, Kathy struggled with her loss, after so many years of training. "Parents must feel like this when their children leave home, but at least they get emails and phone calls," she said. "My split with Hubble was cold turkey."[35]

Kathy and the crew triumphantly returned five days later, on April 29, 1990, landing at Edwards a few minutes before seven in the morning. After traveling over two million miles and completing eighty laps around Earth, they had deployed their all-important astronomical jewel. As she tried to find the words to thank her support team, Kathy choked up.

She did not have long to relax. She was immediately assigned as the payload commander for STS-45, where she would lead a four-member scientific team on a Spacelab mission to measure the human impact on Earth's climate and the upper atmosphere.[36]

Kathy was deep in training for STS-45 when she learned the devastating news: Two months after Hubble's launch, its first images came back blurry. NASA's crisis management team tried to control the story, but within days, headlines decried NASA's $2 billion blunder.[37] Dan Rather reported "Hubble Trouble" on the *CBS Evening News*.[38] CNN's Bernard Shaw called the spyglass "legally blind."[39]

Heartbroken, Kathy and her crew wondered if they were to blame—did they damage the telescope while deploying it? Within months, NASA's optics experts pinpointed the issue: Technicians had misground the telescope's massive mirror by 1/50th the width of a human hair, causing a "spherical aberration."[40] While Kathy, Steve, and everyone involved with Hubble were pleased the issue could be fixed by installing an additional camera—a contact lens, essentially—they also understood that this egregious error should have been caught during testing.

After NASA disclosed the malfunction, Congress called for hearings and publicly grilled agency officials about the massive waste of money and time and questioned whether NASA would be able to accomplish other high-tech projects, like a mission to Mars.[41] Kathy caught bits and pieces of the testimony between simulations and payload briefs for STS-45. "Two days ago, a very dark veil fell on the world of astronomy and astrophysics research," Senate Science Committee chairperson Al Gore intoned in his opening remarks. "Years of hopes and dreams were crushed."[42] Senator Barbara Mikulski, who headed the Senate Appropriations Committee that funded NASA, was more blunt, asking, "Are we going to keep ending up with techno-turkeys?"[43]

Several years and millions of dollars later, NASA would be able to repair Hubble, but Kathy and the original Hubble deployment crew would not be part of that effort. In October 1990, the Navy Reserve—which Kathy had joined in 1988 after *Challenger* grounded shuttle flights—called her into active duty to support the Persian Gulf War. Kathy ran a specialized unit based at the Naval Air Station in Dallas that provided weather and logistics data to the Oceanography Command Center in

Guam.[44] Stationed in Dallas, Kathy worked "virtually" on the war effort and continued training in Houston for her Spacelab mission.[45]

Kathy left NASA in 1993, not because she did not love the work, but because she disliked the politics. The agency did not have a consistent track record of promoting women through its administrative ranks.[46] When an old marine biologist friend called her to become the chief scientist of the National Oceanographic and Atmospheric Administration (NOAA), she took the job.[47]

On December 2, 1993, more than three years after Hubble's deployment, all eyes were on NASA as it launched STS-61, NASA's first complex and daring attempt to fix their big mistake.[48] Even though Kathy was settling into her prestigious new appointment at NOAA, she longed to be space-bound as she watched the mission on television.[49] After rendezvousing with Hubble, the crew, commanded by New Guy Dick Covey, gently berthed the telescope into the payload bay. Over the course of five painstaking spacewalks, a record number for a single shuttle mission, astronauts Story Musgrave, Kathy Thornton, Tom Akers, and New Guy Jeff Hoffman replaced two of Hubble's gyroscopes, added a second wide-field and planetary camera, and upgraded the telescope's solar panels, magnetometers, and onboard processor. Hoffman and Musgrave performed the second longest spacewalk in NASA history, triaging the telescope for almost eight hours on the flight's fourth day.

By day seven, Hubble could finally see clearly, thanks to the delicate installation of the telephone-booth-sized COSTAR instrument.[50] To extend the telescope's lifetime, the crew ferried Hubble up to an altitude of 370 miles and rereleased the telescope.[51] Five days after *Endeavour* returned to Earth, Hubble sent down its first light images. Nearly fifty astronomers and institute engineers gathered around a screen, as the first picture came through. The image, that of a simple star, was crystal clear. The room erupted in shouts and hurrahs. After collecting images and other data from Hubble's full suite of instruments, NASA unveiled the results to the public during a January 13, 1994, news conference at Goddard.

"'Spectacular' was the only word that came to mind," said Kathy, feeling nostalgic.[52] She had helped design the tools, tethers, and caddies

Covey's crew used. "We had indeed laid a good foundation," she smiled.[53] Senator Mikulski, who had famously asked whether NASA was building "techno turkeys," joined deputy administrator Dan Goldin at a news conference and declared, "The trouble with Hubble is over."[54] Kathy said, "The twinges of envy I had felt at launch were long gone, replaced by joy and the kind of warm pride that alumni feel when their old college team wins a national championship."[55]

Once fixed, Hubble lived up to its promise. The telescope identified the first organic molecule seen outside of our solar system when it spotted methane in the atmosphere of a distant exoplanet the size of Jupiter.[56] It glimpsed a new moon orbiting Pluto.[57] It proved that nearly every large galaxy harbored a supermassive black hole.[58] "It has looked into the stellar nurseries where stars are born," Kathy said. "And revealed thousands of galaxies, where there seemed to be only empty patches of sky, found dwarf planets and protoplanetary disks—the clouds of matter from which planets form."[59] Perhaps most importantly, Hubble defined Earth's place in the expanding universe. Using Hubble's fine-grained measurement of distances to faraway galaxies, humans today understand how fast the universe is expanding—which indicates that the universe is 13.7 billion years old.[60]

Although Kathy was not able to watch Hubble take its first look at the universe, her contributions are clear. She helped educate the next generation of astronauts on how to use the tools she had helped designed. She transferred her hard-earned wealth of knowledge to those that would fly on a total of five service missions for the telescope. "Foundations are not ends unto themselves, of course," Kathy said. "The structures that get built upon them are what really matter . . . indeed, a veritable cathedral [was] built upon our foundation stone."[61]

CHAPTER 24

CLOSER TO GOD

Houston. Friday, August 12, 1994.

"You interested in flying on *Mir*?" Shannon heard through her phone.[1] Hoot Gibson, now Astronaut Office chief, barked from the other end.

Mir? Shannon thought. *The Russian space station?*

Three years earlier, the Soviet Union had collapsed. The Cold War was officially over, and the United States, ostensibly, had won. The American mission to *Mir* would usher in a new era of cooperation. More importantly for NASA, its astronauts would gain invaluable experience with long-duration spaceflight and space station operation—critical for finally launching the US space station.[2]

The partnership was the result of years of careful planning by the presidentially appointed National Space Council, which was chaired by Vice President Dan Quayle and which counted among its highly influential members Shannon's old boss: George Abbey.

While exiled in Washington, DC, in the early 1990s,

George capitalized on his proximity to new NASA administrator Dan Goldin. Goldin, who had been a middle manager at TRW Inc., an aerospace and defense contractor, relied on George to show him the ins and outs of the agency.[3,4] Soon George took the reins as "special assistant" to the administrator. "Goldin's Rasputin" was how one *Washington Post* reporter described him. "A shadow never far from his side and the subject of intrigue—whether . . . in Washington, Houston or Moscow."[5]

George had long thought that, despite political differences, reinvigorating cooperation between the two greatest spacefaring nations would spur scientific progress. What better way to symbolize the end of the Cold War than a new US-Russia space partnership? Beyond pure symbolism, it would be practical. Even before the official dissolution of the Soviet Union on December 26, 1991, the White House had feared that cash-starved Russia would sell off its nuclear stockpile, or the expertise of its engineers. Adding Russia to the space station partnership would keep scientists and engineers occupied on civil space projects. As a bonus, the US aerospace industry would get access to Russian space technology.

George privately met with Yuri Koptev, the head of the brand-new Russian space agency Roscosmos, at Dan Goldin's apartment in DC. He pushed Koptev to join the International Space Station project, the name of which had changed from Space Station *Freedom* to Space Station *Alpha* when the Clinton administration took office in 1993. Koptev balked at first. "Why would we do that? We already have *Mir*!"[6] Undeterred, George managed to get Koptev to agree to a partnership, so long as America involved Russia in its plans to return to the moon and fly to Mars.[7]

In October 1993, George; John Young, who was still working as a special assistant to the Johnson director; and a team of engineers flew to Moscow to finalize the deal.[8] In Phase 1 of the partnership, NASA would send American astronauts to *Mir*, the Russian space station already on orbit. Astronauts and cosmonauts would learn how to work together, and the Americans would gain experience with long-duration spaceflight and space station operation. Then, in Phase 2, Roscosmos, the Russian space agency, would join NASA and the space station partners Beggs had signed on back in the 1980s—the Europeans, Japanese, and Canadians—to build a larger, modernized space station. In the third and final phase of the project, the space station would become operational. George, Young, and other planners hoped to reach Phase 3 by the end of the decade.[9]

On December 13, 1993, Goldin sent George back to his old stomping grounds: Houston. His old colleague, Carolyn Huntoon, would be the new head of Johnson, and George would be her deputy director.[10] George and Carolyn found a partnership that worked. Where George played the political game with aplomb, Carolyn kept her focus operational, as she had after *Challenger.* George enlisted all his old bubbas as his new deputies. Steve Hawley helped lead flight crew operations as deputy director, under astronaut David Leestma. Hoot Gibson continued to oversee crew selections as head of the Astronaut Office.[11] TFNG Norm Thagard and his back up Bonnie Dunbar, a member of the 1980 astronaut class, would train for the first long duration stay aboard *Mir.* After Thagard returned from his mission, Shannon would fly to *Mir.* Her stay aboard the Russian vessel, the second of four long-duration missions, would test the strength of the new US-Russian partnership and help determine the future of the new space station project.

Still—*Mir* was Russian, and Russia had been America's sworn enemy for nearly fifty years. The Soviet Union was half the reason there *was* a space race—a race that, according to the Russians, they had won. Barring America's lunar first, the Soviets had launched the first satellite, the first man, the first woman, and the first person of color into space, and established the first space station. Now, with what felt like whiplash to some, Shannon's mission would not be a competition, but a compromise.

Shannon was the unlikeliest of heroes. A sturdy, fifty-one-year-old mother of three, Shannon had been the last to fly of the six women chosen in 1978 and had yet to find her way into the history books. She was not singularly focused like Sally or hypercompetitive like Kathy, but she was resilient.

Born in Shanghai to Baptist missionaries during World War II, she and her parents were rounded up by the Japanese army, among other prisoners of war, and forced into one of over 450 internment camps scattered throughout Asia. Shannon survived her year at the Chapei Civil Assembly Center because of her parents' sacrifices. A *New York Sunday News* story featured a photo of her emaciated parents disembarking from the Swedish motorship *Gripsholm*. "There's my father and my mother, and they both weighed a little under ninety pounds," Shannon explained. "I'm this huge chunky baby . . . It's obvious I had not any deprivation."[12]

After the war ended, Shannon and her parents returned to China. "I'd been around the world twice before I was five years old," Shannon bragged. It was in China that Shannon took her first plane ride and was hooked.[13] Her next flight, at age nineteen, would see her in the pilot's seat, taking her first lesson in the open skies of Oklahoma.[14]

After seeing the all-white, all-male Mercury 7 on her television as a high school junior, Shannon wrote a letter to *Time*, decrying the injustice. Letter mailed, she got to work, saving up the fifty cents an hour she earned from babysitting and housekeeping gigs to pay for $10-an-hour flying lessons.

Shannon came of age in rural Oklahoma, where the ambitions of an inquisitive young woman were not always welcome. By eighth grade, she was questioning authority: "If the science shortage is in such dire status as they claim," she wrote in an essay on space scientists, "they'd let women come in on the same ground as men."[15] Shannon was not going to wait for an invitation. Later that year, she sold her bicycle to purchase a telescope and drew her own map of the moon, imagining she was a real astronomer.[16]

In college, Shannon majored in chemistry, hoping to get a job in a lab. She was sorely disappointed when her professor told her she would never get a job because she was a woman. He was right. Shannon's first job out of college was the night shift at a retirement home.[17]

A year later, the Civil Rights Act passed and employment opportunities for women began to change. The five other TFNG women, she said, "were on one side of the divide and I was on the other."[18] The divide was a gaping chasm that was nearly impossible to bridge.[19]

"I used to tell myself that I only wanted a job because I was qualified, not because I was a woman," Shannon said. By the time she left the University of Oklahoma, Shannon had tempered her expectations: "I began to wonder whether it might not be nice to be somebody's token woman."

When she finally did get a job at an energy company, the Kerr-McGee Corporation, she was fired as soon as she became a new mom.

While she sat at home sinking into a deepening depression, her husband, Michael, encouraged her to return to school to get her doctorate in biochemistry. "You need all the help you can get," he said, acknowledging the discrimination Shannon faced.[20]

Graduate school was no cakewalk. Shannon managed to snag a job as a teaching assistant, but it came with strings. "You certainly wouldn't think about getting pregnant again, would you?" her professor asked.

"I certainly would not," Shannon said soberly, looking him directly in the eye. She was already pregnant with her second daughter, Shani.[21] Shani was due close to finals, and Shannon had asked another professor if she might schedule a make-up exam, in the event she had to miss the final.

"And for what reason would you not be in the final?" he asked, pointedly ignoring her growing belly.

"I might be having a baby," Shannon said.

"If you're not in the room on the final," he said crisply, "you get no credit for the class."[22]

Despite labor contractions, Shannon finished her lab exercises on Friday before rushing to the hospital. She gave birth to daughter Shani at nine that evening, then dashed off her lab report as soon as she awoke the next morning. On Monday morning, after lying to her doctors to secure an early release, she jetted off to complete the final exam. Juggling two toddlers at home, Shannon graduated with her doctorate in biochemistry in 1973 at the age of thirty.

In 1978, after enduring a series of post-doctoral fellowships with a professor who "didn't like females," Shannon became the first mother to enter the astronaut corps.[23] Her missionary father recalled her joking, "The Baptists wouldn't let women preach, so I had to become an astronaut to get closer to God than my father."[24] Though Shannon does not remember making the comment, she was indeed in heaven at NASA—for the first time in her life Shannon was appreciated for her skills, not punished for her gender.

Despite Shannon's previous hard-knock life, or perhaps because of it, she was relentlessly cheerful, balancing her career as an astronaut with her duties as mother of three. She was positive, funny, and "nice" in the classic midwestern sense. Perhaps that was why Shannon had not won a place on any of the more glamorous early shuttle missions. Her open, friendly demeanor and innate gentleness may not have fit the profile of an old-school astronaut: competitive, hyperfocused, and young. Maybe, as happened to many other middle-aged women, her employers had taken her reliability for granted. She did not demand attention, so she became easy to overlook.

Yet those same traits—the kindness, the sense of humor, the quiet resolve—enabled Shannon to shine in one of NASA's unlikeliest missions—a mission defined not by flashy acts of bravery and derring-do, but by patience, humility, and an unapologetic passion for Jell-O. A mission that was ultimately the record-breaking, crowning achievement of the New Guys' class.

"Interested in flying to *Mir*?" she answered Hoot. "You bet! When can I go?"[25]

Thirty Miles North of Moscow. February 1995.

Shannon soared high over Star City, Russia's cosmonaut training center, where she was spending a year away from family and friends preparing for her mission aboard *Mir*. The secretive center, really more of a self-contained town, was surrounded by concrete walls topped with barbed wire and nestled in a dense, snow-covered forest. Its main entrance guarded by a checkpoint, the city had only recently begun appearing on maps.[26]

Even here, up in the sky doing zero-G training, Shannon did not feel at home. She turned to the cosmonaut she was training with, Nadezhda Kuzhelnaya. Maybe Shannon could strike up a conversation? Sure, she had only taken six months of Russian at America's fabled Defense Language Institute, whereas most diplomats studied the language full-time for two years before they were released into the wild.[27] Still, Shannon loved to talk. She was a people person, a natural communicator. *How hard could it be?*

"*Where are you from?*" Shannon asked, in what she thought was flawless Russian. Nadezhda smiled politely, boosting Shannon's confidence.

"*How did you become a cosmonaut? Do you like zero-G training?*" she chattered on.[28]

"WHAT LANGUAGE ARE YOU SPEAKING?" Nadezhda finally shouted over the din of the plane.[29]

Shannon sighed. Without the vocabulary to express herself, she felt like "a partial person," stripped of her personality.[30]

She was not the only American in Star City; her mission director and flight surgeon were there, plus astronaut John Blaha, her backup from the class of 1980. Still, Shannon felt lonely. Her best friend became a solitary orange canary she had acquired at a flea market. Every evening after training, despite Shannon's constant worries that the bird would freeze in her frigid dorm room, it welcomed her back with a cheerful song.[31]

Training with the Russians was far more regimented, and more accelerated, than what Shannon was used to. She had spent a year and a half training for her last shuttle mission, on a vehicle with which she was already familiar. "For *Mir*, I would have just a year . . . to train on two totally new vehicles," Shannon said. "*Mir*, where I would be living, and the Soyuz, our lifeboat." In addition, she would be tasked with mastering NASA scientific experiments for the trip. Shannon was often in back-to-back lectures from nine in the morning to six at night—many of which were in Russian.[32]

Training outside the classroom posed a new physical challenge as well. If something went wrong on *Mir*, their only option for escape was via the Soyuz "lifeboat" capsule that might land in an ocean or in Siberia. The astronauts had to run emergency evacuation drills preparing for any outcome. Water survival training took Shannon, Blaha, and their instructors to the algae-clogged Noginsk pond, an hour's drive from Star City. There, a replica of the Soyuz capsule floated, docked to a dinky pier.

To simulate an evacuation, Shannon, Blaha, and their Russian instructor had to put on their long underwear and launch-and-entry suits, strap themselves into the capsule, and, while inside, dress in their winter survival gear: wool sweaters, coveralls, jackets, hats, and down-filled coats. Once they changed clothes, they had to put on an additional rubber wetsuit, open the hatch, and jump into the lake. Unfortunately for

Shannon and Blaha the water survival training was scheduled during a May heat wave, with temperatures in the high eighties.[33]

Shannon sweated inside the floating capsule.[34] Or was that John Blaha's sweat? Blaha, she joked, was "the type of person who [looked] like he [had] performed an Olympic feat, from a sweatiness standpoint, after just standing up."[35] Local school children swam up to the Soyuz's portholes, giggling and gawking while Shannon fought to peel off her pressure suit and put on the woolen sweater, jacket, hat, and coveralls.[36] *They were quite a sight.*

In late October Shannon officially met her *Mir* crewmates: commander Yuri Onufriyenko and engineer Yury Usachov, both in their thirties compared to Shannon's now fifty-two. Yuri was clean-shaven with blond bushy hair, while Yury had thinning, dark hair and a dense mustache.[37] During one of the many mandatory group psychological sessions, they both made something crystal clear to Shannon: "If they want us to do something, just remember to work together with us and agree with everything that they say."[38] During the crew's final flight qualification, Shannon was not sure what to expect. She only hoped she would not embarrass Yuri and Yury.[39] Although the crew passed with flying colors, Shannon noted that "positive comments seemed to be against protocol."[40] Yuri told her not to worry. Unbridled criticism was simply the Russian way.

Even with all those hurdles, Shannon's biggest issue was not with the cosmonauts—it was with NASA. Her home space agency consistently left her in the dark about even the most basic elements of her mission. She did not even receive her flight assignment until deep into her training in Star City. It fell to the Russians to tell her that she would be a flight engineer on the mission, rather than a visiting researcher. She only learned the nature of the experiments she would be conducting from a PR officer who was prepping her for an interview with ABC News: "Great news, I found out what you'll be doing! You'll be the main engineer to reconfigure *Priroda*, the new [earth sciences] module that will arrive when you're there!"[41]

Administrative disarray at NASA had spilled over into Shannon's Russian mission. On August 4, 1995, in the middle of Shannon's training,

George became acting director at Johnson, replacing Carolyn Huntoon less than two years into her tenure.[42] Goldin claimed he moved Carolyn to headquarters (despite George's protests) because he was dismayed with the speed of progress on the space station project, which, with the addition of the Russians, had officially been renamed the International Space Station (ISS). Within NASA, the party line was that Carolyn had been "pulled away to build the Institute of Biomedical Research," but, in reality, she was shuffled to the side in Washington.[43] After more than twenty-five years of working together, George valued his relationship with Carolyn. He sent her a message assuring her that he had no hand in her exit but doubted that she believed him. Their friendship was never the same.[44]

With George at the helm, communication, at least from Shannon's point of view, did not improve—not that one would expect anything different from the enigmatic Godfather. Amid the turmoil, NASA's training agreement changed hands from one contractor to another. In the transition, no trainer had been assigned to Shannon's mission.[45] To put it simply, she had slipped through the cracks. Toward the end of her training, Russian instructors grilled Shannon on the experiments she would be doing. Shannon told them she still was not certain exactly what was planned.

"I can't believe you work for NASA," one examiner responded.[46]

Sometimes Shannon could not believe it either.

In January 1996, a mere two months before she was set to join *Mir*, Shannon sat across from a NASA contractor who had finally arrived in Russia to walk her through her experiments.

"Why are you telling me this stuff now?" she said. "Where the hell were you nine months ago?"[47] A knot formed in her throat. She jabbed at experiments in the mission plan as tears welled in her eyes. "Well, I don't know anything about this," she said, pointing at one experiment. "And I've never heard of this," she said looking at another. "There's too much for me to do in these last weeks."[48] Shannon all but flipped the table.[49]

It was an unexpected outburst from a woman who almost never complained about anything. Shannon had reached the end of her rope. Within a few days, Shannon's positive attitude returned to her as she came to accept that she might have to live with a calamitous mission. "If things fell apart and your career is over, that's fine. I've had four good flights. I was going to have a good time."[50]

On Orbit. March 24, 1996.

Shannon floated through the middeck of *Atlantis* and caught a glimpse of *Mir*, her home for the next 140 days, through the aft windows. Yuri and Yury had reached *Mir* a month earlier, via a Russian rocket. Shannon was now ferrying water, supplies, and science experiments via the shuttle for the space station. With its solar panels splayed at different angles, the station looked like a lopsided dragonfly with two extra sets of wings. To Shannon, *Mir* gleamed like a space oasis. In that instant, all her anxieties and troubles melted away.

Shannon's uncanny ability to stay positive, no matter what the circumstances, proved to be crucial aboard *Mir*. She did not face a shuttle mission with jam-packed schedules and barely enough time to get everything done. Her time on *Mir* would be more like living with college roommates in space: *The Odd Couple* meets *The Right Stuff*.

After the shuttle docked, Yuri and Yury welcomed her to their home, their cluttered dacha in space. Thick forest green carpeted the "floors," even though up or down did not exist in microgravity.[51] The walls and the consoles were mint green, with dense thickets of fat ventilation tubes and wires sprouting everywhere. A long bungee cord ran the length of the module for ease of navigation, though there was usually something nearby from which to push. Every now and then, a critical system would shut down—like the oxygen-generator or the carbon-dioxide scrubber—but no one ever seemed too concerned; after all, they stored three days of oxygen in the station. If worst came to worst, they could always bail out in the capsule in which the Russians arrived.[52] The whole place had the feel and smell of an old garage.[53]

Life revolved around Base Block, a forty-three-foot-long core module that had been in orbit since 1986. Yuri and Yury's private crew cabins, where they kept personal effects, lay behind two curtains. Each crew member was allotted a porthole, a hinged chair, and a sleeping bag. Shannon got to sleep in the *Spektr* module, where during the day the crew performed atmospheric research and monitored background radiation. Unprotected by Earth's atmosphere, astronauts suffered radiation equivalent to about eight chest X-rays a day.[54]

Each day, Shannon ran in the treadmill's orange harness for forty-five minutes. Even though she scorned the practice, she knew that if she lost too much muscle, she would need help walking when she landed back on Earth, and she loathed that idea even more.[55] *Mir* also featured the latest and greatest in space lavatory technology, boasting a suction toilet and sink.

On the space station, Shannon maintained regular contact with her husband and kids through email and glitchy video conferences. Shannon's youngest now attended college, so none of them needed hands-on parenting. With her friends and family so far away, she turned to her crewmates for more immediate human connection; Yuri and Yury quickly became her space family. Each evening, after signing off with Moscow's control center, the three would take a tea and cookie break and Yuri Onufriyenko would ask his crewmates, "Now is a good time for what?"

"It is a good time to say good night," they would reply in unison, like obedient children.[56]

When the Summer Olympics began in Atlanta, Shannon and Yury Usachov were inspired to invent their own cosmic sporting events. They hurtled through the module at breakneck pace until the risk-taking became too much even for a Russian commander.

"You might kick through the hull and cause a leak!" Yuri cried.[57]

They even developed their own language. Shannon had never quite mastered Russian, so they created unique vocabulary and nonverbal forms of communication. A Yury eyebrow raise meant "flip on the TV downlink."[58] Neither of her crewmates ever seemed to judge Shannon for how poorly she spoke their native tongue. Junk food cut through any language barrier.

Back on land, after learning of a coveted fridge aboard *Mir*, Shannon had requested Jell-O rations from the food science team in Houston. They complied. On her first Sunday aboard *Mir*, she tentatively gave the Jell-O packs a whirl, adding hot water and then throwing the bag into the fridge. A few hours later, she removed the jewel-like bag and took a small spoonful. Savoring it, she passed the bag to Yury, who accepted his spoonful before offering the Jell-O to Yuri.[59]

Shannon counted her remaining Jell-O rations and limited the crew to one precious bag a week, a treat they shared every Sunday, when Shannon would slip on her traditional Sunday pink socks.[60] Chocolate pudding, too, ranked as a hot commodity. Nothing, however, compared to the Yuris' love for mayonnaise. The fatty, creamy substance alone justified the presence of an American aboard a Russian space station.[61] Shannon was partial to the Russians' greasy canned beef stew. Though after she mentioned in a radio interview that she missed M&Ms most, she found bags of them waiting for her in every Russian resupply vessel.[62]

Mir Space Station. April 23, 1996.

Shannon's heart swelled with anticipation as she watched the *Priroda* module grow larger in the window of her spacecraft. Three days earlier, *Priroda* had launched from the Russian spaceport Baikonur Cosmodrome on a Proton rocket. The thirty-foot-long, twenty-ton behemoth, *Mir*'s seventh and final module, carried the earth science laboratory for her experiments. After a month in orbit, it was finally her turn to get to work.[63]

As the module approached the space station, something went very wrong. For reasons Russia's mission control could not discern, *Priroda* suddenly lost half its power supply. One of its battery systems died. Had an electrical fire broken out? A simple connection failure? They had no way of knowing. Since the module's power supply had been halved, they would only have one chance at docking and most dockings failed their first attempt. If the crew could not maneuver the module into position, *Priroda* would be lost, adrift in the void.[64] Shannon's whole purpose on the station would be kaput.

"*Korabel!*" Yuri Onufriyenko shouted at Shannon, the Russian word for "spacecraft." Not quite understanding the circumstances, Shannon smiled and nodded at her commander. *Yes, I see the spacecraft*. Yuri, growing frustrated, barked again "*Korabel!*" pointing to the Soyuz capsule. "Little light bulbs flicked on in my head," Shannon said.[65] Yuri was telling her to get into the Soyuz in case *Priroda* crashed into *Mir* and they needed to escape. Shannon obeyed, flipping in midair and gliding into the Soyuz as fast as she could.[66]

Inside the dim capsule, Shannon waited for a few tense minutes and listened. Soon she felt a small jolt. The Yuris had successfully docked the module, but a potential electrical fire still threatened the craft. Though any fire would have likely burned out, the Yuris had to assume the worst case scenario—that there had been one, and that the 168 batteries bolted to the floor of *Priroda* were leaking hazardous sulfur dioxide. The toxic chemical could burn their skin, sear their lungs, or blind them.[67] They donned surgical masks and goggles to bag up the batteries. "We will get it all ready for you, and then you can start to carry out the American science program," they told Shannon. They would need six days.[68]

Shannon had a choice to make. She could follow instructions—stay out of the way and wait—or she could do something. The American who preceded her on *Mir*, Norm Thagard, had waited around, expecting to be assigned tasks. He had been fine. He had also been bored out of his mind.

Shannon put on a mask and goggles and made herself useful. She had tools ready when the Yuris needed them. She saw that batteries need to be bagged up and so began tying them off. The trio floated through a garbage cloud of loose batteries, bags, equipment, and scrap metal. To Shannon, the gentle clink of tools and metal became a new form of "cosmic music."[69] With Shannon's help, they finished the task in two days, earning her the respect of her colleagues.

Even with all the camaraderie, Shannon's indefatigable good humor, and the full slate of science experiments for her to run—fun stuff like testing how candles burned in microgravity—time began to drag. The classics she had packed, like *The House of Mirth* and *Middlemarch,* began frustrating her with their antiquated portrayals of women.[70] Shannon missed home. She felt lonely.

The days ticked down slowly. She was celebrating her hundredth day—a huge milestone, only forty more days to go—when she checked her email and saw a message from her daughter Shani. Shannon's heart dropped. Shani had learned that NASA was delaying her mother's return for six weeks. The reason? On a recent shuttle mission, NASA had discovered evidence of hot gas leaks between booster segments—the same issue that had doomed *Challenger* a decade earlier. The mission that was slated to pick up Shannon was postponed from July to September, while NASA technicians outfitted her shuttle with new boosters.[71] Safety being paramount, Shannon was not going to argue. If anything, she appreciated NASA's diligence. *But an extra six weeks?*

On day 115, Shannon hit a new record—the most continuous time in space by any American astronaut. Shannon relished the fact that "woman" did not qualify the descriptor.[72] With five flights under her belt and having spent a total of 223 days of her life in space, Shannon would also break the record for most time in space of anyone in her TFNG class.[73]

While Shannon was pleased about her achievement, she was even more thrilled about the arrival of the next Russian resupply vessel, which would include the treasures of capitalism Shannon cherished most: Twinkies, Pringles, gummy bears, M&Ms, and mayonnaise for the Yuris. When Shannon opened her supply bag, not a Twinkie or potato chip was to be found. She was devastated, somehow even more so than when she had learned of the delay. She bit her lip and went to bed early, trying—and failing—to focus on the adventure of it all.[74]

The next day, Shannon strained to hide her homesickness. As her frustration boiled over into hot tears during yet another glitchy family phone call, she heard Yuri say, "Why do I have this? I can't read English."

Wait. A book in English?

Shannon sailed over to him. To her relief, Shannon found the treats had ended up in Yuri Onufriyenko's bag. They had made it to *Mir*; she had not been forgotten after all.

On August 19, with a month to go before Shannon's journey home, Yuri and Yury's replacements arrived at *Mir*. Like Shannon's, the Yuris' departure had also been delayed, but two weeks later, her Russian family members had to return to Earth.

Shannon hugged them tightly and presented each with cardboard Olympic medals. Two "golds" for #1 Commander of the Universe, Yuri Onufriyenko, and #1 Board Engineer of the Universe, Yury Usachov.[75]

On September 7, 1996, Shannon reached day 169 on *Mir*, shattering the women's record for long duration spaceflight previously held by a cosmonaut.[76]

"What do you think of Shannon Lucid taking the spaceflight record from a Russian?" a reporter asked Roscosmos deputy commander Yuri Glaskov in Star City.

"I don't think you've *taken* the record from us," he said after a pause. "We have offered this record to you."[77] Still, even the proud and taciturn Russian commander could not repress his praise for Shannon. "As far as Dr. Shannon Lucid is concerned, I would like to extend my sincerest thanks to the management of the program for making such a selection," he continued, smiling. "Everybody loves her."[78]

On September 26, 1996, eight days after *Atlantis* had docked at *Mir*, the shuttle returned screeching to a halt on Kennedy Space Center's runway. Shannon was finally home.

Scientists and press had long speculated about her condition when she returned to terra firma. They shared concerns that Shannon could have lost 25 to 30 percent of the muscle mass in her legs. Her heart muscle may have atrophied; her bones may have lost calcium.[79] Maybe she would even black out if she tried to stand.[80] No one knew what to expect, not even Shannon. She had exercised on that annoying treadmill religiously, but she would not know if it worked until she set foot on land. All she could do was hope and try.

Shannon's flight surgeon found her lying in a cocoon of drink bags, salt tablets, and "strategic cookies," as she called them.[81] After suit techs wrangled Shannon's stuck helmet off, they readied to carry her off *Atlantis*.

"No, I can stand up," Shannon said.

Defying all expectations, Shannon walked off by herself.

Over the next two days, Shannon would finally get to embrace her husband and children. She landed in Houston to a cheering crowd of hundreds, including elementary students wearing T-shirts that said WELCOME HOME above a picture of her smiling face.

Shannon was shocked. For months, her only human contact had come from a pair of stoic Russians. Her only outside communication had been with NASA contacts and her family. Who knew the rest of the world even cared? President Bill Clinton personally welcomed her home, calling her "a monument to the human spirit."[82,83] George Abbey stood by his side. Shannon's mission had vindicated the National Space Council's entire approach to building the ISS with their Russian partners. He and NASA were in her debt.

And Shannon? Shannon had made history. Her entry may not have been as momentous as Neil Armstrong's moonshot or as well publicized as Sally Ride's first woman mission, but she had surely stepped into history in her own way. Her 188 days in space showed that humans could survive long durations in microgravity, paving the way for humankind to live and work in space. The question was no longer "How long can we stay in space?" The question now was "What can we accomplish while we're up there?"

In Houston, President Clinton followed Shannon to the stage where she greeted her waiting fans. After a lengthy introduction—for which Shannon, out of respect, also stood—Shannon spoke. "All I can say is, Houston never looked so good!" Shannon shouted, gripping the podium, fighting to stay upright. As she walked off to greet her fans, George Abbey shadowed her, keeping a careful eye in case she fell or needed help walking. But Shannon did not stumble, not once.[84]

CHAPTER 25

EVERYTHING THAT RISES MUST CONVERGE

Johnson Space Center. 1996.

Anna sat outside George's new office in Building 1, her stomach turning somersaults. She felt like she had all those years ago, when she waited for George to grill her during the selection week interview. Now a forty-six-year-old mother of two, Anna had not worked full-time at NASA for eight years. Her meeting with George, Johnson's new director, would decide whether she would return to the office and her old title: astronaut.

After *Challenger*, Anna contributed to the return to flight effort along with many of her colleagues. When STS-26 finally flew in 1988, she took a leave of absence to have a second child. Then thirty-eight years old, she knew she would rather grow her family than wait to fly again. Still, the decision to leave NASA, and the job that had meant so much to her, was gut-wrenching. After the birth of her second daughter Kara in January 1989, Anna knew she had made the right choice. She could not care for a newborn, ferry Kristin to kindergarten, *and* train for a shuttle mission.[1] Something had to give. Instead of returning to the Astronaut Office, Anna remained on leave for the next two years.

Meanwhile, Bill started picking up evening shifts at local emergency rooms. He had only flown one mission, STS-51I, and had no role in the return to flight effort. As a mission specialist, Bill started to resent that NASA managers only promoted pilots to leadership positions. Feeling like his career at the agency was stagnating, he transitioned back to the emergency room part-time, where he could maintain a semblance of control over his schedule. Although he tried to balance his hours in the hospital with his full-time job at NASA, he was chronically tired and stressed.[2] His long hours were tough on the marriage.[3]

As a former spacewalker who had performed an EVA to repair the Syncom IV-3 satellite on STS-51I, Bill had been tasked with outlining a maintenance plan for the space station in 1990.[4] His study would affect the number of astronauts NASA hired, the number of spacewalkers it trained, and the station's operating budget. Bill estimated the station's maintenance would require over 3,000 hours—or 273 spacewalks—per year. Congressional staffers were flabbergasted.[5] His estimates conflicted wildly with the mere 130 hours of maintenance NASA had previously advertised. Bill pushed for a whole new station design, but NASA administrators privately decided that the results of Bill's study proved he was "not a team player" and sidelined him.[6,7] Feeling ignored and unheard, Bill left the agency in January 1991 and returned to full-time work in the emergency room.[8]

That same month, after Kara's second birthday, Anna returned to work at NASA part-time.[9] She picked up where Bill left off, diving headfirst into training procedures for the space station. She walked into a minefield. The international agreements that Beggs had negotiated allowed each partner country to conduct their training locally, which meant astronauts would have to fly halfway around the world to Tokyo to train on the new Japanese robotic arm, or to Cologne to learn the ins and outs of the European Space Agency's science modules. *Hadn't anyone in leadership considered the logistical nightmare?* Anna insisted the training simulators be centrally located in Houston.

"That's not something that we're going to be able to change," her counterparts in Washington warned.[10]

"You better hire a bunch of divorce attorneys," she shot back. "This is going to be a nightmare for people training."[11]

Of course, Anna had learned firsthand the toll astronaut life could take on a marriage. Several years earlier, in 1988, with Bill dissatisfied at NASA and working overtime in the ER, and Anna adjusting to life on maternity leave, the Fishers' marriage began to show signs of strain. Having little time to spend together, the two started to grow apart. Still, they held out hope for a less hectic, more harmonious future.

In 1992, a year into Anna's second act at NASA, the Fishers began constructing their dream home on a half-acre along a winding finger of the abundant Armand Bayou, a protected wetland that hugged the Texas oilfields. The floor plan laid out six bedrooms and six bathrooms, sprawled generously across seventy-five hundred square feet.[12] It was the home they had envisioned when they first arrived in Houston in 1978—a place in which they could entertain, a refuge that their daughters could return to, even after they had families of their own. For Anna, who had grown up in two-bedroom military barracks, this was a dream come true.

Meanwhile, Anna's work at NASA became a cycle of endless station redesigns and budget winnowing. The space station project began to feel like an exercise in bureaucratic futility. "If I'm going to come back and work on [the space station], I want it to be real," Anna realized.[13] *Was this busywork worth being away from Kristin and Kara?*

Anna left NASA again to focus on building her perfect home and spending time with her girls. Even as she was prioritizing her personal life, her marriage continued to dissolve. Bill continued to work long hours at the emergency room. Sometimes he came home late, sometimes not at all.[14]

The two had survived medical residency together. They had been the "first Mr. and Mrs. Astronaut," living in the sharp glare of the public eye.[15] They had endured harassing questions from the press. *Are you a good mother? What kind of man lets his wife go to space?* They had weathered the risks of space flight and the demanding schedule of being a double-astronaut family. After enduring so much together, now the foundation upon which their marriage was built was crumbling. Still, Anna and Bill went through the motions of family life for the sake of their daughters.

On Valentine's Day 1996, the Fishers moved into their newly finished house.[16] What should have been a joyful moment, a new start, felt empty. The grand home echoed as hollow as their marriage. In its unfurnished state, Anna wondered if her family would ever be able to fill the cavernous rooms with happy memories.

Anna realized that if her marriage disintegrated, she would need a job to support herself, Kristin, and Kara.[17] She considered a second residency, a return to medicine, but that would mean impossibly long hours. Her identity was intertwined with being an astronaut. "I wasn't ready to leave the space program," Anna said. "I loved it too much."

Who could she turn to?

As Anna mulled over the past eight years, George opened the door to his office and waved her in. She took a seat across from her old boss and said simply, "I want to come back."[18]

George did not even pause. "I never really considered you gone from NASA," he said warmly, handing back her astronaut ID.[19] The man who had started it all, who had walked Anna over to meet Dr. Kraft all those years ago, once again welcomed her into the familiar embrace of the astronaut corps.

In 1996, Anna resumed full-time work with the operations planning branch of the Astronaut Office, supporting astronaut training for the newly christened International Space Station (ISS). The station had been renamed but to Anna, it still meant freedom.

Anna had not worked full-time in an office for eight years. *Web browsing? Microsoft PowerPoint? Email?* They might as well have been a foreign language. Across America, women like Anna, who had entered the workforce in droves with the passage of the Civil Rights Act in 1964 and had taken leave to raise families, now grappled with the unique challenges of blazing a trail back.[20] She buckled down and powered through. Within a year, she was promoted to branch chief of the operations planning office, assuming oversight responsibility for ISS astronaut training.

On a ten-day technical interchange trip to Moscow, the longest Anna had been away from home since her flight to space, she found herself in the warm company of her new work family. As November temperatures outside dipped below freezing, she was invited to the home of a Russian colleague for drinks, which really meant shots of vodka. Lots of vodka. As "international diplomacy" became weeknight revelry, Anna remembered how much fun work could be. She had not felt this uninhibited in nearly a decade.[21]

On her last day in Moscow, Anna ventured out alone to visit Vladimir Lenin's grave. Navigating past the iconic, ice cream–swirled St. Basil's Cathedral, she stepped into the queue outside the glass-walled mausoleum. Surprisingly small, it still encompassed an entire edge of Red Square, a sacred Russian landmark that had been the site of countless Soviet parades and coronations of tsars past. Inside, Lenin's carefully preserved body rested, solemn and cold.

When she returned to Houston, Anna walked through the front door of her home and felt suffocated. The Texas-sized mansion she had created with Bill felt like a mausoleum of her own making. When the front door clicked shut behind her, she collapsed into sobs. She resolved then and there to finally find a way out of her marriage.[22]

Baikonur Cosmodrome, Kazakhstan. November 20, 1998.

Fat gray clouds hung over the elliptical Soviet-age spaceport. Next to Pad 1, from which humanity first escaped Earth's gravity, a crowd of tightly bundled American and Russian spectators in long overcoats and fur caps stomped their feet to keep warm.[23,24] Among them stood George Abbey, NASA administrator Dan Goldin, and Roscosmos head Yuri Koptev, all squinting across the lonely, inhospitable Kazakh steppe toward Pad 81.[25]

There, a black-tipped Proton rocket waited, stamped for the first time with the logos of both NASA and Roscosmos. Strapped inside was the 42,600-pound *Zarya* cargo module—the first piece of the ISS. Thought of as the space station tugboat, the module consisted of a bare-bones living quarter, radios, propulsion, and power.[26] In Russian, *zarya* means "dawn"—signifying a new era in space exploration.[27]

In a crowded Johnson conference room, Anna watched the launch via NASA TV. *Zarya* was the linchpin of the ISS. If it failed to reach orbit, NASA's own mission to launch *Unity*, the first laboratory module of the ISS, would be, at best, delayed. At worst, the dream of the station might implode before their eyes. Already billions over budget, the ISS was losing favor in Congress. Another setback might land the station in the dustbin of history.[28]

Anna watched the screen as the six engines of the Proton lit, delivering 1.9 million pounds of thrust and sending the 180-foot-long rocket hurtling toward the stars.[29] "Lift off, lift off!" cried the NASA announcer. "The International Space Station is under way." Anna whooped and clapped. Tears welled in her eyes. Ten minutes into launch, the Proton's third and final stage separated, leaving *Zarya* to do its work. The module immediately deployed its radio antennas and unfurled eighty-foot-long solar arrays.[30] At Bob Cabana's house, the crew of the upcoming STS-88 exploded in cheers. Their mission was a go.[31] From Baikonur, George Abbey gave a rare grin as he spoke to *Los Angeles Times* reporter Richard Paddock, hailing the launch as "the space program of the future."[32]

Two short weeks later, on December 4, 1998, Anna watched commander Bob Cabana and STS-88 lift off aboard *Endeavour*—the new orbiter NASA built after the *Challenger* accident.[33] On the second day of their mission, the crew docked *Unity* with *Zarya*. Though the two station modules never met on Earth, the duo made a seamless match on orbit.[34] Together, *Zarya* and *Unity* towered seven stories above the orbiter's payload bay. Over the next several days, the crew's spacewalkers connected power cables and cords between the two modules before finally powering *Unity* up on December 10.[35] At 1:54 PM, Bob Cabana and Russian cosmonaut Sergei Krikalev opened the hatch and floated into the new station together.[36]

Back on solid ground in Houston, Anna reflected on the beginning of humanity's new thirty-five-thousand-square-foot home in the sky. In her zeal to be the perfect wife and mother, to build the perfect home for her family, she had neglected a part of herself. Now, with the construction of the space station, she was starting to rebuild. In June 1999, Anna became chief of the space station branch of the Astronaut Office, managing fifty astronauts and coordinating with the program's international partners. "It was like having fifty children," Anna joked, when asked about having such a large team depend on her. "No one wanted to do it."[37]

Still in its infancy, the ISS was the neglected stepsister at NASA, at least as far as astronauts were concerned. Those assigned to the office's space shuttle branch felt sure they would fly soon, whereas an assignment to the space station branch meant a much later flight. However, Anna saw that NASA's future was the ISS: There was no purpose for the shuttle without the station. "If you don't get the space station to work," she warned the incoming class of AsCans, "you guys are going to be out of a job."[38]

Anna was committed to the success of the ISS and loved firing up new blood. She threw herself into negotiations with the Russians to establish common design requirements and standards for displays and procedures. Working across time zones, language barriers, and cultural differences to develop and assemble the ISS proved orders of magnitude more complex than the first moon landing.

For Anna, the newness of the space station harkened back to the early days of the shuttle when she had joined NASA. Building something from the ground up was hard, creative work, and few at NASA had the experience required. Of the original thirty-five, only three New Guys remained in the Astronaut Office: Anna, Shannon Lucid, and Steve Hawley.[39] Anna was exactly where she wanted to be, helping to create a new program from scratch. She was good at this. She was good at beginnings—less so endings.

While Anna was stitching together the ISS partnership, her own union was ripping apart. "Home" no longer meant Bill. In 2000, she filed for divorce.[40]

In February 2001, Anna said goodbye to George, too.

George had been lobbying on the Hill—for Goldin to remain head of NASA under the incoming Bush Administration—when Goldin called him into headquarters and asked him to give up his post as head of Johnson and to stay in DC permanently. Of course, the ever-savvy George had seen this coming. He had been reluctant to accept the Johnson Space Center directorship for this very reason: It would put him in Goldin's crosshairs. As director, George had clashed with Goldin over the usual things—time and money. The station was already behind schedule when George inherited it. Now his projected budget of $30 billion exceeded Congress "hard cap" by nearly $5 billion. Goldin had long branded his own management style as "faster, better, cheaper," so when he needed to explain the space station's slow progress and budget overage to Congress, he made George his scapegoat.[41,42]

George could not deny that his once-symbiotic, ten-year-long partnership with Goldin had irretrievably broken. Mark Albrect, who worked with both men, likened the relationship to that between Henry II and Thomas Becket. Like Becket, George had been given power and used it. "George changed. From being the guy who would never openly disagree with his boss, for the first time in his career, he went head-to-head with Dan," said Albrect.[43] The fact that George had been advocating for Goldin was an additional twist of the knife.[44] For the moment, George demurred, protesting that the timing was too delicate. He needed to stay for the all-important shuttle launching of the *Destiny* module, the future home of American experiments on the ISS.[45]

But George could only stave off the inevitable for so long.

On February 23, 2001, NASA issued a press release publicly reassigning him as a "special assistant to international programs."[46] This time, he was out for good. George would not return to the heart of manned space flight. His secretary Mary Lopez, with the help of Ellison Onizuka's widow Lorna, still a close friend, cleared out his Houston office. George and his old deputy, Jay Honeycutt, with whom he had selected the New Guys, marked the occasion with a dinner at Frenchie's.[47] George remained in his special assistant role at NASA until 2003, when he became a fellow at Rice University's James B. Baker Institute. It was a whimper of an end to an otherwise storied career. George had worked at NASA for nearly four decades.

On November 2, 2000, the ISS officially began its life as an orbiting home for humankind when astronaut Bill Shepherd and cosmonauts Yuri Gidzenko and Sergei Krikalev docked their Soyuz spacecraft with the station, unlocked the hatch, and floated aboard the three-module-long configuration to begin the first long-duration stay on the ISS. Bill Shepherd, who had been recruited by George in 1984 and was part of the *Challenger* recovery crew with Jim Bagian and Bob Crippen, served as the commander. He was only the second American to launch to space aboard a Russian spaceship and the first to live and work on the ISS.[48]

Three days earlier, Anna, chief of the space station branch, had watched Shepherd's launch via NASA TV from a seat in Mission Control Houston.[49] Even as Shepherd and his cosmonaut crewmates hurried to make the ISS habitable, Anna was already planning the next manned expedition to the station. In four and a half months, the three-person Expedition 2 crew would replace Shepherd's team, providing for a continuous human presence on the ISS for decades to come.

In August 2001, the crew of Expedition 2 passed the torch to their own replacements, the three-man crew of Expedition 3. Astronaut Frank Culbertson, Judy Resnik's former boyfriend, commanded the 117-day mission, which was largely overshadowed by the attacks of September 11, 2001. As the ISS passed over New York City that morning, Culbertson photographed smoke clouds billowing from the World Trade Center towers and noted that "tears don't flow the same in space." The only American not on Earth during the attacks, Culbertson penned an open letter to the country days later, reflecting on the tragedy and the anger, fear, and uncertainty that it had unleashed upon the country. "It's horrible to see smoke pouring from wounds in your own country from such a fantastic vantage point," Culbertson wrote. "The dichotomy of being on a spacecraft dedicated to improving life on the earth and watching life being destroyed by such willful, terrible acts is jolting to the psyche, no matter who you are." Culbertson closed the letter with a sentiment of determined optimism: "Life goes on, even in space. We're here to stay."[50]

Indeed, the ISS was there to stay. By the end of 2002, six major modules had been connected to the orbiting space station, three Russian and three American—*Zarya*, *Zvezda*, and *Pirs*; and *Unity*, *Destiny*, and *Quest*, respectively—along with solar arrays to supply power and a Canadian-built robotic arm like the one aboard the shuttle. Construction on the space station was expected to finish by 2006, with continuous habitation planned until at least 2015.[51]

Even though George was no longer in Houston, the shuttle program he helped create continued to propel the building and inhabitation of the ISS, the future of NASA's human spaceflight program. Anna would help ensure its success. NASA was where she belonged. She did not need Bill or seventy-five hundred square feet of prime Texas real estate—humanity's home in space dwarfed her former dream house. The space station marked the start of a new era, both for NASA and for Anna. As 2002 rolled into 2003, the future looked spectacularly bright.

CHAPTER 26

YESTERDAY IN TEXAS

Sabine County, Texas.
February 1, 2003, 8:05 AM CDT.

The early morning fog clung to the pine forests of East Texas. At nearly freezing temperatures, small ice crystals glazed the grasses of the bayou.[1] Since it was Saturday, many residents of Sabine County still slumbered. Just as the sun began to warm up the day, a deafening boom obliterated the peace of the quiet morning. A distinct rumbling sound followed the blast, growing louder and louder over the next few minutes. The air popped and crackled. This was not thunder—thunder did not produce this kind of metallic clanging. Houses shook so violently that people feared their windows would shatter. *Was it an earthquake?* That would be unheard of in these parts. One after another, tectonic booms echoed across town.

For the next half hour, *things* started falling from the sky.[2] A casting sheared a backyard trampoline as a stunned mother watched from her window. A metal shard struck a moving car, sending the driver careening off the road. An aluminum I-beam impaled a carport roof and broke the concrete floor below. Objects torpedoed into the water around a fisherman on the Toledo Bend Reservoir. Then something the size of a small car plummeted into the lake, creating a swell so large it nearly flooded his boat.[3,4]

Four-H kids presenting their livestock at the Sabine County agricultural show witnessed the roof of the building start to quake. Pigs squealed. Chickens took off flying. Horses bucked. The kids ran outside to see what caused all the commotion. Looking up they saw smoke trails streaming through the sky. *Get back inside,* the teachers shouted.[5]

At his desk, Sabine County Sheriff Tom Maddox felt the whole office convulse. When the noise stopped, his phone lines lit up. Everyone had a different theory. *A gas pipeline broke. Terrorists were attacking. An airliner had crashed. A meth lab exploded. Aliens were invading.* The sheriff tried to keep up. How many disasters could be happening at the same time?[6]

In counties stretching from Dallas, Texas, to Shreveport, Louisiana—a near three-hundred-mile band—thousands of panicked calls jammed the landlines of emergency operators and police offices. Like most Americans, the residents of East Texas had not known the space shuttle *Columbia* was on its way back to the Cape that morning. After more than a hundred workmanlike launches, a shuttle landing did not garner coverage on the network news.

Columbia was a SPACEHAB flight—a new carrier module similar to Spacelab. At a time when NASA was desperately trying to make its deadline to complete the International Space Station, the science mission received little attention. NASA's radars at Vandenberg Air Force Base and White Sands Missile Range were not even tracking the orbiter's reentry.[7]

Still, space enthusiasts waited and watched for *Columbia*'s descent. Stationed on hilltops, rooftops, and cliff edges across the country, their cameras captured the only recordings of *Columbia* as a blaze of light and flame—brilliant and piercing and violent—engulfed the orbiter. While traveling at more than eleven thousand miles per hour—181,000 feet above Corsicana and Palestine, Texas—*Columbia,* the oldest shuttle in

NASA's fleet, was torn asunder. The many sonic booms the residents of Sabine heard that morning were the sounds of tens of thousands of pieces of shuttle debris falling from the clear blue sky.[8]

Washington, DC. February 1, 2003.

Fred Gregory was at home finishing breakfast with his wife, Barbara, when he got the call. *We've lost Columbia.*[9] "What do you mean we've lost it?" Fred asked. Then he switched on his television and saw the pink streaks in the sky.[10]

At sixty-two years old, Fred had raised his children and moved to a wooded neighborhood on Maryland's South River, where he had an acre of land, a dock, and a boat on a rivulet that fed into the Chesapeake Bay. Since flying his last shuttle mission twelve years ago, Fred had climbed NASA's administrative ranks to become deputy administrator of the agency—second only to NASA administrator Sean O'Keefe.[11] George W. Bush had appointed O'Keefe, formerly a deputy at the Office of Management and Budget, to put a stop to NASA's cost overruns, largely from the over-budget space station. O'Keefe was a level-headed money man, not a science guy. He leaned heavily on Fred's quarter century of experience to understand NASA's inner workings. The two often took turns being the onsite NASA representative for launches and landings. That day, Fred had stayed home while O'Keefe went to the Cape for the return of STS-107.

When NASA lost transmission from *Columbia,* Mission Control went into lockdown. National news networks, grasping for whatever information they could, showed eyewitness video clips of the shuttle breaking apart, like a shooting star splintering across the sky. Fred understood almost immediately that the orbiter had incinerated on reentry. There would be no survivors. He grabbed his coat and car keys and sped into central DC.

When he got to headquarters, his staff had set up a conference room for contingency operations. "The first thing I noticed was that the Mishap Investigation Document was open," Fred said. After *Challenger,* he had created the document with fellow TFNG Loren Shriver to detail procedures in the event of a disaster. He had spent fourteen long months writing the treatise but prayed he would never have to use it. Yet here it was, open to a checklist. Item number one jumped out at him: "Assemble an investigation board."[12]

NASA had delayed STS-107 thirteen times in favor of missions for the ISS. Experiments in the backlog built up.[13] To "clear out the garage," Congress earmarked appropriations for a SPACEHAB mission, a sixteen-day flight crammed with life-science experiments and weather observations.[14] *Columbia,* the oldest, largest, and heaviest of the orbiters, was the only shuttle that could accommodate SPACEHAB.

During the crucible of training, the crew members developed a camaraderie that owed much to commander Rick Husband, a former Air Force pilot and deeply religious family man. *Columbia* would be Husband's second flight, but his first as commander. It would also be the second mission for Kalpana "KC" Chawla, an aerospace engineer and the first Indian American to fly, and for Michael Anderson, a US Air Force officer and NASA's ninth African American in space.[15] Willie McCool, the mission's pilot; Laurel Clark, a doctor, Navy captain, and mission specialist; Dave Brown, a naval aviator and flight surgeon; and Ilan Ramon, an Israeli Air Force pilot who would be the first of his country to go to space, rounded out the crew. For these four, STS-107 would earn them their golden pins.[16]

Columbia launched on January 16, 2003. The air was cold and dry, a foretaste of the dark chill that lay beyond the atmosphere's edge. Notwithstanding a terrorist scare—a cluster of party balloons in the airspace that caused a brief fracas—the liftoff appeared to be perfect.[17] Later that day, however, the imagery analysis team at Kennedy was knocked flat when they reviewed the grainy launch videos. At the 81.7 second mark, a briefcase-sized chunk of orange foam, the external tank's insulation, fell off, smashed into *Columbia*'s left wing, and crumbled into dust. "We watched it on the big screen again and again and again," an analyst recounted, "trying to understand where the foam impacted the orbiter."[18]

NASA had long worried about launch debris damaging the shuttle during liftoff. Before the shuttle's inaugural flight, more than twenty years prior, Anna Fisher had been assigned to a team to determine if NASA could fix the shuttle's fragile heat tiles while in orbit, if such an event were to occur. Anna's team found repair nearly impossible without doing additional damage to the orbiter. NASA would have needed two years and millions of dollars to engineer a solution. Managers could not stomach the hefty price tag, so they dropped the project. STS-1 flew successfully anyway. Crip and Young gritted their teeth on the way back, profoundly aware of the orbiter's tile damage, but lived to tell the tale. Since then, the shuttle had flown 111 successful launches, and tile damage had never caused a failure.

The imagery analysts noted, however, that nothing this big had ever hit the orbiter before.[19] Because one of the tracking cameras that was supposed to film the launch had not operated correctly, and another camera was out of focus, Kennedy's imagery analysis team could not see precisely where the foam struck the orbiter or the severity of the damage.[20] Engineers feared that the foam might have impacted and compromised the shuttle's heat shield, which protected the vehicle from the nearly 3,000°F temperatures upon reentering the Earth's atmosphere.[21]

NASA formed an ad hoc Debris Assessment Team to analyze the incident over a three-day period. The engineers came back worried. Nevertheless, the on-orbit Mission Management Team met the engineers' concerns with various levels of indifference. *This has happened dozens of times with no effect,* went the logic. Moreover, the foam had hit the leading edge of the left wing, which was sheathed in durable reinforced carbon-carbon (RCC), not the fragile silica tiles that covered much of the orbiter. *It would be like a Styrofoam cooler hitting a car on the freeway,* the managers figured. *It may cause a scratch, but it wouldn't kill you.*[22] Still, some of the engineers continued to insist that, although the reinforced carbon-carbon was harder than the silica tiles, it was also brittle, and therefore susceptible to damage.[23] "Their objections largely went unappreciated," Kennedy's launch director Michael Leinbach later said. "While someone could almost crush tiles in their hands, RCC felt like an extremely tough and capable material."[24]

At 2:04 PM on the day of the accident, President George W. Bush addressed the nation from the White House cabinet room. "My fellow Americans, this day has brought terrible news and great sadness to our country," he began. "Mission Control in Houston lost contact with our space shuttle *Columbia.* A short time later, debris was seen falling from the skies above Texas . . . *Columbia* is lost; there are no survivors."[25] The news sent ripples of shock and pain across the country. How did an accident like this happen *again*?

After fielding questions from members of Congress all morning, Fred turned his attention to the creation of the Columbia Accident Investigation Board (CAIB). "By noon of that day we had already pulled together the main group," Fred said. As for the head of the investigation, O'Keefe gave Fred five names.[26] Fred homed in on one: Navy admiral Harold Gehman. Gehman had the breadth of experience for the task. He had been the vice chief of the Navy, the NATO Atlantic commander, and head of the US Joint Forces Command. Gehman also led the difficult investigation into the USS *Cole* incident, in which a suicide attack nearly sank an American destroyer in Yemen in October 2000.[27]

The next morning, Fred was up at 5:00 AM. He boarded the NASA Gulfstream, which zipped along the eastern seaboard to pick up members of the new investigation board and ferry them to Barksdale Air Force Base in Shreveport, Louisiana, which had become the hub of the search for its proximity to the wreckage.

"By late afternoon, we had a team in place at Barksdale," said Fred. "Admiral Gehman and I opened the gate and introduced ourselves to the press. This was significantly different from what happened after *Challenger.* Congress and the administration were watching, and they said, *NASA has this under control.*"[28] Thanks to Fred's initiative, President Bush did not request an independent commission as President Reagan had for *Challenger.*

The debris had fallen on a swath of land that stretched from Dallas, Texas to Fort Polk, Louisiana—two thousand square miles.[29] Given the scope of the search, and the fact that much of the debris was either toxic or in flames, President Bush brought in the Federal Emergency Management Agency (FEMA) and called a state of emergency. The FBI cornered scavengers hawking debris on eBay.[30] The Environmental Protection

Agency (EPA) vacated schoolyards and playgrounds before sending in hazmat teams to clear noxious materials. Helicopters surveyed the area for anything that looked out of place. Anyone available to lend a hand—police agencies, fire departments, and emergency services—chipped in, forming "an alphabet soup of agencies," Fred said.[31,32]

A command post went up in Hemphill, Texas, a small town of roughly 1,100 people near the Texas-Louisiana border.[33] Although rural, Hemphill ranked as the largest town in East Texas's sparsely populated Sabine County. The town nearly buckled from the throngs of volunteers that flocked in to help. Sheriff Maddox instructed the utility company to install thirty new phone lines at the local firehouse, which previously only had one.[34] They upgraded the electrical grid to handle the new population and all the new devices. The wastewater treatment facility nearly overflowed.[35] Winter in Hemphill challenged workers—freezing rain, buggy swamps, dense thickets. Communal food service, cramped sleeping quarters, and poor restroom facilities produced illness and dysentery among the workers. Despite the conditions, volunteers kept showing up. Over twenty-five thousand people came to recover *Columbia*.[36]

Three hours after the accident, the firehouse received a call from a jogger who spotted what he had assumed was the body of a dead animal—a deer or a wild boar—on Beckcom Road, a few miles southwest of Hemphill.[37] Sheriff Maddox, who answered the call, found human remains—a member of *Columbia*'s ill-fated crew.[38] A trooper who accompanied Maddox to the site draped a raincoat over the fallen crew member. Overhead, an unscrupulous reporter nudged a helicopter pilot to dip low enough to blow off the coat and reveal the body. The trooper with an FBI agent stood on the corners of the raincoat to secure it and gave the finger to the pilot.[39]

With the first body as a guidepost, workers traversed the territory along a grid system, usually reserved for airline crashes and crime scene investigations. Ten thousand National Guard troops were dispatched to look for the bodies. Over the next five days, searchers found the rest of the crew's remains along a fifteen-mile track in Sabine County. Some of the bodies lay deep in the swampy Sabine Forest, surrounded by heavy mud and briar thickets. Extracting them proved physically grueling and emotionally draining.[40] Grief crippled the astronauts who came to help when they came upon the personal effects of their classmates or the parts of the wrecked shuttle that they had worked on.[41]

Astronaut Mark Kelly requested that a member of the clergy say a few words over crew remains before they were moved. Brother Fred Raney, a local pastor, fulfilled that sad duty, providing short memorial services in the woodlands and pastures.[42] *Columbia*'s commander Rick Husband had recited Joshua 1:9 to his crew: "Be strong and of good courage; do not be afraid, nor be dismayed, for the Lord your God is with you wherever you go." Brother Fred added the scripture to his services and found fitting words for the crewmembers of Hindu and Jewish faiths.[43,44]

On February 4, President Bush led a memorial service at Johnson as Reagan had for *Challenger*. "The grief is heavy, our nation shares in your sorrow and in your pride," President Bush said, also reiterating his commitment to the space program. "This cause of exploration and discovery is not an option we choose; it is a desire written in the human heart."[45]

Two days later, firefighters found a videotape cassette near Palestine, Texas, that was, remarkably, in working condition.[46] Its contents revealed the crew in the last minutes before the breakup, happy and eager to come home. Eerily, the crew commented on the glow of the plasma as they entered Earth's atmosphere. "Looks like a blast furnace," commander Husband marveled. "You definitely don't want to be outside now."[47] In the last seconds of the video, gravity tugged their equipment down, as they neared home. Four minutes later, Mission Control detected a problem with the sensors, the first sign of trouble.[48]

As winter gave way to spring, searchers braved alternating days of cold sleet and muggy rain. Especially for those unfamiliar with the heavily forested terrain, the conditions were perilous—muddy, difficult to traverse, rife with insects and snakes. Naturally occurring swamp gas caused some workers to faint.[49] A few turned back, but most crews pressed ahead, regardless of the peril. On March 27, a search helicopter hit the crown of a large oak tree and crashed in the swampy Ayish Bayou, killing two of its passengers.[50]

Despite the challenges, searchers found deep meaning in their work. Volunteers gave their hearts and souls to the effort. Fire crews from nearly every Native American tribe and nation in the Midwestern and Western United States—including the Apache, Blackfoot, Caddo, Cherokee, Choctaw, Creek, Iowa, Kiowa, Pawnee, Sioux, among others—contributed to the recovery effort.[51] Fred remembered the Native American firefighters gathering around the fire at night, singing songs, drawing pictures, and writing poems about the shuttle.[52] They reminded Fred of the deep, spiritual nature of their search.

In practical terms, uncovering the mystery of *Columbia* was the only way NASA would get back to space. NASA needed to understand what had caused the accident.[53] The dream of the International Space Station—humanity's newest home in the sky—lay in the balance. "Billions of dollars' worth of space station modules were sitting in a processing facility at Kennedy waiting to be launched," Leinbach said.[54]

In all, the workers ferreted out eighty-four thousand pieces of *Columbia* weighing over forty-two tons, or nearly 40 percent of the orbiter.[55] Fred designated a hangar at Kennedy to house the recovered debris. Truckloads of wreckage were driven cross country to the Cape throughout February, March, and April. As Fred flew back and forth between Barksdale, the Cape, and headquarters, he watched the giant jigsaw puzzle being put back together.

"That's when we could confirm that the reinforced carbon on the leading edge of the wing had been broken," said Fred. "We were now able to understand how the heat came through that leading edge of the wing and began taking out systems."[56] The effects of the heat of reentry were visible everywhere. Heat-resistant tiles were warped beyond recognition. Globs of metal spattered the interior and exterior surfaces of the recovered left-wing pieces and inside the crew module.[57] "Then it became very obvious," Fred said. "This is where the failure occurred."[58]

On March 19, firefighters found a metal box on the ground next to a small crater where it had impacted Earth and bounced. Astonishingly, they had found the OEX, a close analog of the flight data recorder found on all commercial airlines.[59] The OEX had recorded hundreds of channels of data from the ascent and reentry.[60]

The last few seconds of telemetry suggested that *Columbia's* crew knew something was wrong in the moments before the orbiter broke apart. As heat breached the orbiter's shield, the sensors would have set off alarms and the crew would have reacted. A recovered control panel from the crew capsule showed that a crew member pulled switches to try to control the hydraulic system. Just as in *Challenger*, even in their last moments, the crew had tried to save themselves.[61]

The next day, the Bush administration began Operation Iraqi Freedom with a massive bombardment of Baghdad, and *Columbia* slipped from front-page news. Still, to those following the investigation, the findings confirmed the earliest suspicions about what had caused *Columbia*'s demise. *The debris spoke for itself.*

Fred turned over the agency's reports to Gehman and the CAIB.[62]

Many in Congress and the press argued that Gehman's board did not have enough independence from NASA. To win their confidence, Gehman established a congressional-liaison office, and, on March 5, 2003, he added five new civilian members to the board.[63] They included Nobel-laureate physicist Douglas Osheroff, former Air Force Secretary Sheila Widnall, and former astronaut Sally Ride.[64]

Orlando, Florida. February 1, 2003.

On the morning of the accident, Sally woke up in a Florida hotel room to an urgent call from her sister. *NASA's lost contact with Columbia,* Bear Ride told her over the hotel phone. Sally flipped on her TV. Having arrived from California the prior evening, she was suffering from jet lag and had slept in.

"Oh, that's not good," Sally said.[65] She watched as the tragic news unfolded, detail by somber detail. The *Challenger* disaster had occurred almost seventeen years to the day. The loss of *Columbia* and her crew hit her like an uppercut to the jaw.[66]

Sally had traveled to Florida to lecture a group of eight hundred fifth- through eighth-grade girls, and their parents, at the University of Central Florida.[67] Even though she was working in academia at the University of California San Diego, Sally was also building Sally Ride Science, a non-profit that encouraged girls to study the sciences.

Sally and her organizers considered whether to cancel the conference in light of the tragedy. After much discussion, they agreed it was important to give the girls a place to grieve, time to ask questions, and an opportunity to give them perspective on why, despite the dangers, it still mattered to take risks in life.[68]

"Why wasn't there a backup plan?" the girls asked Sally the next day. "How do astronauts feel about going up into space now?" and "What good has come out of yesterday's tragedy?" Sally told the girls that risk went hand in hand with exploration.[69]

"I think that although yesterday was a horrible day for the space program, the space program will go on," Sally consoled. "We'll pick up the torch the astronauts carried and carry it forward." With emotion straining her voice, she said, "We need kids to look to the stars."[70]

"Didn't you investigate *Challenger*?" another asked.

Yes, I did, she told them, but no, she would not be investigating the *Columbia* accident.[71] Sally had left NASA sixteen years ago. Even though she still believed in exploration, she did not think she would ever go back. Then, a month after the accident, she got a call from Admiral Gehman.

"It took forty-five minutes to convince her," Gehman recalled, "but with Sally we got a scientist, a veteran of the *Challenger* investigation, and we got an astronaut. She exceeded our expectations."[72] Sally was the only astronaut on the board and the only person to serve on both the *Challenger* and *Columbia* investigation committees.

As the inquiry spun up, Sally, once again, found herself working eighty-hour weeks.[73] She flew to Louisiana's Barksdale Air Force Base, where the Mishap Investigation Team had based its command center, only one hundred miles north of Hemphill.[74] She traveled to her old stomping grounds in Houston and to NASA Headquarters in DC. She reviewed the launch video that showed the briefcase-sized block of foam hitting *Columbia*'s left wing. Although the shuttle had broken up on reentry, she believed *Columbia*'s fate may have been written on liftoff. "Follow the foam," she jotted down in her notebook.[75]

For perspective, Sally called up her old friend Hoot Gibson, who was now a pilot for Southwest Airlines. Back in 1988, Hoot had experienced a significant foam strike on STS-27, one of the last of the shuttle's military flights. "The belly looked as if it had been blasted with shotgun fire," he said of *Atlantis*.[76] The damage merited a report but was deemed not significant enough to stop the launch schedule.[77] In 1992, an even larger piece of foam, twenty-six-by-ten inches, hit *Columbia* on takeoff. When the orbiter returned, techs discovered the worst tile damage in the shuttle's history.[78]

While these flights were notable for their tile damage, foam strikes had become commonplace over the years.[79] Evidence of foam loss appeared on sixty-five of the seventy-nine missions for which imagery was available.[80] Since the issue never caused a serious problem, NASA managers believed it to be an "inevitable" and "acceptable risk."[81] "We had this mindset that this was a nuisance," said Wayne Hale, the shuttle program manager. "It wasn't a safety-of-flight issue."[82] In truth, the risk had not disappeared, but the fear of it had.[83]

Despite NASA's complacency, Kennedy engineers who had seen the video of STS-107's launch were jolted by the size of debris they saw and the speed with which it hit the orbiter.[84] Engineers surfaced the issue to their superiors, who dismissed their concerns out of hand. Linda Ham, the shuttle program deputy director and chair of the Mission Management Team responsible for investigating in-flight anomalies like the foam strike, believed that given the material properties of the foam, the debris was unlikely to cause much damage. In a meeting, five days after *Columbia*'s launch, Ham then referred to the impossibility of tile repair on orbit anyway even if there were a problem, saying: "I really don't think there is much we can do. So it's not really a factor during the flight because there isn't much we can do about it."[85]

Unable to get their managers' attention, engineers on the Debris Assessment Team contacted their connections at the Department of Defense, asking to use spy satellites to take images of the damage, as they had when STS-1 launched. On January 22, the seventh day of the mission, the Air Force started to honor the engineers' request, but then Ham interceded.

High-level managers like Ham not only preferred for employees to follow the appropriate chain of command, but also for all analysis to be kept in-house at NASA, rather than involving outside agencies. When Ham learned that the engineers had made a request for images directly to the DoD, completely bypassing her required approval, she reached out to their superiors—high-level Johnson managers on the Mission Management Team—to see if *they* wanted to make a *formal* request for the satellite images.[86] The managers backed down, saying that they believed they did not have enough data to justify involving the DoD. Of course, the reason they did not have "enough data" was because they did not have the satellite images of the orbiter, which the DoD would have happily provided.[87] Ham then terminated the request for images with the Department of Defense.[88] The action later came under scrutiny as a move that prioritized office politics over the safety of the crew.[89]

Absent imagery from the DoD, engineers had to rely on a computer modeling program developed by the Boeing Corporation known as "Crater" to assess the damage. Boeing had designed the software to model damage from small debris striking the orbiter, not debris the size of a "briefcase."[90] STS-107 was also the first time that Crater was used to analyze potential damage from a foam hit to a shuttle that was already in orbit.[91] Crater predicted the damage to the orbiter would not cause issues on reentry.

To check the veracity of the Crater model post facto, the CAIB hired a San Antonio aeronautics organization (a group that normally simulated birds hitting the windshields of airplanes) to reproduce foam hitting the left wing's reinforced carbon panels. Commissioners gaped when they saw the test produce a hole big enough for someone to put their head through.[92] Some engineers and managers who had still been living in denial were close to tears.[93]

"Their whole house of cards came falling down," Gehman said.[94]

Sally pored through emails between NASA managers and engineers and interviewed shuttle personnel. When presented with boxes of information that spanned the length of the conference room, she popped Goldfish crackers in her mouth and carefully thumbed through the files, absorbing it all. All the while, she penciled her thoughts into another notebook with the mission patch of the lost crew on the cover, making an annotation to "call A^2," aka Steve Hawley.[95] The pattern crystallizing in her mind was chilling in its similarity to *Challenger*: a rush to flight, a culture that stifled dissent, and a persistent engineering problem. This time, managers explained away tile damage, instead of O-rings, to save on cost and time. Again, the safety of the crew had been fatally compromised.

"I'm hearing a little bit of an echo here," Sally said at a Houston press conference in April 2003.[96] Gehman was floored by the parallels she made between *Challenger* and *Columbia*, as were the other attendees.

"We were kind of sniffing around this idea," said Gehman, "but we hadn't zeroed in on it. Hadn't made it a priority. We were still looking for widgets and brackets and foam. That single statement significantly reinforced in us the need to change the direction of the investigation. She caused us to reconsider where our priorities were."[97]

It wasn't the foam; it was the culture.

As the Rogers Commission had learned in their *Challenger* investigation, the CAIB was now realizing that interested NASA officials, such as Linda Ham, should not be allowed to be the filter through which the board received information.[98] Gehman told Ham and her colleague, Ralph Roe, the launch director and manager of the vehicle engineering office, that they needed to exit the investigation. Ham and Roe, alternating between anger and tears, refused to leave.[99] O'Keefe backed them. Without recourse, Gehman took his argument public by publishing his request on the CAIB website. O'Keefe seethed, but Gehman won the fight.

Enter Anna Fisher.

The previous fall, Anna's daughter Kristin left for Boston College to begin her freshman year. Kara, her youngest, was finishing high school. With her babies out of the nest and her divorce with Bill finalized, Anna settled into a new home in Houston. Bill got their old manse in the divorce, but Anna finally felt free. She was invigorated by her new role as head of the International Space Station branch of the Astronaut Office. She was even dating again. Then Kent Rominger, head of the Astronaut Office, presented her with the opportunity to fly to the space station to which she had devoted much of her life.

Now, two decades after her first flight, Anna considered a second act in space. She jumped at the chance and began working on updating her training. To accrue flight hours, she would pilot her T-38 to visit her daughter Kristin at Boston College. She took classes with the other younger astronauts and volunteered to be CapCom on upcoming ISS missions. By January 2003, she was ready to jump back in line. If successful, Anna, at fifty-three, would become the oldest woman to fly to space. A month later NASA was grounded.

Anna, out on a date the previous evening, woke up that Saturday morning only moments before *Columbia*'s approach. "I wasn't really following that flight because I wasn't working it . . . I was home."[100] When she saw the disaster unfolding, her heart sank. She got dressed and headed into work. Anna did not realize it then, but the fact that STS-107 was not one of her flights made her the perfect candidate to investigate it.

After Gehman dismissed O'Keefe's Mission Management Team, the director of flight operations tapped Anna to lead a new group to investigate NASA managers. Her team, which would report to the CAIB, would evaluate the managers' decision to ignore the foam strike rather than triage the damage it may have caused. Was it true what Linda Ham had said, that there was nothing NASA could have done to save *Columbia*? To answer that question, Anna played out various rescue and repair scenarios.

"If you don't know much about the program, you would think, 'Well, why not just go to the space station?'" Anna said. "It was in a much lower orbit, and there was just absolutely no way that the shuttle, at this point, had sufficient fuel onboard to make it to the space station."[101] Still, another possibility for rescue existed. The orbiter *Atlantis* was already at Kennedy preparing for its upcoming flight. In theory, engineering crews could have prepped *Atlantis* for flight early while *Columbia* powered down, terminated the SPACEHAB experiments, and put the astronauts on extended sleep schedules to preserve oxygen and prolong their flight. With a crew of two, *Atlantis* would rendezvous with *Columbia*, bringing *Columbia*'s crew onboard with a series of spacewalks. Some thought that the rescue would have been near impossible. Anna pointed out that "NASA is very good at responding to crises," as the improbable return of Apollo 13 proved, but "they opted not to do that."[102]

Anna's team also studied whether the crew could have repaired the shuttle on orbit. The crew "lacked suitable repair materials," Anna said. They also did not have the robotic arm, which could have potentially taken photos of the tile damage and supported the astronauts during a spacewalk. Yet had the crew known of the damage, they may have been able to improvise a solution. "They could have at least cold soaked the wing by orienting it toward deep space to make it really cold before reentry. It might have bought them some time," Anna said.[103] If only they had known the extent of the damage.

Mission Control did not inform the crew about the foam strike until one week into the mission, when they received this upbeat email:

> Rick and Willie, You guys are doing a fantastic job staying on the timeline and accomplishing great science. . . . There is one item that I would like to make you aware of for the upcoming [Public Affairs] event . . . this item is not even worth mentioning other than wanting to make sure that you are not surprised by it in a question from a reporter.[104]

The email went on to explain the foam hit and attached a video clip.

> Experts have reviewed the high-speed photography and there is no concern for RCC or tile damage. We have seen this same phenomenon on several other flights and there is absolutely no concern for entry. That is all for now. It's a pleasure working with you every day.[105]

The lede-burying communications to the crew sent Anna into a rage: "If it were me and I were on orbit, and there was a chance that I was going to die, I would have liked to have known that so that I could talk to my family."[106] Anna suggested the managers should have given each crew member a chance to speak with their loved ones, even if they thought the reentry risk was marginal.

Anna's final report was damning. NASA deprived the crew of the knowledge of the risk they faced and forestalled their chance to survive by not acting. Sally and the other CAIB investigators knew that NASA might not be receptive to their findings, so Sally volunteered to leak the report to the public using a middleman, as she had in the *Challenger* investigation.[107] She called Todd Halvorson of *Florida Today,* with whom she had worked at Space.com.[108]

"I was in the media newsroom in Houston," Halvorson said. "And I remember running outside when I saw who was calling on my cell phone."[109] Halvorson took meticulous notes and churned out the copy. The story ran the next day: "DARING RESCUE MAY HAVE SAVED CREW: ATLANTIS RESCUE FLIGHT POSSIBLE, STUDY SAYS."[110] The article quoted a "senior investigator"—Sally—as saying, "It would have been high drama, but there was a realistic chance . . . of returning the crew."[111] The story laid out details of a bold rescue mission that might have been attempted had NASA chosen to do so.

Before publishing the final report, Sally's colleagues on the CAIB asked her to write a summary of the Rogers Commission's findings on the *Challenger* accident, as context for the *Columbia* disaster. "I've been putting this off because it's so irritating to go back and read that report," Sally wrote in an email to her colleagues. "It's definitely Deja Vous [*sic*] all over again."[112] NASA knew about the O-rings like they knew about the foam, but "they never stopped to fix the problem, never stopped and understood it."[113]

The seventeenth anniversary of the *Challenger* accident occurred as *Columbia* orbited Earth. Commander Rick Husband and his crew paused to remember the crews of *Challenger* and the Apollo 1 mission. "They made the ultimate sacrifice, giving their lives in service to their country and for all mankind," said Husband. "Their dedication and devotion to the exploration of space was an inspiration to each of us and still motivates people around the world."[114] In an inexplicable byproduct of fate, all three of NASA's fatal spacecraft accidents occurred within one week of each other, between January 27 and February 1.[115] These great tragedies unfolded not merely because of engineering mistakes or scientific unknowns, they arose out of human error, miscommunication, politics, and hubris.

Sally penned the final lines of the CAIB report on the twentieth anniversary of her historic first flight to space. On August 26, 2003, seven months after the accident, Gehman hand-delivered an official copy of the report to NASA administrator Sean O'Keefe, who accepted the report somberly. The document served as a stern rebuke of NASA's managers like Mission Management Team chair Linda Ham and shuttle program manager Ron Dittemore, saying that "organizational practices detrimental to safety were allowed to develop," "reliance on past success [was used] as a substitute for sound engineering practices," and professional differences of opinion were stifled. The commission found that "NASA's culture and structure had as much to do with this accident as the external tank foam."[116] The report cast a particularly unflattering light on Linda Ham. Was she a villain or a scapegoat? No matter, she had already been reassigned within NASA a month before the report's publication.[117]

On the same day the CAIB released its report, Sally gave an unvarnished interview to the *New York Times* about the accident. "There wasn't any of that quality that Mission Control is almost famous for," she explained, "which is grabbing onto the pants legs of a problem and not letting go until it understands what the problem is and what the implications are. And that didn't happen in this case."[118]

As for NASA administrator Sean O'Keefe and deputy Fred Gregory, the report was largely silent, saying simply that O'Keefe had come to NASA as a managerial and financial leader, but not as a scientist.[119] Fred lamented that the shuttle program managers had not been more open to dissent. "For making leadership decisions, you must understand the importance of listening, not tossing off an idea because it does not fit into the way you were thinking," he said. After *Columbia*, Fred stayed on at NASA, rising from deputy to acting administrator, becoming the first African American to lead the agency. As the new administrator, Fred did not want to dwell on blaming specific individuals for the accident: "My role was to learn from what mistakes we had made so that we would not repeat those mistakes."[120]

Given the shuttle's track record of failures and near misses, lawmakers wondered whether NASA could learn from its errors. The CAIB delivered a grim answer: "Because of the risks inherent in the original design of the Space Shuttle, because that design was based in many aspects on now-obsolete technologies, and because the Shuttle is now an aging system but still developmental in character, it is in the nation's interest to replace the Shuttle as soon as possible as the primary means for transporting humans to and from Earth orbit."

While in the near-term the CAIB supported the shuttle returning to space, the committee ultimately recommended that NASA develop a new generation of spacecraft.[121] The Bush administration reviewed the CAIB's report and, in January 2004, President Bush announced that, as soon as the shuttle finished assembly of the International Space Station in 2010, the fleet would be retired from service.[122] "The space shuttle has flown more than a hundred missions. It has been used to conduct important research and to increase the sum of human knowledge," the president said. "Yet for all these successes, much remains for us to explore and to learn. In the past thirty years, no human being has set foot on another world or ventured farther up into space than 386 miles, roughly the distance from Washington, DC, to Boston, Massachusetts. America has not developed a new vehicle to advance human exploration in space in nearly a quarter century. It is time for America to take the next steps."[123]

Hemphill, Texas. April 22, 2003.

Nearly three months after the accident, Choctaw firefighters from the state of Oklahoma performed a "victory dance" in Hemphill's VFW outpost.[124] The ceremony commemorated the end of the search for *Columbia* in Sabine and San Augustine counties.[125] The next day, family members of the *Columbia* crew came to East Texas to thank the recovery teams in person and see where their loved ones had been found.[126] Brother Fred Raney told the widows of Willie McCool and Ilan Ramon that he had provided a "chapel in the woods" for their loved ones.[127] "I wanted them to know that they were being thought of during that whole time," he said.[128]

On April 29, NASA administrator Sean O'Keefe hosted a memorial dinner at Lufkin's Civic Center, about sixty miles west of Hemphill. Sabine County Sheriff Tom Maddox as well as other East Texas government leaders and dignitaries from the Native American tribes who contributed to the search attended. The *Columbia* crew families thanked the people of East Texas for returning their loved ones to them. Via a live video message from the ISS, the crew of Expedition Six thanked the searchers.[129] NASA's first woman shuttle commander, Eileen Collins, who would return the shuttle to space, concluded the ceremony at the civic center on behalf of herself and her crew.[130]

Almost two years later, in 2005, Collins docked at the ISS and sent a message to families of *Columbia*. "For all our lost colleagues, we leave you with this prayer often spoken for those who sacrifice themselves for all of us"—quoting Laurence Binyon's poem "For the Fallen."

> They shall grow not old, as we that are left grow old:
> Age shall not weary them, nor the years condemn.
> At the going down of the sun and in the morning
> We will remember them.[131]

Nearly twenty-five thousand workers, contributing over a million hours of labor, had played a part in NASA's return to space.[132] Once more, NASA had risen from the ashes, but now the shuttle was entering its last phase of service: finishing the ISS.

NASA, along with its international partners, planned to complete the station by 2010.[133] While the shuttle had been grounded, Russian Soyuz rockets continued to ferry supplies and two-person crews to and from the station, allowing work and research to carry on virtually uninterrupted. Now, with the shuttle back in service, ISS program managers planned to add over a dozen major components to the station over the next half decade. A 2005 NASA directive read that ISS research would be shifted toward "development of countermeasures against space radiation and the long-term effects of reduced gravity," in order to "prepare human explorers to travel beyond low Earth orbit."[134] NASA and the world readied for the next stage in human space travel: conquering new worlds.

Back in Houston, Anna looked at herself in the mirror. A quarter of a century earlier, she had started at the Astronaut Office, a young, wide-eyed New Guy. She had hardly known what the shuttle was. They had come so far together. She now had to ask herself, *Do I want to stay? Do I want to finish this?* Anna's faith in NASA had cracked but not broken. She believed in the promise of human spaceflight. She trusted that completing humanity's home in space was a necessary step toward deep space exploration, and she still wanted to guide that voyage as head of the International Space Station branch of the Astronaut Office. So Anna dried her tears, rolled up her sleeves, and went back to work.

CHAPTER 27

GOD BLESS

Marriott Hotel, Orlando, Florida. July 7, 2011.

In the high-ceilinged lobby, decorated with Florida palm fronds and pink marble, orchestral tunes, cascading water, and buzzy bar talk echoed gently, suffusing the cavernous hall with the ambient soundscape common in malls, airport lounges, and mega-hotels.

"You are a whiskey delta," Anna's voice cut through the hum, teasing Hoot as he threatened to retire for the evening. This epithet, military slang for "weak dick," was hurled at astronauts who tried to go to bed rather than stay out drinking.[1] They were both in their sixties, and while Hoot still had his handsome swagger, his tolerance for partying was not what it was in his flaming-hooker days. Tomorrow *Atlantis* would launch, bookending the space shuttle program after 134 missions.[2] The event was a de facto class reunion, a bittersweet coda for the New Guys who had grown up and grown gray in the shuttle's shadow.

Hoot and Rhea were mindful of the clock. They had an early start the next morning, meeting up with their now-adult children. Rhea left NASA in 1997, after completing two more shuttle missions, STS-40 in 1991 and STS-58 in 1993, both dedicated to life sciences research, and serving as assistant to the director of flight crew operations for shuttle/*Mir* payloads. "We were sort of told you had to be willing to go travel to Russia to train for two years," Rhea said. "Train with cosmonauts in a little tin can. I had three little kids at home—that wasn't going to work for me."[3] With two decades at NASA, Rhea had a difficult time finding employment that measured up. *There is no career quite like being an astronaut.* She returned to medicine as the assistant chief medical officer at Vanderbilt University. After a decade at Vanderbilt, she became a healthcare coach and national speaker on science and leadership.[4]

Hoot served as chief of the Astronaut Office and deputy director of flight crew operations before retiring from NASA in 1996—a year after his fifth and final shuttle mission, STS-71. He took a job at Southwest Airlines.[5] He would never get tired of flying: In his spare time, he still set speed records. Four years earlier, he had flown his red-and-white Sea Fury plane *Riff Raff* 437 miles per hour, a mere sixty feet off the ground at the Reno Air Races.[6]

"Whiskey delta?" Hoot repeated, shaking his head and laughing. He used to lead the bar crawls during their AsCan years. The student clearly had learned from the teacher. They ribbed each other as only time-forged companions could.

"See you tomorrow," Rhea said, hugging Anna goodnight.

Anna and Fred stayed up, much like they had done during their astronaut days. Anna wanted to savor every minute. Her NASA career spanned from the birth of the shuttle to the completion of the ISS, and she was not planning on throwing in the towel yet. Though the shuttle program was ending, Anna had more road to run in the Astronaut Office. Rich in wisdom and hard-won lessons, she was training a new class of AsCans, destined to fly *Orion*, the shuttle's successor.[7] *Orion* looked nothing like the shuttle but was instead a return to a capsule design like the Apollo module, albeit sleeker and more sophisticated.[8]

The next morning, Anna, Hoot, and Rhea met Fred and his new wife, Annette, in the lobby. Fred remarried a year earlier after Barbara, his wife of forty-four years, passed away.[9] After three years of serving as NASA's deputy administrator, Fred resigned from NASA in 2005.[10] As second-in-command at NASA, he had been the agency's chief operating officer, responsible for several of its largest programs, including the completion of the ISS.[11]

Boarding the bus that would take them out to the Cape, Anna recalled the trips they had taken together as astronaut candidates. This time, there would be no food fights, no scandalous sleepovers, no pranks on George Abbey.

As they approached Kennedy Space Center, their bus worked its way through the enormous crowds. Cars and vans packed the lots. Nearly a million spectators lined the beaches and causeways to see *Atlantis* take off, almost double the turnout for Sally's first flight and triple that for STS-1.[12] For many onlookers, the shuttle was an intergenerational emblem, its history strung from the immortal words of Neil Armstrong to the dawn of the third millennium.

Opposite the crowded stands, *Atlantis* sat perched on Launch Pad 39A—the same pad that sent the Apollo astronauts to the moon, the first shuttle into orbit, and the first American woman and African American to space. *Atlantis* and her sisters *Columbia*, *Challenger*, *Discovery*, and *Endeavour* propelled many of the technological breakthroughs of the era. The shuttle launched and maintained Hubble, whose unobstructed view helped to unmask the origins of the universe; deployed Galileo, the first spacecraft to orbit Jupiter; and installed scores of satellites that thrust telecommunications, weather monitoring, and national security systems into the twenty-first century.[13,14]

This fleet of shuttles carried into orbit the modular segments of the ISS. Constructed over forty missions and a hundred-plus spacewalks, the orbiting laboratory boasts living quarters, docking ports, propulsion and control modules, power-generating solar arrays, and airlocks for spacewalks.[15] In addition, the shuttle and its crews installed three science modules: the US laboratory *Destiny*, the European *Columbus* astrophysics module, and two Japanese modules. *Destiny* featured ultrasound imaging machines, medical equipment for life sciences research, and a microgravity glovebox that created a sealed environment for smaller experiments.[16]

Now, in summer of 2011, the station was complete—a huge triumph for NASA. Though human space exploration was born out of the space race, the twenty-first century saw cooperation between the US, Russia, and more than fifteen other nations to live and learn in space. Blazing a trail through the night sky, the ISS was a testament to what humanity's peaceful collaboration could achieve, and a reminder that, viewed from space, Earth had no borders.

Prior to liftoff, the New Guys and their companions gathered under a tent to say a few words about the shuttle. Astronauts and NASA staffers spoke when moved to do so, reminiscing about missions gone by and mourning those that would no longer come. After decades of routine press conferences, the astronauts now spoke without filter and from the heart.[17]

Bob Crippen recounted his experience on the first shuttle mission thirty years before, then singled out Sally Ride and Kathy Sullivan, with whom he had flown. "Both of them proved that women can do anything they want to, as far as the space program is concerned," he said. The crowd cheered.[18]

Kathy, now fifty-nine, had arrived at the Cape several days earlier to speak on climate change. A few months prior, the Senate confirmed her as assistant secretary and deputy administrator of the National Oceanic and Atmospheric Administration (NOAA).[19] The job would be her second stint at NOAA—her first began in 1993 as the agency's chief scientist. She spent the intervening decade in Columbus, Ohio, running the Center of Science and Industry (COSI), one of the first "hands-on" science museums in America.[20] Kathy also lent a hand when Sally's science camps came to town, connecting Sally's students with local scientists.[21]

Sally was conspicuously absent for Crip's praise.

Now sixty, Sally was living with Tam in San Diego.[22] In 2007, after eighteen years as a professor at the University of California, San Diego, Sally retired to devote all her time to running her company, Sally Ride Science, with Tam. Though Sally and Tam did not speak about their relationship, their coworkers noted the way they cared for each other, stepping in when one sensed the other needed support.[23] They wore matching bands on their left ring fingers, moved into a beautiful Spanish-style home in the San Diego hills, and co-parented a floppy-eared bichon.

Sally's refusal to make their relationship public was not easy on Tam.[24] She almost left Sally, a few times. Once, Sally declined to go on a cruise with Tam, then immaturely gave Tam the silent treatment when Tam dared to go without her. Tam said Sally "turned into a fourth-grader."[25] "We didn't talk about it, but she knew that I was thinking of leaving her," Tam said. "But we loved each other passionately . . . No one is a perfect mate, but Sally came close enough."[26]

In 2008, Sally received a call from the Obama transition team offering her the NASA administrator job. She could finally implement the many recommendations she had made throughout the years. Despite the influence she could have had, Sally could not tolerate the attention and disliked the politics. She loved running Sally Ride Science. She told the transition team *not* to have the president ask her, but Sally did agree to join the 2009 US Human Spaceflight Plans Committee to sketch out the future of NASA.

The goals of what would become known as the Augustine Committee were not only to examine new potential targets for exploration, but also to discuss the viability of NASA's current manned spaceflight program, including the space shuttle and George W. Bush's Constellation Program. The Constellation Program was to build spacecraft to replace the space shuttle, in order to ferry astronauts and cargo to the space station, return to the moon, and ultimately launch a voyage to Mars. Sally spent the summer working day and night, sometimes sending the committee chairman 2 AM emails about crucial issues.[27]

At the committee's final public hearing in DC, she gave a two-hour presentation, with slides and statistics illustrating the unrealistic nature of the Bush-era strategy: Constellation's spacecraft and rockets were not only impractical, but they were behind schedule and overbudget, unable even to launch during the lifetime of the space station.[28] Unless the United States spent a lot more money, Sally said, NASA could not achieve manned spaceflight to any destination beyond low Earth orbit, and certainly could not reach Mars or establish a base on the moon.[29]

Sally recommended shuttering the Constellation Program with an eye toward building cheaper, more streamlined spacecraft.

As the "major architect of the report," Sally found that the shuttle program—due to its cost and complexity—stifled other ventures.[30] In order to finish the ISS, however, she lobbied to extend the life of the shuttle for one more year and push out the retirement of the space station to 2020. After the shuttle was decommissioned, the US could purchase seats on Russian rockets to launch its astronauts to the station, while private industry built new spacecraft to reach low Earth orbit.[31]

The committee's recommendations did not sit well with many of NASA's older astronauts, including Neil Armstrong, Apollo astronaut Jim Lovell, and Sally's two-time mission commander Bob Crippen. Crip considered it misguided to relinquish America's only vehicle for human spaceflight. The *Challenger* tragedy, the *Columbia* accident, and the many near misses before, between, and after, however, proved the shuttle too risky a vehicle to operate. For Sally, ending the shuttle program was an act of love. There were safer spacecraft to build and new worlds to explore.

Five months before *Atlantis*'s final launch, in July 2011, Sally said to Tam, "I want to rest. I don't feel well."[32] That's when Tam noticed Sally's skin had a yellowish tinge. She convinced Sally to see a doctor the next day, and an ultrasound revealed a golf ball–sized tumor in Sally's abdomen. A CT scan confirmed the worst—pancreatic cancer. Because of the location of the tumor, her prognosis was grim; the survival rate was 14 percent.[33] Sally was determined to beat those odds and approached her treatment with the same poise and unflappability as she did her astronaut missions.

Sally kept her diagnosis a secret from everyone, even her closest friends. Her rapidly deteriorating condition torpedoed her busy calendar, but she still flew to meetings and, failing that, Skyped in to greet children at her conferences. She wore a wig to hide her hair loss. No one was fooled—"The wig looked like a wig," said her biographer, Lynn Sherr—but no one knew the details of her condition.[34]

Tam focused all her energy on making Sally comfortable. She bought a hospital bed to put next to the one they shared so that Sally could watch the beautiful San Diego sunsets from their bedroom window.[35] Sally's nurses and surgeon Dr. Lowy stopped by to care for her and saw a side to Sally that precious few people got to see. Sally, despite her brave persona, was terrified—not unlike she was during her first shuttle launch. *It was the only time in my life I remember feeling I wasn't in control of something,* Sally told her doctor. Sally then pointed to her hospital bed, the IV bag, and herself, and added, *Kind of like this.*[36]

Back at the Cape, *Atlantis*'s crew waited as the seconds ticked by. Like legions of astronauts before them, adrenaline coursed through their veins. This final flight was a gift. *Discovery* and *Endeavour* flew their last missions in February and May 2011, officially completing the ISS by bringing the final equipment and storage modules. Four decades after its creation, the shuttle program had fulfilled its purpose. Even though the fiscal taps had all but shut off after the 2008 financial crisis, President Barack Obama signed off on one last shuttle flight: STS-135 would transport the supplies needed to keep the ISS functional.[37]

NASA kept the crew lean with four space veterans: commander Chris Ferguson, pilot Doug Hurley, and mission specialists Sandra Magnus and Rex Walheim.[38] The small crew was largely a safety measure.[39] By 2011, *Atlantis* was the only shuttle that had not been decommissioned or—in the case of *Challenger* and *Columbia*—destroyed. If something went awry in orbit, NASA would need to transfer *Atlantis*'s crew to the ISS and rely on Russian rockets and personnel for a thorny rescue mission.[40] The absence of a rescue vehicle illuminated the sudden void in the American space program. Private contractors like Boeing and SpaceX would take over low Earth orbit flights while NASA refocused on deep space exploration, but the realization of those goals was still more than a decade away.

The final mission also meant that over eight thousand people who worked on the shuttle were losing their jobs. Ferguson and his three crewmates made frequent visits to the workers at Kennedy, to try to boost their spirits. Even though layoffs loomed, NASA employees put their shoulders to the wheel for the final mission.[41] Emotions ran high.

Meetings at Kennedy now ended with tearful goodbyes as colleagues realized they would never work together again.

As the launch time approached, Anna, Rhea, Hoot, and Fred joined the other members of their class, as well as shuttle astronauts from the other twelve classes that had followed them. They migrated to the VIP stands, greeting their old friends with smiles, hugs, and tears. Anna tried to absorb the moment. So much had changed since they first joined the shuttle program. Their group was the first to send a Black American to orbit; now a Black man was in the Oval Office. Since Sally Ride's historic flight, more than sixty women had "slipped the surly bonds" of Earth. The final class of eleven astronauts chosen to train for the shuttle program included two women, an African American, a Mexican American, and the first Puerto Rican astronaut. One of the women, Dorothy Metcalf-Lindenburger, was the first Space Camp alumna to become an astronaut; her flight to space on STS-131 marked the first time four women were together in space. Now it was a footnote, not a headline, that one of *Atlantis*'s crewmembers was a woman—spaceflight veteran Sandra Magnus.

Even though it was no longer news to see a minority or woman astronaut, parity remained a goal rather than a reality. The overwhelming majority of the 355 shuttle astronauts were white men. Nevertheless, the program flew forty-nine women, fourteen African Americans, six Asian Americans, and nine Hispanic Americans.[42] Fred said of the process, "Sometimes it's slow coming, sometimes it's a revolution, sometimes it's an evolution."[43] For all its sunspots, the story of the shuttle astronauts remains a triumphant chapter in America's troubled history on race and gender. The New Guys forever changed the face of the astronaut—the quintessential American hero.

Shannon Lucid, sixty-eight, did not join her friends in the stands at Kennedy because she was working as CapCom at Mission Control in Houston. Almost a decade earlier, she had been called to DC as NASA chief scientist but wheedled her way back to Houston within a year. "Who wanted to work at headquarters? You want to work where the action is happening, not in offices," she said. "I liked doing real things."[44] Since 2003, Shannon had been overnight CapCom for fifteen shuttle missions. With five space flights, 223 days in space, and multiple records to her name—many of them logged in her forties and fifties—Shannon's storied career was a beacon of light for many women who came after her.[45]

Though many New Guys traveled great distances to make the reunion, some decided not to attend. Guy Bluford watched the launch on television from his home in Cleveland. After flying two more shuttle missions—STS-39 and STS-53, both of which deployed classified DoD payloads—Guy departed from NASA in 1993. He moved on with his life. "When I left," he said, without emotion, "I left."[46] After holding leadership and board positions at various aerospace companies in the 1990s and 2000s, he founded a consulting firm for public and private aerospace technology developers.[47]

George Abbey did not attend the launch as a protest to the shuttle's termination. His stated reason masked a more personal motive. George simply could not stand to watch the end of the program to which he had given all of himself—his best years, his passion, his every waking moment.[48] After retiring from NASA in 2003, George returned to Houston joining Rice University as a senior fellow in space policy.[49] He became a proselytizer for the US space program, appearing in the public sphere like never before. He spoke at space conferences, corporate events, and universities around the world, from the UK to Russia to China.[50] Through it all, he remained a regular at Frenchie's.

George had come a long way from the talented test pilot who the inflexible Chuck Yeager denied a ticket to space. Rather than succumbing to bitterness, George spent a lifetime correcting that injustice by bringing people of all backgrounds into the fold—"the whole family of humankind," as Nichelle Nichols once said. A maxim of his distilled worldview: *Space doesn't just belong to the warriors; it belongs to all of us.*[51] George was not universally beloved—far from it—but even his critics conceded that nobody was as passionate about the shuttle program, or as successful at bending the agency to his will.[52] Though George Abbey is not a household name, like Neil Armstrong or Sally Ride, his imprint on space travel is everywhere.

A thick fog hung listlessly in the air. The odds of launching *Atlantis* were dismal: 30 percent by NASA's predictions. Heavy rain showers and thunderstorms had inundated the Cape during preflight preparations. "We all thought the mission would be canceled or delayed," Fred said.[53] *If there wasn't a delay, would it really be a shuttle launch?* Then a heaven-sent gap in the weather allowed *Atlantis* to make its move skyward.

"Yippee, woo!" cried Anna with the same enthusiasm she brought to every launch.[54] As she watched *Atlantis* rise into the sky, her eyes glinted with nostalgia and heartache. This chapter of space travel was coming to an end.[55]

STS-135's mission was to haul *Raffaello*, an Italian module, to the ISS. Mission specialist Magnus would transfer almost ten thousand pounds of food, supplies, and equipment and dispose of more than five thousand pounds of trash and outdated equipment.[56] The cargo brought to the ISS included the first smartphones in space—an iPhone and two Androids—with apps like "Spacelab for iOS" intended to help astronauts track experiments.[57]

Atlantis docked with the ISS and Magnus floated through the station's hatch. Back in 1997, Anna had chosen Magnus as a "Russian Crusader," her own version of George's Cape Crusaders.[58] Magnus learned Russian procedures, supported their product testing and development, and "made friends" as Anna put it.[59] Magnus had already flown to space twice, including a 133-day stay aboard the ISS in 2008. "I felt like I was coming home," she said.[60] She found notes in her handwriting around the station and dug up a pair of fuzzy pink-blue-and-white socks that she had forgotten on her previous mission a few years back.

During her record-setting stay aboard *Mir* in 1996, Shannon, too, had worn a pair of pink socks every Sunday to help mark the passage of time.

"Here it is, another Sunday on *Mir*!!!" Shannon emailed mission controllers. "And how, you might ask, do I know that it's Sunday? Easy!!! I have on my pink socks and Yuri, Yury, and I have just finished sharing a bag of Jell-O!!!"[61] Now, fifteen years later, Magnus floated over to a camera and wiggled her toes in her own pink socks so Shannon could see. "Just a little nostalgia there for a moment," Magnus purred over the radio.[62]

On day four of the mission, Shannon dedicated the wake-up song to Magnus: "Tubthumping" by the British anarcho-communist punk group Chumbawamba.

He sings the songs that remind him of the good times
He sings the songs that remind him of the better times
I get knocked down, but I get up again
You are never gonna keep me down

The song had resonance for the shuttle program as a whole: Disheartening as the shuttle's end was, Shannon trusted that America's yen for exploration would bring the country back to space.[63]

When the *Atlantis* crew prepared to return home, the astronauts held a small and somber departure ceremony in the ISS's *Harmony* module. Flanked by the crews of STS-135 and ISS Expedition 28, shuttle commander Chris Ferguson presented the station with an American flag that Bob Crippen and John Young had flown aboard STS-1.

"This flag represents not just a symbol of our national pride and honor, but in this particular case, it represents a goal," said Ferguson. "This flag . . . [is] to be returned to Earth once again by an astronaut that launches on a US vehicle."[64]

"I understand it'll be kind of like a capture the flag moment for commercial spaceflight," President Obama joked.[65] It would be nine years before STS-135 pilot Doug Hurley would return in a SpaceX Crew Dragon capsule to make good on that promise.[66]

On the final morning of the mission, Shannon woke the STS-135 crew with a more traditional number, Kate Smith's 1938 rendition of Irving Berlin's "God Bless America." She and the other mission controllers solemnly placed their hands over their hearts. "It's for all the men and women who have put their heart and soul into the shuttle program for all these years," Shannon trilled proudly at the song's conclusion.[67]

The station now complete, *Atlantis* detached from the ISS. "When a generation accomplishes a great thing, it's got a right to stand back and for just a moment admire and take pride in its work," Ferguson said to ground controllers, marveling at the space station through his window. "As the ISS now enters the era of utilization, we'll never forget the role the space shuttle played in its creation. Like a proud parent, we anticipate great things to follow from the men and women who build, operate, and live there . . . Make us proud."[68]

Orbiting in the sky above, the ISS lit a way forward, toward humanity's collective future in space. A mother ship of scientific discovery, the ISS has enabled in-space research that has broken new ground in medicine, yielding improvements in vaccine development, breast cancer detection, ultrasound technology, and inventing new drugs only manufacturable in microgravity.[69] Scientists are paving the way for deep space exploration by studying astronauts' reactions to long-duration stays, monitoring muscle mass and cardiovascular stress. As humans imagine living off the planet, scientists are learning how to grow our food in space.

Physics research on the ISS has led to the discovery of "cool flames," fire that, in microgravity, burns two and a half times cooler than the flame of a typical candle and releases an entirely different set of chemical products. The station has collected data on over a hundred billion cosmic particles, unavailable on Earth, and allowed researchers to better understand dark matter and the origins of our universe. In 2018, astronauts on the ISS produced a fifth state of matter, a Bose-Einstein condensate, in which a substance is cooled so close to absolute zero, that its atoms lose their free energy and begin to act as a single atom. The event provided researchers insight into the fundamental laws of quantum mechanics.

As humanity looks for ways to preserve our own planet, the ISS helps forecast and track natural disasters, monitors Earth's natural resources from space, and produces ground-breaking research on smart agriculture, water purification, and climate science.[70] Every night, since November 2, 2000, the ISS passes overhead, a tiny light in the sky, visible to the naked eye, and a constant reminder that the human spirit of exploration is still alive.

The ISS astronauts watched as *Atlantis* departed and then reentered Earth's atmosphere, blazing a plasma path over the Yucatan.[71] Even though he protested the launch, George showed up at Johnson Space Center for landing. No longer on NASA's payroll, he was simply an invited guest. Forty-two years and a day from when he had sat in Mission Control watching Neil Armstrong take his first stride on the moon, he now watched the end of the program that had defined him.

Atlantis, its exterior cast in amber by the rising sun, touched the tarmac in Florida right before six o'clock that morning. "Having fired the imagination of a generation, a ship like no other, its place in history secured," said NASA mission commentator Rob Navais, as *Atlantis* pulled into port.[72] The STS-135 crew looked at each other. No one wanted to move. This was the last time anyone would crew a shuttle. "Mission complete, Houston," Ferguson radioed. "After serving the world for over thirty years, the space shuttle found its place in history, and it's come to a final stop."[73]

NASA now decommissioned its last orbiter *Atlantis*. The retired shuttle was moved to the Kennedy Space Center Visitor Complex.[74] The two other remaining orbiters, along with *Enterprise*, flew to museums across America on the back of the Shuttle Carrier Aircraft.[75] Around the country, thousands of onlookers came out to watch the shuttles make their journeys to their final resting places. Wall Street types pushed into corner offices for a prime view of *Enterprise* as it flew over the Hudson River heading to the Intrepid Sea, Air and Space Museum in New York City. Tech analysts and software engineers set alarms to avoid missing *Endeavour*'s fly-by of San Francisco Bay before landing at the Los Angeles airport.[76] Throngs of locals flooded out of their apartments and craned their necks to watch the orbiter squeak down Crenshaw Boulevard to the California Science Museum.[77]

School children flooded the National Mall to watch *Discovery* soar over the White House, the Washington Monument, and the Capitol building. Having once ferried Judy Resnik to space and twice returned NASA to flight after the *Challenger* and *Columbia* disasters, *Discovery* now headed for the National Air and Space Museum to take its place among the artifacts of human space flight: the Wright brothers' first powered plane; Alan Shepard's *Freedom 7*; and the Apollo 11 lunar landing module.[78] For the millions who visit the museums, these monuments from the past enliven the future and animate the next generation of scientists, engineers, and explorers.[79]

Anna and Shannon were the last New Guys to work for NASA.

Shannon retired shortly after *Atlantis*'s final flight to care for her husband, who was struggling with dementia. It reminded her of her first job working at a nursing home. "Look, the full circle of life, I started out working changing diapers, I'm ending my career changing diapers," she told her children. Shannon's remark had that mix of humor and pathos for which she was known and loved.[80] When asked why America should continue the space program, Shannon said: "Human beings learn things by going out there and doing things. That's how we thrive, otherwise we stagnate."[81]

After wheelstop, Anna set her sights on the moon and Mars, helping train astronauts for new spacecraft. The design of the display system for the *Orion* capsule, a four-person module intended for deep space exploration, owed much to her vision. NASA will launch *Orion* on the newly designed Space Launch System, which boasts 13 percent more thrust and can carry at least five more tons of cargo than the shuttle.[82]

In the year following the final shuttle mission, Sally and Tam came face-to-face with the challenges of end-of-life care. For the first months of her battle with cancer, Sally and those around her believed she would be one of the few to survive. As her condition deteriorated, she and Tam needed to plan for the inevitable. The same summer that STS-135 flew, Sally and Tam quietly registered as domestic partners, a practical measure that helped them deal with the endless forms and questions from healthcare professionals.

"She's my partner," Sally would snap at nurses when they questioned

Tam.[83] For the first time, Sally had openly acknowledged that she and Tam were a couple.[84] But Tam wanted more.

"Who am I going to be to the world?" she asked Sally.[85]

"Being open about us might be very hard on NASA and the astronaut corps. But I'm okay with that," Sally finally said. Facing death, Sally was finally liberated from her secret. "Whatever you think is right is fine with me," Sally reassured Tam.[86]

On the night of July 21, 2012, exactly one year after *Atlantis* landed on Earth one last time, Sally fell into a coma.[87] She passed away two days later and, after her death, the final line of her obituary stated plainly that Sally and Tam had been partners for twenty-seven years.

"I wish I had another twenty-seven years with you," Sally had told Tam in the last weeks of her life.[88] Sally's obituary shocked the world, coming at a time when no astronaut from any nation was openly gay. Sally's posthumous admission sparked much debate about whether she should have come out sooner, or at all.

One queer journalist called Sally a "gay trailblazer," but for other commenters, Sally became "the absent heroine" whose life in the closet was a "missed opportunity."[89] Some wrote letters complaining that the story was being covered in papers at all, wishing for a world where sexuality was not newsworthy.[90] "Sally was a national hero and a powerful role model," President Obama said. "She inspired generations of young girls to reach for the stars."[91] As her sister, Bear, noted, Sally "hated labels of every kind—including 'hero.'"[92]

At Johnson Space Center, where Saturn Lane intersects NASA Road 1, there is a grove of oak trees. When President Ronald Reagan came to speak after the *Challenger* disaster, his advance team chopped down the oaks to clear the view behind his podium. A decade later, new trees were planted, each one memorializing a deceased astronaut, director, or NASA employee.[93] Plaques bearing the names of McAuliffe, Scobee, Resnik, Smith, Onizuka, Jarvis, and McNair rest at the center of the Astronaut

Memorial Grove. Surrounding those are many more with familiar names: Armstrong, Grissom, Low. For Steve Hawley, the hallowed grove is a reminder of life and renewal. "This is our personal memorial," he said. "It's ongoing, and it lives."[94]

On September 18, 2012, a group of New Guys including Anna, Rhea, and Shannon gathered to plant an oak tree bearing the name of Sally Ride. Lorna Onizuka, Bob Crippen, and George Abbey left roses at the tree's base. High above Earth, aboard the ISS, astronaut Sunita Williams delivered a few words of remembrance. "I don't think I'd be here without her having done the things she did," Williams reflected.[95]

Now in his nineties, George Abbey still visits the grove.[96] Many of the plaques bear names that first crossed George's desk decades ago. It was he who brought them into the NASA family. Though most of the trees in the main cluster are dedicated to astronauts, a few lining the path to the grove bear the names of former center directors: Gilruth, Cohen, Estess. A fourth tree stands reserved for a center director but has yet to be dedicated.[97]

In the aftermath of the *Challenger* accident, June Scobee marshaled the crew's families together to create a lasting memorial to their loved ones. Together they built the Challenger Center—a global network of learning hubs where space-themed simulations help millions of students bring their classroom studies to life and cultivate the skills needed for STEM careers.[98]

Lake City, South Carolina, rededicated the library where a clerk once called the police on young Ron McNair: Children of all races can now visit the Ronald E. McNair Library for story time, where volunteers rap Dr. Seuss's *Green Eggs and Ham* and encourage the next generation of inquisitive minds.

After Ellison's death, Lorna Onizuka worked for NASA as a Japanese American liaison for the International Space Station. She lives in Houston and remains close with George Abbey. "We were so fortunate to have worked, lived, and celebrated NASA's phenomenal Shuttle era," Lorna said, reflecting on her and El's time at NASA. "We worked hard physically and with heart. We celebrated the highs and glories of success. We huddled together as one large family to pull ourselves back up from tragic losses. With everlasting friendships and indelible pride, we now get to witness a new era of commercial space flight."[99]

Guy lives in Philadelphia with his wife, Linda. They have two grown children. In addition to leading his aerospace technology consulting firm, Guy still finds time for active hobbies like handball, racquetball, and golf.[100] He occasionally gets to suit up for some scuba diving. Gliding through currents in his neutral buoyancy suit, Guy can close his eyes and remember what it felt like to float in zero gravity.

Kathy's intrepid spirit endures. In 2020, at sixty-eight years old, Kathy became the first woman to reach the deepest known spot in the ocean, called Challenger Deep.[101] She wrote a poignant memoir chronicling her experience with the Hubble Space Telescope. "I think I will be exploring until they put me in a little wooden box," Kathy pronounced.[102]

Rhea and Hoot live in Nashville and have three grandchildren. Rhea recently ended operations at her healthcare coaching company and wrote a memoir about her time as an astronaut. She credits the shuttle program with inspiring a generation of students in the sciences. "If we want the kids now to go into STEM programs, we need to have exciting things for them to do," she expressed.[103]

Anna, also a grandmother, retired from NASA in 2017. Her thirty-six-year history at the agency made her the longest-serving astronaut on record.[104] She remains close with the current generation of astronauts, including Victor Glover, who became the fifteenth African American in space on the second crewed SpaceX flight.[105] She appears at speaking engagements and travels with her boyfriend. Her ex-husband, Bill, still lives in their old house, but Anna has moved on.

Sally Ride's shuttle checklists, T-38 helmet, and crew jacket all became part of the space shuttle's collection at the Smithsonian National Air and Space Museum—the same museum where a plucky young Judy Resnik once barged into Michael Collins's office with nothing more than a hand-drawn map and a dream to be an astronaut.[106] Sally blushed when her friend Susan had suggested there would one day be an exhibit on her here.[107] Although Sally struggled with the attention she received in 1983, Susan's prophesy came to pass. From Sally's childhood telescope to Judy Resnik's name tag and Guy Bluford's flight suit, the New Guys now have an enduring place in the museum, their history a part of humanity's journey to the stars. *Ad astra.*

Fred Gregory lives in Annapolis. He keeps himself plenty busy in retirement. When he is not water skiing, hunting, or fishing, Fred collects stereo equipment and drives his vintage specialty cars, all yellow, to automobile shows. When asked whether he would go to space again, he supplied a terse no. The missions are different now: long stays aboard the space station and other incidental challenges like learning Russian—sacrifices Fred does not want to make. "That's not what I signed up for," he explained. "I signed up to fly this thing that had wings."[108] *Besides, there's a new generation now.* The new astronauts are an increasingly diverse and exceedingly capable group of pilots, scientists, and engineers, eager to shine a light into the vast expanse of our universe and sew up the pockets of the unknown. As Fred put it, "God bless the new guys."[109]

Acknowledgments

To the astronauts and NASA employees who so generously shared their personal recollections with me: Without you, this book simply could not exist.

Among them, a special thanks to the New Guys Fred Gregory and Anna Fisher, who were the first to speak with me, spending dozens of hours, some over skinny margaritas on the oceanside patio at the Marina del Rey Cheesecake Factory, but many (many!) over video conference. To Guy Bluford, Mike Coats, John Fabian, Rick Hauck, Steve Hawley, Shannon Lucid, George "Pinky" Nelson, Rhea Seddon, Brewster Shaw, and Kathy Sullivan, all of whom spent countless hours speaking with me as well.

To the NASA brass who shepherded this class of astronauts through their first flights and the media frenzy that followed: George Abbey, Jay Honeycutt, Carolyn Huntoon, and Rick Nygren; thank you for generously lending your time to this project. A special thanks to Mr. Abbey, who took me to some of his—and the New Guys'—favorite Clear Lake haunts, including at least two interviews over rigatoni at Frenchie's, an illustrious Italian restaurant run by the Camera brothers.

To the astronauts from the classes before and after the New Guys who also aided: Jim Bagian, Charlie Bolden, Mary Cleave, Bob Crippen, Bill Fisher, Mike Fossum, and Jerry Ross. Thank you for helping fill in gaps and rounding out scenes. To Frankie Camera, Estella Gillette, Karen Ross, and Sylvia Salinas Stottlemyer, who all contributed to the shuttle program: Thank you for offering a generous picture of your Astronaut Class 8 friends.

To Sylvester James Gates Jr., Charlie Justiz, and Susan Okie, for sharing their recollections of friends who have passed on. A special thank you to Carl McNair, Michael Oldak, and June Scobee, who were gracious with their time and memories of loved ones who were lost in the *Challenger* accident. To Sally Ride, whose life has always served as an inspiration.

To the NASA history department, I learned so much from the in-depth interviews conducted for the Johnson Space Center's Oral History Project, and especially to Jennifer Ross-Nazzal and Sandra Johnson, for your careful guidance and feedback—it was invaluable.

This book was born out of a collective effort, beginning with my cousins Bob Adams and Gary Ouzts (who worked as a NASA contractor), with whom I had a conversation that sparked the idea for a book about the New Guys; Tom Sarko (my fifth-grade science teacher, who competed in the Teacher in Space Project and first introduced me to NASA); and to space writer Rick Houston, who generously arranged for my first trip to NASA's Johnson Space Center.

To my HarperCollins editor, Mauro DiPreta, who believed in me as a writer and supported the vision of this book from the nascent idea through its final evolution. Your depth of intelligence, crucial feedback, and steady guidance helped shape the text and made me a better writer.

I would be remiss not to mention my agent, Anthony Mattero, of CAA, for championing this project from the get-go. To Chris Farrah, for his editorial input and generous spirit.

To my business partners, Kyra Sedgwick and Valerie Stadler, for giving me the space and time to disappear into this project; I am forever grateful.

To Giselle Cheung, Danielle DiSpaltro, Kirsten Johnson, Tom Waddick, and Erin Williams, for their brilliant research, editorial wizardry, and endless curiosity, and for always going above and beyond. To Alexander Czarnecki and Sam Richman, for their excellent research and thoughtful contributions.

To my mother, Martha, whose passion for literature and gift for storytelling inspired my love for a tale well told.

To Ralph Greco, my first English teacher, who stood ready with his red pen to teach me firsthand that old adage, "writing is rewriting."

To Yvonne Cheng, my first and final reader, whose intelligence and wisdom I rely upon daily—thank you.

Appendix

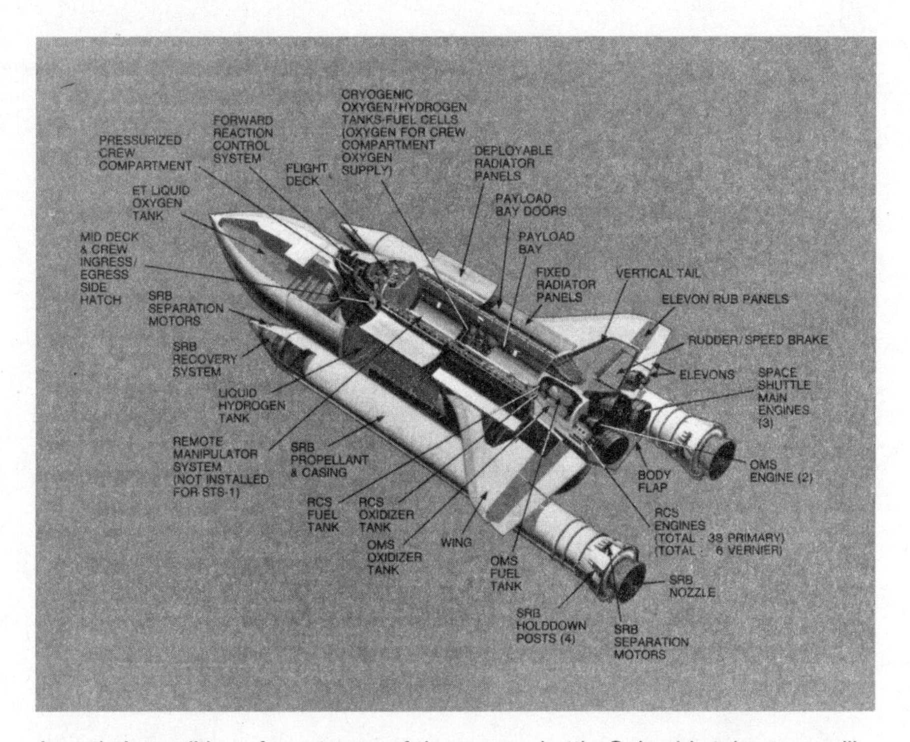

An artist's rendition of a cutaway of the space shuttle *Columbia* takes some liberties to reveal major components of the vehicle. Note the external liquid oxygen tank, top left, and twin solid rocket boosters (SRBs) flanking the vehicle.

Notes

Chapter 1: Ad Astra

1. Mark Jones, "NASA Picks Two Southland Women," *Los Angeles Times*, January 17, 1978, https://www.newspapers.com/image/384114242/.
2. "Astronaut Bio: Judith A. Resnik," NASA, https://er.jsc.nasa.gov/seh/resnik.htm.
3. Scott Spencer and Chris Spolar, "The Epic Flight," *Esquire*, December 1, 1986, https://classic.esquire.com/article/1986/12/1/the-epic-flight-of-judith-resnik.
4. Jones, "NASA Picks Two Southland Women."
5. Jones, "NASA Picks Two Southland Women."
6. "Architectural History of the National Air and Space Museum, 1972," Smithsonian Institution Archives, https://siarchives.si.edu/collections/siris_sic_14442.
7. Pamela Scott, *Buildings of the District of Columbia*. New York: Oxford University Press, 1993, p. 91.
8. The artist Richard Lippold created this abstract sculpture, which was installed in front of the museum in 1976. "Installation of *Ad Astra* sculpture," Smithsonian Institution Archives, https://siarchives.si.edu/collections/siris_arc_391956. Also, "Ad Astra, an Abstract Sculpture by Richard Lippold," *What Is Abstract Sculpture?* blog, March 1, 2011, https://abstractsculpture.wordpress.com/2011/03/01/ad-astra-an-abstract-sculpture-by-richard-lippold.
9. Peter Jakab, "Leonardo da Vinci and Flight," Smithsonian National Air and Space Museum, August 22, 2013, https://airandspace.si.edu/stories/editorial/leonardo-da-vinci-and-flight.
10. Carl McNair, *In the Spirit of Ronald E. McNair, Astronaut: An American Hero*. Atlanta, GA: MAP Publishing Associates, 2005, pp. 107–08. Also, author interview with Carl McNair, April 1, 2022.
11. McNair, *In the Spirit,* 108.
12. "The First Laser," HRL Laboratories, https://www.hrl.com/about/laser.
13. McNair, *In the Spirit,* 107, 110.
14. McNair, *In the Spirit,* 15.
15. Dudley Clendinen, "Two Pathes to the Stars: Turnings and Triumphs; Ronald McNair," *New York Times,* February 9, 1986, https://www.nytimes

.com/1986/02/09/us/two-pathes-to-the-stars-turnings-and-triumphs-ronald -mcnair.html.

16. McNair, *In the Spirit,* 15–16.
17. McNair, *In the Spirit,* 8.
18. McNair, *In the Spirit,* 8–10.
19. McNair, *In the Spirit,* 10.
20. McNair, *In the Spirit,* 10.
21. Clendinen, "Two Pathes to the Stars."
22. "Tobacco Harvesting," George Washington's Mount Vernon, December 29, 2015, https://www.youtube.com/watch?v=C6rq4X_zWzQ.
23. McNair, *In the Spirit,* 11.
24. McNair, *In the Spirit,* 12.
25. Clendinen, "Two Pathes to the Stars."
26. McNair, *In the Spirit,* 13.
27. Clendinen, "Two Pathes to the Stars."
28. Author interview with Carl McNair, April 1, 2022.
29. McNair, *In the Spirit,* 29.
30. Clendinen, "Two Pathes to the Stars."
31. McNair, *In the Spirit,* 29.
32. Clendinen, "Two Pathes to the Stars."
33. McNair, *In the Spirit,* 28.
34. Author interview with Carl McNair, April 1, 2022.
35. McNair, *In the Spirit,* 57.
36. McNair, *In the Spirit,* 57.
37. McNair, *In the Spirit,* 59.
38. Michael Graff, "Star Trek: 15 Things You Didn't Know About Uhura," Screenrant, September 13, 2017, https://screenrant.com/star-trek-uhura-facts -trivia.
39. *Black in Space: Breaking the Color Barrier,* directed by Laurens Grant, performance by David Harewood, Smithsonian Channel, 2020, https://www .youtube.com/watch?v=I7jJ8jEh608.
40. McNair, *In the Spirit,* 106.
41. McNair, *In the Spirit,* 106.
42. Author interview with Anna Fisher, December 9, 2020.
43. Billie Jean King, "Publisher's Letter," *womenSports,* February 1978, vol. 4, issue 2, p. 4, https://archive.org/details/sim_womensports_1978-02_4_2 /page/2/mode/2up?q=molly+tyson.
44. Lynn Sherr, *Sally Ride: America's First Woman in Space.* New York: Simon & Schuster, 2014, p. 84.
45. Sharon Rosenthal, "Taking Off!" *Daily News,* August 8, 1982, p. 3, https://www .newspapers.com/image/490828510.
46. Sherr, *Sally Ride,* 13.
47. Rosenthal, "Taking Off!"
48. Sherr, *Sally Ride,* 15.
49. Author interview with Susan Okie, April 8, 2022.
50. Rosemarie Wittman Lamb, "Astronaut Sally Ride: America's First Woman in Space," *El Paso Times,* May 29, 1983, p. 4, https://www.newspapers.com /image/435803670/.

51. Author interview with Susan Okie, April 8, 2022.
52. Author interview with Susan Okie, April 8, 2022.
53. Susan Okie, "America's First 'A Remarkable Woman,' Friend Says," *Florida Today,* May 8, 1983, p. 8A, https://www.newspapers.com/image/125264104/.
54. Sherr, *Sally Ride,* 52.
55. Sherr, *Sally Ride,* 51.
56. Sherr, *Sally Ride,* 58.
57. Sherr, *Sally Ride,* 84.
58. David Shayler and Colin Burgess, *NASA's First Space Shuttle Astronaut Selection: Redefining the Right Stuff.* Germany: Springer, 2020. Also, author interview with Jay Honeycutt, December 16, 2020, author interview with Anna Fisher, November 12, 2020.

Chapter 2: Light This Candle

1. Henry C. Dethloff, "Chapter 3: Houston—Texas—USA," *Suddenly Tomorrow Came . . . A History of the Johnson Space Center.* Houston, TX: NASA History Series, 1993, https://historycollection.jsc.nasa.gov/JSCHistoryPortal/history/suddenly_tomorrow/chapters/Chpt3.pdf.
2. John E. Riley, "NASA to Interview Astronaut Applicants," Lyndon B. Johnson Space Center, Press Release No. 77-42, July 29, 1977, https://www.nasa.gov/centers/johnson/pdf/83129main_1977.pdf.
3. Milton E. Reim, "Tenth Group of Astronaut Applicants Report to JSC November 14," Lyndon B. Johnson Space Center, Press Release No. 77-75, November 11, 1977, https://www.nasa.gov/centers/johnson/pdf/83129main_1977.pdf.
4. Mike Mullane, *Riding Rockets: The Outrageous Tales of a Space Shuttle Astronaut.* New York: Scribner, 2006, p. 2 (Kindle ed.).
5. Mullane, *Riding Rockets,* 23.
6. Mullane, *Riding Rockets,* 29–30.
7. Shayler and Burgess, *NASA's First Space Shuttle Astronaut Selection,* 75.
8. Ben Evans, *Space Shuttle Columbia: Her Missions and Crews.* Germany: Springer New York, 2007, p. 5.
9. Rhea Seddon, *Go for Orbit: One of America's First Women Astronauts Finds Her Space,* United States: Your Space Press, 2015, p. 27.
10. Carmen Bredeson, *Shannon Lucid: Space Ambassador.* Minneapolis, MN: Lerner Publishing Group, 2000, p. 18.
11. "NASA Picks Six Women Astronauts with the Message: You're Going a Long Way, Baby," *People,* February 6, 1978, https://web.archive.org/web/20160426212619/http://www.people.com/people/article/0,20070132,00.html.
12. Linda Gillan, "A Space-Age Equation: First Female Astronauts = X," *Los Angeles Times,* August 30, 1977, https://www.newspapers.com/image/383728479/.
13. Seddon, *Go for Orbit,* 31.
14. Susan Okie, "NASA Appeal Gave a Physicist Wings," *Washington Post,* May 9, 1983, https://www.washingtonpost.com/archive/politics/1983/05/09/asa-appeal-gave-a-physicist-wings/565c3c99-0cac-4356-b8bf-fa11c7fae288.
15. Sherr, *Sally Ride,* 86.
16. Mullane, *Riding Rockets,* 2.

17. Shayler and Burgess, *NASA's First Space Shuttle Astronaut Selection,* location 900. Also, Margaret Rhea Seddon transcript, NASA Johnson Space Center Oral History Project, May 20, 2010, https://historycollection.jsc.nasa.gov/JSCHistoryPortal/history/oral_histories/SeddonMR/SeddonMR_5-20-10.htm.
18. Sherr, *Sally Ride,* 86, 114.
19. Author interview with Anna Fisher, November 12, 2020.
20. Shayler and Burgess, *NASA's First Space Shuttle Astronaut Selection,* 30.
21. Author interview with Jim Bagian, April 2, 2021.
22. Melanie Wallace, "Video Interview: Remembering Sally Ride," interview with *Nova, PBS,* 1984, https://www.pbs.org/wgbh/nova/insidenova/2012/07/remembering-sally-ride.html.
23. Author interview with Jim Bagian, 2021. Also, Sherr, *Sally Ride,* 91.
24. Author interview with Jay Honeycutt, December 14, 2020.
25. Seddon, *Go for Orbit,* 32–33.
26. Author interview with Steve Hawley, 2018.
27. Author interview with Bob Crippen, July 30, 2018.
28. Carolyn L. Huntoon transcript, NASA Johnson Space Center Oral History Project, June 5, 2002, https://historycollection.jsc.nasa.gov/JSCHistoryPortal/history/oral_histories/HuntoonCL/HuntoonCL_6-5-02.htm.
29. Lloyd Leigh, "Our Men in the Spacecraft Center," *Chicago Daily Defender,* December 7, 1965.
30. Michael Cassutt, *The Astronaut Maker: How One Mysterious Engineer Ran Spaceflight for a Generation.* Chicago, IL: Chicago Review Press, 2018, p. 181 (Kindle ed.).
31. Author interview with Anna Fisher, June 22, 2020.
32. Anna L. Fisher transcript, NASA Johnson Space Center Oral History Project, February 17, 2009, https://historycollection.jsc.nasa.gov/JSCHistoryPortal/history/oral_histories/FisherAL/FisherAL_2-17-09.htm.
33. Thomas Wolfe, *Look Homeward, Angel.* United Kingdom: Prabhat Prakashan, 1929.
34. Jones, "NASA Picks Two Southland Women."

Chapter 3: Ten Interesting People

1. Shayler and Burgess, *NASA's First Space Shuttle Astronaut Selection,* 46.
2. Karlyn Bowman, "The Decline of the Major Networks," *Forbes,* July 27, 2009, https://www.forbes.com/2009/07/25/media-network-news-audience-opinions-columnists-walter-cronkite.html?sh=516228e847a5.
3. "NASA News Conf to Announce Names of New Astronauts Including Some Women," *CBS Evening News,* January 16, 1978, https://commerce.veritone.com/search/asset/43705534.
4. "NASA Astronaut Group 8," Wikipedia, https://en.wikipedia.org/wiki/NASA_Astronaut_Group_8.
5. "NASA Bias Against Blacks and Women," *Sacramento Observer,* November 14, 1973.
6. "NASA Bias Against Blacks and Women," *Sacramento Observer.*
7. Jane Van Nimmen, Leonard C. Bruno, and Robert L. Rosholt, "NASA Historical Data Book, 1958–1968 Vol. I: NASA Resources," NASA Scientific and Technical Information Office, 1976, https://history.nasa.gov/SP-4012v1.pdf.

8. Melanie Whiting, "60 Years Ago, Eisenhower Proposes NASA to Congress," NASA, April 2, 2018, https://www.nasa.gov/feature/60-years-ago-eisenhower-proposes-nasa-to-congress.
9. Joseph D. Atkinson and Jay M. Shafritz, *The Real Stuff: A History of NASA's Astronaut Recruitment Program*. New York: Praeger, 1985, p. 37.
10. Atkinson and Shafritz, *The Real Stuff,* 37.
11. Tom Wolfe, *The Right Stuff*. New York: Farrar, Straus and Giroux, 1979, p.17.
12. Kristy N. Kamarck, "Women in Combat: Issues for Congress," Congressional Research Service, December 13, 2016, https://fas.org/sgp/crs/natsec/R42075.pdf, p. 2.
13. Kamarck, "Women in Combat: Issues for Congress."
14. Hal Sider and Cheryl Cole, "The Changing Composition of the Military and the Effect on Labor Force Data," US Bureau of Labor Statistics, July 1984, https://www.bls.gov/opub/mlr/1984/07/art2full.pdf, p. 13.
15. Matthew H. Hersch, *Inventing the American Astronaut*. New York: Palgrave Macmillan, 2012, pp. 29–34.
16. Neil Armstrong had previously been a naval aviator.
17. "Civilian Pilots Are Eligible Says Webb," *NASA Roundup*, April 4, 1962, vol. 1, no. 12, https://historycollection.jsc.nasa.gov/JSCHistoryPortal/history/roundups/issues/62-04-04.pdf, p. 1.
18. John F. Kennedy first announced his call to action to land a man on the moon in May 1961 to a joint session of Congress. "Address to Joint Session of Congress May 25, 1961," John F. Kennedy Presidential Library and Museum, https://www.jfklibrary.org/learn/about-jfk/historic-speeches/address-to-joint-session-of-congress-may-25-1961.
19. "John F. Kennedy Moon Speech—Rice Stadium," NASA, September 12, 1962, https://er.jsc.nasa.gov/seh/ricetalk.htm.
20. Planetary Society, "How Much Did the Apollo Program Cost?" https://www.planetary.org/space-policy/cost-of-apollo.
21. Richard Hollingham, "Apollo in 50 Numbers: The Workers," BBC, June 19, 2019, https://www.bbc.com/future/article/20190617-apollo-in-50-numbers-the-workers.
22. "NASA Langley Research Center's Contributions to the Apollo Program," NASA, https://www.nasa.gov/centers/langley/news/factsheets/Apollo.html.
23. Emily Ludolph, "Ed Dwight Was Set to Be the First Black Astronaut," *New York Times,* July 16, 2019, https://www.nytimes.com/2019/07/16/us/ed-dwight-was-set-to-be-the-first-black-astronaut-heres-why-that-never-happened.html.
24. *Black in Space: Breaking the Color Barrier*, Smithsonian.
25. Steven L. Moss, "NASA and Racial Equality in the South, 1961–1968," master's thesis, December 1997, Texas Tech University, pp. 52–53.
26. "Ed Dwight Describes His Selection to Train as the First Black Astronaut, Pt. 1," interviewed by Larry Crowe, HistoryMakers Digital Archive. June 19, 2002, session 1, tape 4, story 6.
27. Atkinson and Shafritz, *The Real Stuff,* 101.
28. J. Alfred Phelps, *They Had a Dream: The Story of African Astronauts*. Novato, CA: Presidio Press, 1994, p. 7.
29. Ludolph, "Ed Dwight Was Set to Be the First Black Astronaut."
30. Yeager's comment was recalled by Ed Dwight in *Black in Space: Breaking the*

Color Barrier, 10:00–11:07. Yeager denied racism against Dwight. While those allegations linger, no evidence one way or another has surfaced. Dwight has not changed his recollections. See also, Betty Kaplan Gubert, Miriam Sawyer, and Caroline M. Fannin, *Distinguished African Americans in Aviation and Space Science*. United Kingdom: Oryx Press, 2002, p. 115.

31. "I Was Poised to Be the First Black Astronaut. I Never Made It to Space." "Almost Famous" by Op-Docs, *New York Times*, https://www.youtube.com/watch?v=Xj1sJQW98nE.
32. Phelps, *They Had a Dream*, 10–11.
33. Ludolph, "Ed Dwight Was Set to Be the First Black Astronaut."
34. Charles L. Sanders, "The Troubles of 'Astronaut' Edward Dwight," *Ebony*, June 1965.
35. "I Was Poised to Be the First Black Astronaut. I Never Made It to Space," *New York Times*.
36. *Black in Space: Breaking the Color Barrier*, Smithsonian, 8:46–12:26.
37. Shareef Jackson, "Ed Dwight Was Going to Be the First African American in Space. Until He Wasn't," *Smithsonian Magazine*, February 18, 2020, https://www.smithsonianmag.com/history/ed-dwight-first-african-american-space-until-wasnt-180974215/.
38. Mark Garcia, "Robert Lawrence: First African-American Astronaut," NASA, February 21, 2018, https://www.nasa.gov/feature/robert-lawrence-first-african-american-astronaut.
39. *Black in Space: Breaking the Color Barrier*, Smithsonian.
40. Katherine Q. Seelye, "Geraldyn M. Cobb, 88, Who Found a Glass Ceiling in Space, Dies," *New York Times*, April 19, 2019, https://www.nytimes.com/2019/04/19/obituaries/geraldyn-m-cobb-dead.html. Also, Kelli Gant, "Women Involved in Aviation," https://www.ninety-nines.org/women-in-aviation-article.htm.
41. "Women Who Reach for the Stars," NASA, March 22, 2005, https://www.nasa.gov/missions/highlights/f_mercury13.html.
42. Swapna Krishna, "The Mercury 13: The Women Who Could Have Been NASA's First Female Astronauts," Space.com, July 24, 2020, https://www.space.com/mercury-13.html.
43. "Women Who Reach for the Stars," NASA. Also, Gant, "Women Involved in Aviation." Also, Krishna, "The Mercury 13."
44. Amy E. Foster, "Robert Gilruth to Jerrie Cobb, 17 Apr. 1962, Women in Space Program File, NASA HQ" from *Integrating Women into the Astronaut Corps: Politics and Logistics at NASA, 1972–2004*. Baltimore, MD: Johns Hopkins University Press, 2011.
45. NASA required "a bachelor's degree or equivalent" and Glenn argued that he had "equivalent." *Qualifications for Astronauts: Report of the Special Subcommittee on the Selection of Astronauts*, Committee on Science and Astronautics, US House of Representatives, 87th Congress, 2nd Session, July 17–18, 1962. United States Government Printing Office, 1962, p. 56.
46. "Honoring the Mercury 13 Women," biography of Geraldyn "Jerrie" Cobb, May 22, 2012, University of Wisconsin Oshkosh, http://www.uwosh.edu/mercury13/index.php.
47. "'Astronauttes' Beg for Place in Space," UPI Archives, July 17, 1962, https://

www.upi.com/Archives/1962/07/17/Astronauttes-beg-for-place-in-space/9691512607445.

48. *Qualifications for Astronauts: Report of the Special Subcommittee on the Selection of Astronauts,* Committee on Science and Astronautics, US House of Representatives, 87th Congress, 2nd Session, July 17–18, 1962. United States Government Printing Office, 1962, p. 67.
49. Foster, "Women in Space Program File," 60–61.
50. "Valentina Tereshkova: Cosmonaut," National Air and Space Museum, https://airandspace.si.edu/people/historical-figure/valentina-tereshkova.
51. Some believed Tereshkova's flight was another Soviet stunt to embarrass the American space program: Indeed, after her mission, the Soviet Union dismantled the women cosmonaut program. It would be almost twenty years before the next female cosmonaut went to space. "Mission Monday: The First Women in Space," Space Center Houston, June 15, 2020, https://spacecenter.org/mission-monday-the-first-women-in-space.
52. "History of Executive Order 11246," Office of Federal Contract Compliance Programs, US Department of Labor, https://www.dol.gov/agencies/ofccp/about/executive-order-11246-history.
53. Allen Fisher, "Women's Rights and the Civil Rights Act of 1964," National Archives, https://www.archives.gov/women/1964-civil-rights-act. Also, "Founding: Setting the Stage," National Organization for Women, https://www.archives.gov/women/1964-civil-rights-act.
54. Max Frankel, "Johnson Signs Order to Protect Women in US Jobs From Bias," *New York Times*, October 14, 1967, https://timesmachine.nytimes.com/timesmachine/1967/10/14/90410926.pdf. Also, "Executive Order 11375—Update," National Organization for Women, https://350fem.blogs.brynmawr.edu/1967/10/13.
55. The earnings gap is defined as the difference between median earnings of these two groups. Patrick Bayer and Kerwin Kofi Charles, "Divergent Paths: Structural Change, Economic Rank, and the Evolution of Black-White Earnings Differences, 1940–2014," November 2016, https://www.nber.org/system/files/working_papers/w22797/w22797.pdf.
56. Deborah Michaels, "Shirley Chisholm: 1924–2005," Womenshistory.org, 2015, https://www.womenshistory.org/education-resources/biographies/shirley-chisholm.
57. "Chisholm, Shirley Anita," History, Art & Archives, United States House of Representatives, https://history.house.gov/People/Listing/C/CHISHOLM,-Shirley-Anita-(C000371).
58. "Chapter Three: NASA Personnel," SP-4012 NASA Historical Data Book: Vol. IV, NASA Resources 1969–1978, https://history.nasa.gov/SP-4012/vol4/ch3.htm.
59. Hearings before the Subcommittee on Civil Rights and Constitutional Rights of the Committee on the Judiciary, House of Representatives, 93rd Congress, 2nd Session on NASA's Equal Opportunity Employment Program, March 13–14, 1974, https://www.govinfo.gov/content/pkg/CHRG-93hhrg44377/pdf/CHRG-93hhrg44377.pdf.
60. Elizabeth Howell, "NASA's Real 'Hidden Figures,'" Space.com, February 24, 2020, https://www.space.com/35430-real-hidden-figures.html.
61. Brynn Holland, "Human Computers: The Women of NASA," History.com,

August 22, 2018, https://www.history.com/news/human-computers-women-at-nasa.

62. Atkinson and Shafritz, *The Real Stuff,* 134.
63. William Sims Bainbridge, "Chapter 1: The Impact of Space Exploration on Public Opinions, Attitudes, and Beliefs," from *Historical Studies in the Societal Impact of Spaceflight*, Washington, DC: NASA, 2015, ed. Steven J. Dick, pp. 18–19.
64. Imani Perry, "For the Poor People's Campaign, the Moonshot Was Less Than a Triumph," *New York Times,* July 16, 2019, https://www.nytimes.com/2019/07/16/us/for-the-poor-peoples-campaign-the-moonshot-was-less-than-a-triumph.html.
65. Alexis C. Madrigal, "Gil Scott-Heron's Poem, 'Whitey on the Moon,'" *Atlantic,* May 28, 2011, https://www.theatlantic.com/technology/archive/2011/05/gil-scott-herons-poem-whitey-on-the-moon/239622.
66. Department of Housing and Urban Development, Space, Science, Veterans, and Certain Other Independent Agencies Appropriations for Fiscal Year 1975, Hearing Before a Subcommittee of the Committee on Appropriations, United States Senate, 93rd Congress, 2nd session, on H.R. 15572, https://books.google.com/books?id=6BgAjXoTb5cC.
67. Eric Fenrich, "Détente and Dissent: Apollo-Soyuz, Ruth Bates Harris, and NASA's Rhetoric of Cooperation," *Quest: The History of Spaceflight Quarterly,* vol. 22, no. 1, 2015, p. 7, https://www.academia.edu/18619282/D%C3%A9tente_and_Dissent_Apollo_Soyuz_Ruth_Bates_Harris_and_NASA_s_Rhetoric_of_Cooperation.
68. Kim McQuaid, "Chapter 22: 'Racism, Sexism, and Space Ventures': Civil Rights at NASA in the Nixon Era and Beyond," *Societal Impact of Spaceflight,* Washington, DC: NASA, 2007, p. 431, https://www.history.nasa.gov/sp4801-chapter22.pdf.
69. July 19, 1972, memorandum from Ruth Bates Harris to associate administrator for manned space flight Todd Groo, quoted in Atkinson and Shafritz, *The Real Stuff,* 135.
70. Jack Anderson, "White House Kept Lid on Nixon Homes' Costs," *Gadsden Times,* November 3, 1973, https://news.google.com/newspapers?nid=1891&dat=19731103&id=83otAAAAIBAJ&sjid=ptcEAAAAIBAJ&pg=793,258371.
71. "NASA Bias Against Blacks and Women," *Sacramento Observer.*
72. Hearing before the Committee on Aeronautical and Space Sciences, US Senate, 93rd Congress, 2nd session, on Review of NASA's Equal Employment Opportunity Program, January 24, 1974, https://hdl.handle.net/2027/mdp.39015078060459.
73. Shayler and Burgess, *NASA's First Space Shuttle Astronaut Selection,* 51.
74. "SP-407 Space Shuttle: Space Shuttle System and Mission Profile," NASA, https://history.nasa.gov/SP-407/part1.htm.
75. "President Nixon's 1972 Announcement on the Space Shuttle," NASA, January 5, 1972, https://history.nasa.gov/stsnixon.htm.
76. Harry W. Jones, "NASA's Understanding of Risk in Apollo and Shuttle," NASA, September 17, 2018, https://ntrs.nasa.gov/archive/nasa/casi.ntrs.nasa.gov/20190002249.pdf.
77. Atkinson and Shafritz, *The Real Stuff,* 141.
78. Atkinson and Shafritz, *The Real Stuff,* 142–143.

79. Atkinson and Shafritz, *The Real Stuff,* 154.
80. Atkinson and Shafritz, *The Real Stuff,* 151.
81. *NASA Astronaut Recruitment Final Report: Women in Motion, Inc.,* NASA Contract NASW-3049 Final Report by Women in Motion Inc. for the National Aeronautics and Space Administration, August 10, 1977, http://womaninmotionmovie.com/wp-content/uploads/2019/03/WIM-Final-NASA-Report-Execitive-Summary.pdf.
82. "Nichelle Nichols—NASA Recruitment Film (1977)," https://www.youtube.com/watch?v=Lca9_EDMcX0.
83. "Astronaut Selection Is Delayed," *Johnson Space Center Roundup,* December 23, 1977, https://historycollection.jsc.nasa.gov/JSCHistoryPortal/history/roundups/issues/77-12-23.pdf. Also, Cassutt, *The Astronaut Maker,* 185.
84. Robert H. Williams, "PostScript," December 19, 1977, *Washington Post.*
85. Donald K. Slayton and Michael Cassutt, *Deke! An Autobiography.* New York: Macmillan/Tom Doherty Associates/Forge Books, 1995, p. 404 (Kindle ed.).
86. Sherr, *Sally Ride,* 91.
87. Mullane, *Riding Rockets,* 29.
88. Kathy Sullivan transcript, NASA Johnson Space Center Oral History Project, May 10, 2007, https://historycollection.jsc.nasa.gov/JSCHistoryPortal/history/oral_histories/SullivanKD/SullivanKD_5-10-07.htm.
89. Sullivan, NASA Oral History Project, May 10, 2007.
90. Kathy Sullivan, *Handprints on Hubble.* Cambridge, MA: MIT Press, 2019, p. 23. Also, "Kathryn D. Sullivan Biographical Data," NASA, https://www.nasa.gov/sites/default/files/atoms/files/sullivan_kathryn.pdf.
91. Phelps, *They Had a Dream,* 73.
92. Bruce Nichols, "Astronauts Feel No Difference," *Longview News-Journal,* February 22, 1978.
93. Stephen J. Lynton, "Frederick Gregory, Frederick Hauck," *Washington Post,* January 17, 1978.
94. "For All Mankind . . . Space Agency Names First Women, Blacks and Oriental to Orbit Earth," UPI, *Modesto Bee,* January 17, 1978.
95. *NBC Nightly News,* January 16, 1978.
96. Author interview with Kathy Sullivan, October 14, 2020.
97. Sullivan, *Handprints on Hubble,* 26–27.

Chapter 4: Baptism by Fire, Water, and Air

1. Mullane, *Riding Rockets,* 58–59.
2. David Dismore, "July 9, 1978: Feminists Make History with Biggest-Ever March for the Equal Rights Amendment," Feminist Majority Foundation, July 9, 2014, https://feminist.org/news/july-9-1978-feminists-make-history-with-biggest-ever-march-for-the-equal-rights-amendment.
3. Author interview with Sylvia Salinas Stottlemyer, July 20, 2020.
4. Milton E. Reim, "NASA Selects 35 Astronaut Candidates," Lyndon B. Johnson Space Center, Press Release No. 78-03, January 16, 1978, https://www.nasa.gov/centers/johnson/pdf/83130main_1978.pdf.
5. Reim, "NASA Selects 35 Astronaut Candidates."
6. "50 Years Ago: NASA Benefits from Manned Orbiting Laboratory Cancellation,"

NASA, June 10, 2019, https://www.nasa.gov/feature/50-years-ago-nasa-benefits-from-mol-cancellation.

7. Mullane, *Riding Rockets,* 47.
8. Mullane, *Riding Rockets,* 48.
9. Rowland White, *Into the Black: The Extraordinary Untold Story of the First Flight of the Space Shuttle Columbia and the Astronauts Who Flew Her.* New York: Atria Books, 2017, p. 198.
10. Manned Spacecraft Center's Roundup, May 13, 1964, p. 8.
11. White, *Into the Black*, 198.
12. Sherr, *Sally Ride,* 98.
13. Interview with Sylvia Salinas Stottlemyer, May 6, 2022. Also, Mullane, *Riding Rockets,* 49–50.
14. Author interview with Jim Bagian, April 13, 2021.
15. This recollection is imagined and constructed from interviews with Rhea Seddon and her book, *Go for Orbit,* and from interviews with Anna Fisher. Also, "NASA Astronaut Ronald McNair 1979," Black STEM Rockets, https://www.youtube.com/watch?v=RqL5ilEIN8M.
16. Seddon, *Go for Orbit*, 47.
17. Seddon, *Go for Orbit*, 49–50.
18. "Ronald E. McNair Biographical Data," NASA, https://web.archive.org/web/20201031225827/https://er.jsc.nasa.gov/seh/mcnair.htm.
19. McNair, *In the Spirit*, 40.
20. Author interview with Carl McNair, April 1, 2022.
21. McNair, *In the Spirit*, 18.
22. McNair, *In the Spirit*, 18–19.
23. Seddon, *Go for Orbit*, 51.
24. Seddon, *Go for Orbit*, 51.
25. Seddon, *Go for Orbit*, 7.
26. "They Doubted Her Sanity, Now the Doubt Is Larger," UPI, *St. Lucie News Tribune,* January 18, 1978, https://www.newspapers.com/image/778676257/.
27. Seddon, *Go for Orbit*, 51–52.
28. "People," *Time*, Vol. 112, No. 7, August 14, 1978, p. 79.
29. Peter Gwynne/Holly Morris, "Sextet for Space," *Newsweek,* August 14, 1978.
30. *Time,* August 14, 1978, p. 79.
31. "Female Astronaut Makes a Big Splash," United Press International, *Tampa Tribune,* August 1, 1978.
32. Gwynne/Morris, "Sextet for Space."
33. Seddon, *Go for Orbit*, 38, 61.
34. Cassutt, *The Astronaut Maker,* 200.
35. Phelps, *They Had a Dream,* 75.
36. Author interview with Guy Bluford, May 25, 2021.
37. Seddon, *Go for Orbit*, 58.
38. Mullane, *Riding Rockets,* 38.
39. Alex Ronan, "The First Mother in Space Is Looking Forward to Seeing a Woman on Mars," *New York*, September 19, 2016, https://www.thecut.com/2016/09/astronaut-anna-lee-fisher-wants-to-see-a-woman-on-mars.html.
40. Seddon, *Go for Orbit*, 59.
41. Author interview with Anna Fisher, April 12, 2022.

42. Seddon, *Go for Orbit*, 105–06.
43. Seddon, *Go for Orbit*, 107.
44. Seddon, *Go for Orbit*, 107.
45. Seddon, *Go for Orbit*, 107.
46. Seddon, *Go for Orbit*, 107–08.
47. Seddon, *Go for Orbit*, 125.
48. Seddon, *Go for Orbit*, 75.
49. Seddon, *Go for Orbit*, 125.
50. Seddon, *Go for Orbit*, 125.
51. Seddon, *Go for Orbit*, 125.
52. Author interview with Rhea Seddon, April 27, 2022.
53. Mullane, *Riding Rockets*, 74.
54. Mullane, *Riding Rockets*, 75.
55. "The Dream Is Alive: NASA's Space Shuttle Program—1985," Smithsonian Institute, an IMAX Documentary Film. Also, Seddon, *Go for Orbit*, 70.
56. Seddon, *Go for Orbit*, 70.
57. McNair, *In the Spirit*, 74–75.
58. McNair, *In the Spirit*, 75.
59. McNair, *In the Spirit*, 76.
60. Staff of the *Washington Post, Challengers: The Inspiring Life Stories of the Seven Brave Astronauts of Shuttle Mission 51-L*. New York: Pocket Books, 1986.
61. Corinne Naden, *Ronald McNair*. New York: Chelsea House, 1991, p. 46.
62. Phelps, *They Had a Dream,* 107.
63. Seddon, *Go for Orbit*, 70.
64. Seddon, *Go for Orbit*, 70.
65. Seddon, *Go for Orbit*, 71.
66. Seddon, *Go for Orbit*, 3.
67. Seddon, *Go for Orbit*, 3.
68. Seddon, *Go for Orbit*, 3.
69. Seddon, *Go for Orbit*, 5.
70. Rhea Seddon, "How Not to Drown," http://astronautrheaseddon.com/how-not-to-drown.
71. Seddon, "How Not to Drown."
72. Seddon, "How Not to Drown."
73. Seddon, *Go for Orbit*, 70.
74. Seddon, "How Not to Drown."
75. Seddon, *Go for Orbit*, 71. Also, author interview with Rhea Seddon, March 3, 2022.

Chapter 5: I'll Be You

1. Peter S. Pritchard, "Pumpkin Center Died Like Al Neuharth Lived—Big and Bold," *USA Today*, March 17, 2016, https://www.usatoday.com/story/opinion/2016/03/17/usa-today-founder-pumpkin-center-al-neuharth-burned-column/81924180.
2. "Pumpkin Center," Historical Marker Database/HMdb.org, https://www.hmdb.org/m.asp?m=183487.
3. Seddon, *Go for Orbit*, 77.

4. George's long time friend was Tom Tate, who was the administrative assistant to US Representative Don Fuqua of Florida, chairman of the House Science and Technology Committee, which had oversight over the shuttle program's budget. Cassut, *The Astronaut Maker,* 311.
5. Author interview with Rhea Seddon, April 27, 2022.
6. Author interview with Kathy Sullivan, December 12, 2020.
7. Author interview with Kathy Sullivan, December 12, 2020.
8. McNair, *In the Spirit*, 85.
9. Seddon, *Go for Orbit*, 75.
10. Malcolm McConnell, *Challenger: A Major Malfunction—A True Story of Politics, Greed, and the Wrong Stuff.* New York: Doubleday, 1986, p. 34.
11. Because Marshall had employed so many of the Operation Paperclip scientists—those Nazi rocketeers who had fled the Russians at the end of the war—the place was nicknamed "Hunsville." McConnell, *Challenger: A Major Malfunction,* 34.
12. Mullane, *Riding Rockets,* 53.
13. Author interview with June Scobee, June 10, 2021.
14. Mullane, *Riding Rockets,* 52.
15. Mullane, *Riding Rockets,* 52.
16. Sullivan, *Handprints on Hubble*, 11.
17. Sullivan, *Handprints on Hubble*, 11–12.
18. Author interview with Kathy Sullivan, December 12, 2020.
19. Sullivan, *Handprints on Hubble*, 9–10.
20. Sullivan, *Handprints on Hubble*, 13–14.
21. Sullivan, *Handprints on Hubble*, 14–16.
22. Sullivan, *Handprints on Hubble*, 18.
23. Sullivan, *Handprints on Hubble*, 21–22.
24. Interview between Kathy Sullivan and David Barclay, December 12, 2020.
25. White, *Into the Black*, 3.
26. Author interview with Fred Gregory, March 31, 2022.
27. Guion S. Bluford transcript, NASA Johnson Space Center Oral History Project, August 2, 2004, https://historycollection.jsc.nasa.gov/JSCHistoryPortal/history/oral_histories/BlufordGS/BlufordGS_8-2-04.htm.
28. Cassutt, *The Astronaut Maker,* 198.
29. Bluford, NASA Oral History Project.
30. Shayler and Burgess, *NASA's First Space Shuttle Astronaut Selection,* 177–78.
31. Daniel C. Brandenstein transcript, NASA Johnson Space Center Oral History Project, January 19, 1999, https://historycollection.jsc.nasa.gov/JSCHistoryPortal/history/oral_histories/BrandensteinDC/brandensteindc_1-19-99.htm.
32. Sullivan, *Handprints on Hubble*, 60.
33. Sherr, *Sally Ride,* 101.
34. Author interview with Kathy Sullivan, May 13, 2022.
35. Donald P. Myers, "'Good Times' for a NASA Neighborhood," *Newsday,* October 8, 1988, https://www.newspapers.com/image/711443401/.
36. Myers, "'Good Times' for a NASA Neighborhood."
37. Myers, "'Good Times' for a NASA Neighborhood."
38. Myers, "'Good Times' for a NASA Neighborhood."

39. Cassutt, *The Astronaut Maker,* 147.
40. Robert T. McCall transcript, NASA Johnson Space Center Oral History Project, March 28, 2000, https://historycollection.jsc.nasa.gov/JSCHistoryPortal/history/oral_histories/McCallRT/RTM_3-28-00.pdf.
41. Author interview with Kathy Sullivan, March 10, 2021.
42. Author interview with Kathy Sullivan, April 14, 2022.
43. Cassutt, *The Astronaut Maker,* 160.
44. Author interview with Pinky Nelson, September 10, 2020.
45. Author interview with Kathy Sullivan, June 14, 2021.
46. Author interview with Guy Bluford, December 5, 2020.
47. Author interview with Kathy Sullivan, April 14, 2022.
48. Author interview with Anna Fisher, May 25, 2021.
49. Seddon, *Go for Orbit,* 95.
50. Seddon, *Go for Orbit,* 79.
51. Author interview with Anna Fisher, May 25, 2021.
52. The notable exception is Kathy Sullivan, with whom Fred and Guy played racquetball and were often bested. Author interview with Kathy Sullivan, October 14, 2020.
53. Author interview with Charlie Bolden, August 24, 2020.
54. Author interview with Fred Gregory, October 23, 2020.
55. Author interview with Kathy Sullivan, March 10, 2021.
56. Mullane, *Riding Rockets,* 59–60.
57. Author interview with Kathy Sullivan, April 14, 2022.
58. Mullane, *Riding Rockets,* 59.
59. Sherr, *Sally Ride,* 118.
60. Sherr, *Sally Ride,* 117.
61. Author interview with Anna Fisher, January 26, 2021, and author interview with Kathy Sullivan, March 10, 2021.
62. Sherr, *Sally Ride,* 117.
63. Sherr, *Sally Ride,* 117.
64. Sherr, *Sally Ride,* 119.
65. Sherr, *Sally Ride,* 120–21.
66. Mullane, *Riding Rockets,* 113.
67. Author interview with Rhea Seddon and Hoot Gibson, October 5, 2020, and Steven Hawley transcript, NASA Johnson Space Center Oral History Project, December 4, 2002, https://historycollection.jsc.nasa.gov/JSCHistoryPortal/history/oral_histories/HawleySA/HawleySA_12-4-02.htm.
68. Sherr, *Sally Ride,* 120.
69. Email with Kathy Sullivan, March 20, 2022.
70. Author interview with Anna Fisher, Rhea Seddon, and Kathy Sullivan, November 2, 2020.
71. Mullane, *Riding Rockets,* 63.
72. Author interview with Anna Fisher, Rhea Seddon, and Kathy Sullivan, November 2, 2020.
73. Author interview with Anna Fisher, Rhea Seddon, and Kathy Sullivan, November 2, 2020.
74. Mullane, *Riding Rockets,* 59.

75. *NASA Roundup*, September 7, 1979, vol. 18, no. 18, https://historycollection.jsc.nasa.gov/JSCHistoryPortal/history/roundups/issues/79-09-07.pdf.
76. Seddon, *Go for Orbit*, 92.

Chapter 6: Get the Son of a Bitch in Space

1. Thomas O'Toole, "Touchdown for Orbiter *Enterprise*: Space Shuttle Responds Flawlessly on Test Flight, 1st Earth Landing," *Washington Post*, August 13, 1977.
2. Terry Lee Rioux, *From Sawdust to Stardust: The Biography of DeForest Kelley, Star Trek's Dr. McCoy*. New York: Pocket Books, 2005, p. 221.
3. "A Spaceship Landed on Earth," Rockwell International, August 1978, https://www.youtube.com/watch?v=Hgk4GskErjQ.
4. "Air Force Plant 42," GlobalSecurity.org, https://www.globalsecurity.org/military/facility/afp-42.htm.
5. "Technical Development of the Space Transportation System," NASA, https://www.nasa.gov/sites/default/files/files/1b.pdf.
6. *Johnson Space Center Roundup*, https://historycollection.jsc.nasa.gov/JSCHistoryPortal/history/roundups/issues/75-10-10.pdf. Also, "How Much Did It Cost to Create the Space Shuttle?" Planetary.org, https://www.planetary.org/space-policy/sts-program-development-cost. Also, T. A. Heppenheimer, *The Space Shuttle Decision, 1965–1972*. Washington, DC: Smithsonian Institution Press, 2002, https://space.nss.org/the-space-shuttle-decision-by-t-a-heppenheimer.
7. White, *Into the Black*, 153.
8. Cassutt, *The Astronaut Maker,* 8.
9. Author interview with Sylvia Salinas Stottlemyer, May 6, 2022.
10. Cassutt, *The Astronaut Maker,* 16.
11. Cassutt, *The Astronaut Maker,* 20–21.
12. Author interview with George Abbey, December 4, 2020. Also, Cassutt, *The Astronaut Maker,* 57. Also, "A Real Buzz about Our George," The Free Library, 2013, https://www.thefreelibrary.com/A+real+Buzz+about+our+George%3B+George+Abbey%27s+career+path+reads+like+a-a0327071001.
13. Cassutt, *The Astronaut Maker,* 29, 57.
14. Cassutt, *The Astronaut Maker,* 422.
15. "Manned Mars Landing," presentation to the Space Task Group by Dr. Wernher von Braun, NASA, August 4, 1969, p. 4, https://www.nasa.gov/sites/default/files/atoms/files/19690804_manned_mars_landing_presentation_to_the_space_task_group_by_dr._wernher_von_braun.pdf.
16. "Beyond the Atmosphere: Early Years of Space Science," statement by President Nixon on the Space Program, Appendix J, March 7, 1970, https://history.nasa.gov/SP-4211/appen-j.htm.
17. Roger D. Launius, "NASA and the Decision to Build the Space Shuttle, 1969–72," *The Historian*, vol. 57, no. 1, Autumn 1994, pp. 18–19, https://www.jstor.org/stable/24449159.
18. James Fletcher, "Where Do We Go from Here in Space?" address before the Antelope Valley Board of Trade, Lancaster, California, October 18, 1974; NASA History Office, Washington, DC, p. 9.
19. Heppenheimer, *The Space Shuttle Decision*, pp. 396–400.
20. George Low, interview by John Logsdon, July 7, 1970, NASA History Office,

quoted in Richard Jurek, *The Ultimate Engineer: The Remarkable Life of NASA's Visionary Leader George M. Low*. Lincoln, NE: University of Nebraska Press, 2019, p. 157.

21. NASA said that they could reduce the cost per launch well below what it cost for most payloads to go up on expendable launch vehicles (ELVs) like Titan, estimated at that point to be about $12 million per launch during the first decade of the shuttle's operational existence. NASA claimed its fully reusable two-stage flyback booster design would reduce per-launch costs to $5.5 million.
22. President Nixon's 1972 announcement on the Space Shuttle, NASA, https://history.nasa.gov/stsnixon.htm.
23. Jeff DeTroye, "National Security" in *Wings in Orbit: Scientific and Engineering Legacies of the Space Shuttle,* ed. Wayne Hale, NASA, p. 44, https://www.nasa.gov/centers/johnson/pdf/584720main_Wings-ch2c-pgs42-52.pdf.
24. Logsdon (in Jurek), *The Ultimate Engineer,* 161.
25. Hans Mark, *An Anxious Peace: A Cold War Memoir,* College Station, TX: Texas A&M University Press, 2019, p. 409 (Kindle ed.), and Logsdon (in Jurek), *The Ultimate Engineer,* 162.
26. Mark Damohn, *Back Down to Earth: The Development of Space Policy for NASA During the Jimmy Carter Administration*. Lincoln, NE: iUniverse, 2001. Also, Logsdon (in Jurek), *The Ultimate Engineer,* 160.
27. Chronological History Fiscal Year 1978 Budget Submission, NASA, https://www.nasa.gov/sites/default/files/atoms/files/o45128943_1978.pdf.
28. Wayne Biddle, "The Endless Countdown," *New York Times,* June 22, 1980, https://timesmachine.nytimes.com/timesmachine/1980/06/22/113945981.pdf.
29. Victor K. McElheny, "Space Shuttle's Timetable Is Set Back," *New York Times,* May 31, 1975, https://www.nytimes.com/1975/05/31/archives/space-shuttles-timetable-is-set-back.html.
30. Judy A. Rumerman, "Human Space Flight: A Record Of Achievement, 1961-1998," Washington, DC: NASA History Division, https://www.hq.nasa.gov/office/pao/History/40thann/humanspf.htm. The thirty-one crewed spaceflights before 1980 included six Mercury flights, ten Gemini, eleven Apollo, three Skylab, and one for the Apollo-*Soyuz* Test Project.
31. "Rogers Dry Lake," National Park Service, https://www.nps.gov/articles/rogers-dry-lake.htm.
32. Molly Tyson, "Women in Space," *womenSports,* February 1978, p. 22, https://archive.org/details/sim_womensports_1978-02_4_2/page/22/mode/2up.
33. John Noble Wilford, "Space Shuttle Makes a Bumpy Landing, But Officials Are Pleased," *New York Times,* October. 27, 1977, https://www.nytimes.com/1977/10/27/archives/space-shuttle-makes-a-bumpy-landing-but-officials-are-pleased.html.
34. "'... Go for Sep'—The Space Shuttle Approach and Landing Tests," NASA, 1977, https://www.youtube.com/watch?v=SovgWyhAMY4.
35. "Chapter Four: Computers in the Space Shuttle Avionics System," NASA, p. 105, https://history.nasa.gov/computers/Ch4-4.html.
36. Steven Siceloff, "Shuttle Computers Navigate Record of Reliability," NASA, https://www.nasa.gov/mission_pages/shuttle/flyout/flyfeature_shuttlecomputers.html.
37. Wilford, "Space Shuttle Makes a Bumpy Landing."

38. "Fifth Shuttle Orbiter Free Flight Set for October 26," NASA News, October 21, 1977, https://ntrs.nasa.gov/api/citations/19770083455/downloads/19770083455.pdf.
39. Lane E. Wallace, *Flights of Discovery: 60 Years at the NASA Dryden Flight Research Center.* Vol. 4318. NASA History Office, 2006.
40. "Approach & Landing Test (ALT-5) Onboard Audio," October 26, 1977, https://www.youtube.com/watch?v=rjPMqtYET_A&t=384s.
41. White, *Into the Black*, 174–77.
42. Tyson, "Women in Space."
43. Howard Benedict, "Reaching for Universe: Space Shuttle Flight Breathes New Life into Space Program," *Monroe News-Star*, August 30, 1977, p. 9, https://www.newspapers.com/image/87192580.
44. UPI, "Space Shuttle Test Runs Flawlessly," *The Bellingham Herald*, August 12, 1977, p. 1, https://www.newspapers.com/image/769573093.
45. T. A. Heppenheimer, *History of the Space Shuttle, Volume Two: Development of the Space Shuttle, 1972–1981.* Washington, DC: Smithsonian Books, 2010, p. 575.
46. Heppenheimer, *History of the Space Shuttle*, 220.
47. "40 Years Ago: The Space Shuttle at Ellington Air Force Base March, 1978," *South Belt Houston Digital History Archive*, March 1, 2018, https://southbelthouston.blogspot.com/2018/03/3118-40-years-ago-space-shuttle-at.html
48. White, *Into the Black*, 27.
49. *NSTS Shuttle Reference Manual*, NASA, 1988, https://science.ksc.nasa.gov/shuttle/technology/sts-newsref/sts-msfc.html.
50. Nola Taylor Redd, "Stennis Space Center: NASA's Largest Rocket Testing Site," Space.com, January 25, 2018, https://www.space.com/39498-stennis-space-center.html.
51. "Four Down, Four to Go: Artemis I Rocket Moves Closer to Hot Fire Test," NASA, https://www.nasa.gov/exploration/systems/sls/four-down-four-to-go-artemis-i-rocket-moves-closer-to-hot-fire-test.html.
52. "Historical Overview: Early Engine Studies," NASA, https://www.nasa.gov/sites/default/files/files/3HO.pdf.
53. "Sen. John C. Stennis Celebrates a Successful Space Shuttle Main Engine Test," NASA/Stennis Space Center, October 23, 1978, https://archive.org/details/78-459-33.
54. "Engineering Innovations" in *Wings in Orbit: Scientific and Engineering Legacies of the Space Shuttle,* ed. Wayne Hale, NASA, https://web.archive.org/web/20220121095917/http://er.jsc.nasa.gov/seh/536823main_Wings-ch4.pdf.
55. "Space Shuttle Main Engines," NASA, 2009, https://www.nasa.gov/returntoflight/system/system_SSME.html.
56. "Space Shuttle Era: External Tank and Boosters," NASA, https://www.nasa.gov/multimedia/podcasting/flyout_et_srb.html.
57. White, *Into the Black*, 145.
58. Calculated in 2022 dollars; about $120 million, with each SSME costing about $40 million; Jeff Foust, "Aerojet Rocketdyne Defends SLS Engine Contract Costs," *SpaceNews*, May 7, 2020, https://spacenews.com/aerojet-rocketdyne-defends-sls-engine-contract-costs/.
59. "Historical Overview: Early Engine Studies," NASA. Also, White, *Into the Black*, 206.

60. Heppenheimer, *History of the Space Shuttle,* 163.
61. "A-1 Test Stand," NASA, https://www.nasa.gov/sites/default/files/atoms/files/a-1_test_stand_v1.pdf.
62. White, *Into The Black,* 206. Also, Heppenheimer, *History of the Space Shuttle,* 286.
63. James Horton, Jeffrey Megivern, et al., "Summary of Results from Space Shuttle Main Engine Off-Nominal Testing," ResearchGate.net, September 2011, https://www.researchgate.net/publication/266461005.
64. Robert E. Biggs, "Space Shuttle Main Engine The First Ten Years," *History of Liquid Rocket Engine Development in the United States, 1955-1980,* American Astronautical Society History Series, Vol. 13, ed. Stephen E. Doyle, pp. 69–122, https://gandalfddi.z19.web.core.windows.net/Shuttle/SSME_MPS_Info/.
65. Michael E. Hampson and S. Barkhoudarian, "Reusable Rocket Engine Turbopump Condition Monitoring," NASA. Marshall Space Flight Center *Advanced High Pressure O2/H2 Technology,* 1985.
66. "SRB Overview," spaceflight.nasa.gov, https://web.archive.org/web/19990421084627/http://spaceflight.nasa.gov/shuttle/reference/shutref/srb/srb.html. Archived from the original on April 21, 1999.
67. "40 Years Ago: Preparations for STS-1," NASA, February 13, 2020, https://www.nasa.gov/feature/40-years-ago-preparations-for-sts-1.
68. "NASA Railroad Keeps Shuttle's Boosters on the Right Track," NASA, https://www.nasa.gov/mission_pages/shuttle/flyout/railroad.html. Also, "The NASA Railroad," NASA, https://www.nasa.gov/sites/default/files/files/NASA-Railroad.pdf.
69. "Critical to the Countdown: The NASA Railroad," TrainMuseum.org, July 20, 2019, https://www.train-museum.org/2019/07/20/critical-to-the-countdown-the-nasa-railroad.
70. McConnell, *Challenger: A Major Malfunction*, 50–52.
71. William H. Greene, Norman F. Knight Jr., and Alan E. Stockwell, "Structural Behavior of the Space Shuttle SRM Tang-Clevis Joint," *Journal of Propulsion and Power*, vol. 4, no. 4, July–August 1988, https://arc.aiaa.org/doi/10.2514/3.23069.
72. "Chapter VI: An Accident Rooted in History," *Report of the Presidential Commission on the Space Shuttle Challenger Accident*, https://history.nasa.gov/rogersrep/v1ch6.htm.
73. McConnell, *Challenger: A Major Malfunction*, 50–52.
74. "Flyback" technology was originally suggested by the rocket engineers at Marshall. Flyback boosters were filled with liquid propellant, and unlike solids they would burn evenly and predictably. Once they got the orbiter to Earth's upper atmosphere, they would disconnect and fly themselves back to Earth. They could land on solid ground and would not need to be fished out of the ocean by huge Naval ships and crews, like the SRBs. Over the long haul, flyback boosters promised to be cheaper, since they would be fully reusable. Sarah Turner, "Maxime Faget and the Space Shuttle," NASA Activities, 21 (November/December 1990): 22, https://history.nasa.gov/708235main_Shuttle_Bibliography_1-ebook.pdf.
75. White, *Into the Black*, p. 99.
76. Edwards Air Force Base, weather on March 9, 1979, https://www.wunderground.com/history/daily/us/ca/edwards/KEDW/date/1979-3-9.
77. Nieson Himmell, "Bolts and Tape Ground Giant Space Shuttle," *Los Angeles Times*, March 19, 1979.

78. Cassutt, *The Astronaut Maker,* 202.
79. White, *Into the Black*, 153.
80. Cassutt, *The Astronaut Maker,* 202.
81. Thomas L. Moser transcript, NASA Johnson Space Center Oral History Project, April 9, 2010, https://historycollection.jsc.nasa.gov/JSCHistoryPortal/history/oral_histories/MoserTL/MoserTL_4-9-10.htm. Also, White, *Into the Black*, 178.
82. John F. Yardley transcript, NASA Johnson Space Center Oral History Project, June 30, 1998, https://historycollection.jsc.nasa.gov/JSCHistoryPortal/history/oral_histories/YardleyJF/YardleyJF_6-30-98.htm.
83. White, *Into the Black*, 132.
84. Rob Stein and Guy Gugliotta, "Ceramic Shuttle Tiles Had History of Glitches," *Washington Post*, February 7, 2003, https://www.washingtonpost.com/wp-srv/articles/A38144-2003Feb6.html.
85. Damond Benningfield, "Shuttle Tiles," *Air & Space Magazine*, May 2006.
86. Richard S. Lewis, *The Voyages of Columbia: The First True Spaceship*. New York: Columbia University Press, 1984, p. 91.
87. National Air and Space Museum, "Shuttle Strain Isolation Pad, STS-1," Smithsonian, https://airandspace.si.edu/collection-objects/shuttle-strain-isolation-pad-sts-1/nasm_A19820050000.
88. Moser, NASA Oral History Project, April 9, 2010.
89. Stein and Gugliotta, "Ceramic Shuttle Tiles Had History of Glitches."
90. Lewis, *The Voyages of Columbia: The First True Spaceship*, 91.
91. National Air and Space Museum, "Shuttle Strain Isolation Pad, STS-1." Also, "Reliable Shuttle Orbiters Need Special Care," United Space Alliance (archived), https://web.archive.org/web/20110522111605/http://unitedspacealliance.com/news/newsletters/issue080/Articles/ReliableShuttleOrbitersNeedSpecialCare.asp.
92. Dennis Jenkins, *Space Shuttle: Developing an Icon, 1972-2013, Volume III: The Flight Campaign*. Forest Lake, MN: Specialty Press, 2017, pp. 1–3. Also, Heppenheimer, *History of the Space Shuttle,* 239.
93. Stein and Gugliotta, "Ceramic Shuttle Tiles Had History of Glitches."
94. Biddle, "The Endless Countdown."
95. Jenkins, *Space Shuttle, Vol. III*, 214, and "Cronology of KSC and KSC Related Events for 1979," NASA, https://ntrs.nasa.gov/api/citations/20060017819/downloads/20060017819.pdf.
96. Cassutt, *The Astronaut Maker,* 127, 232.
97. Lydia Chavez, "Martin Marietta: Missiles to Cement," *New York Times*, August 26, 1982, https://www.nytimes.com/1982/08/26/business/martin-marietta-missiles-to-cement.html.
98. Author interview with Anna Fisher. Also, Jenkins, *Space Shuttle, Vol. III*, 16.
99. Eric Berger, "A Cold War Mystery: Why Did Jimmy Carter Save the Space Shuttle?" Ars Technica, July 14, 2016.
100. Jimmy Carter, *White House Diary*. New York: Farrar, Straus and Giroux, 2010, p. 63.
101. Mark, *An Anxious Peace,* 409.
102. Mark, *An Anxious Peace,* 412.
103. Berger, "A Cold War Mystery."
104. White, *Into the Black*, 87.

105. Hans Mark interview, March 12 1997, https://documents.theblackvault.com/documents/nro/hansmark-haines-interview-jan2019release.pdf.
106. Mark, *An Anxious Peace*, 409 (Kindle ed.).
107. Hans Mark, *The Space Station: A Personal Journey*. Durham, NC: Duke University Press, 1987, p. 96.
108. Mark, *An Anxious Peace*, 411–12.
109. Melvin Croft and John Youskauskas, *Come Fly with Us: NASA's Payload Specialist Program*. Lincoln, NE: University of Nebraska Press, 2019, pp. 161–62.
110. Released classified document, October 10, 1979, National Reconnaissance Office via Freedom of Information Act portal, February 27, 2017, www.nro.gov/Portals/65/documents/foia/declass/FOIA%20for%20All%20-%20Releases/F-2017-00080a.pdf.
111. Robert F. Thompson transcript, NASA Johnson Space Center Oral History Project, October 3, 2000, https://historycollection.jsc.nasa.gov/JSCHistoryPortal/history/oral_histories/ThompsonRF/ThompsonRF_10-3-00.htm.
112. James E. David, *Spies and Shuttles: NASA's Secret Relationships with the DoD and CIA*. Washington, DC: Smithsonian, 2015, p. 232.
113. David, *Spies and Shuttles*, 196.
114. David, *Spies and Shuttles*, 207.
115. Christopher C. Kraft transcript, NASA Johnson Space Center Oral History Project, August 6, 2012, https://historycollection.jsc.nasa.gov/JSCHistoryPortal/history/oral_histories/KraftCC/KraftCC_8-6-12.htm.
116. Heppenheimer, *History of the Space Shuttle*, 242.
117. Dwayne A. Day, "Invitation to Struggle: The History of Civilian-Military Relations in Space," in *Exploring the Unknown*, vol. 2, NASA, pp. 233–410, https://history.nasa.gov/SP-4407/vol2/v2chapter2-1.pdf
118. Author interview with Estella Gillete, June 24, 2020. Author interview with Jim Bagian, April 13, 2022.
119. Cassutt, *The Astronaut Maker*, 210.
120. Author interview with Anna Fisher, December 11, 2020.
121. Jenkins, *Space Shuttle, Vol. III*, 6.
122. Author interview with Anna Fisher, November 12, 2020.
123. Author interview with Anna Fisher, June 8, 2021.
124. Author interview with Bob Crippen, January 28, 2022.

Chapter 7: The Dream Is Alive

1. Biddle, "The Endless Countdown."
2. "Launch Complex 39, Pads A and B," NASAfacts, https://www.nasa.gov/sites/default/files/167416main_LC39-08.pdf, p. 2.
3. Author interview with Fred Gregory, October 16, 2020.
4. Emma Brown, "A Local Life: Nora Drew Gregory, 98, longtime DC teacher and library advocate, dies at 98," *Washington Post*, July 23, 2011, https://www.washingtonpost.com/local/obituaries/a-local-life-nora-drew-gregory-98-longtime-dc-teacher-and-library-advocate-dies-at-98/2011/07/21/gIQAhJmeVI_story.html.
5. Phelps, *They Had a Dream*, 128.
6. Author interview with Fred Gregory, May 17, 2021.

7. Phelps, *They Had a Dream*, 131.
8. Author interview with Fred Gregory, May 17, 2021.
9. Frederick Gregory transcript, NASA Johnson Space Center Oral History Project, April 29, 2004, https://historycollection.jsc.nasa.gov/JSCHistoryPortal/history/oral_histories/GregoryFD/GregoryFD_4-29-04.htm.
10. Author interview with Fred Gregory, May 17, 2021.
11. Gregory, NASA Oral History Project, April 29, 2004.
12. Gregory, NASA Oral History Project, April 29, 2004.
13. Author interview with Fred Gregory, May 21, 2018.
14. "John W. Young," *New York Times*, April 13, 1981, https://www.nytimes.com/1981/04/13/us/john-w-young.html.
15. Terry Burlison, "*Columbia*'s First Victims," 2013, https://www.baen.com/columbia.
16. "The Boldest Test Flight in History," April 10, 2006, NASA, https://www.nasa.gov/mission_pages/shuttle/sts1/sts1_25.html.
17. "Space Shuttle—Solid Rocket Boosters," NASA, https://web.archive.org/web/20130406193019/http://www.nasa.gov/returntoflight/system/system_SRB.html.
18. "The Greatest Test Flight—STS-1," https://www.youtube.com/watch?v=cT4ADwS66X0.
19. Sullivan, NASA Oral History Project, May 10, 2007.
20. Gregory, NASA Oral History Project, April 29, 2004.
21. White, *Into the Black*, 200. In this abort scenario, Crip and Young would jettison the booster rockets and external tank before pitching the shuttle around and attempting to land back on the Kennedy runway. Sally and Scobee would fly along with them, radioing altitude readings to the shuttle to help Crip and Young gauge the distance between runway and landing gear as they brought the bird down.
22. George "Pinky" Nelson transcript, NASA Johnson Space Center Oral History Project, May 6, 2004, https://historycollection.jsc.nasa.gov/JSCHistoryPortal/history/oral_histories/NelsonGD/NelsonGD_5-6-04.htm.
23. "Presentation: Solid Rocket Boosters," NASA, http://www.nasa-klass.com/Curriculum/Get_Oriented%202/Solid%20Rocket%20Boosters/PRES_SRB.pdf, 2.
24. White, *Into the Black*, 276.
25. "Hail *Columbia*! The Way It Was," *NASA Space News Roundup*, April 14, 1981, vol. 20, no. 8, https://historycollection.jsc.nasa.gov/JSCHistoryPortal/history/roundups/issues/81-04-14.pdf, p. 4.
26. White, *Into the Black*, 5.
27. "The Greatest Test Flight—STS-1."
28. White, *Into the Black*, 284.
29. Linda Herridge, "STS-1: Astronaut Bob Crippen Remembers the Ride of His Life," NASA's John F. Kennedy Space Center, April 12, 2021, https://www.nasa.gov/feature/sts-1-astronaut-bob-crippen-remembers-the-ride-of-his-life.
30. Author interview with Anna Fisher, June 8, 2021.
31. White, *Into the Black*, 372.
32. White, *Into the Black*, 281–92. The National Reconnaissance Office (NRO) is an agency of the United States Department of Defense that designs, builds, launches, and operates the reconnaissance satellites of the US federal government, and provides satellite intelligence to several government agencies. NRO is considered,

along with the Central Intelligence Agency (CIA), National Security Agency (NSA), Defense Intelligence Agency (DIA), and National Geospatial-Intelligence Agency (NGA), to be one of the "big five" US intelligence agencies.

33. White, *Into the Black*, 300.
34. "The Greatest Test Flight—STS-1."
35. Robert Crippen transcript, NASA Johnson Space Center Oral History Project, May 26, 2006, https://historycollection.jsc.nasa.gov/JSCHistoryPortal/history/oral_histories/CrippenRL/CrippenRL_5-26-06.htm.
36. In aviation, "Judy" is a radio call issued by a pilot to indicate that they have made visual contact with the correct target.
37. Seddon, *Go for Orbit*, 105.
38. "Remembering John Young—STS-1: First Shuttle Flight: A Remarkable Flying Machine," April 12, 1981, https://www.youtube.com/watch?v=x_yzDOxKVJQ1981.
39. "Whoops, Cheers Greet Return," Associated Press, *Longview Daily News*, April 15, 1981.
40. White, *Into the Black*, 346.
41. John Young with James R. Hansen. *Forever Young: A Life of Adventure in Air and Space*, University Press of Florida, 2012, p. 235.
42. White, *Into the Black*, 369.
43. White, *Into the Black*, 368.
44. "Astronauts Promote Space Station," *Florida Today*, May 21, 1981, p. 1, https://www.newspapers.com/image/125143533/.
45. Steve Nesbitt, "Astronaut Injured in Traffic Accident," Lyndon B. Johnson Space Center, Press Release No. 81-035, September 14, 1981, https://www.nasa.gov/centers/johnson/pdf/83133main_1981.pdf.
46. *A&T Register* staff, "McNair Seriously Injured in Car Accident in Texas," *A&T Register*, September 18, 1981, p. 1.
47. McNair, *In the Spirit*, 122.
48. William E. Rone Jr., "King Urges Negroes to Ballot Box March," *The State*, May 9, 1966, https://www.newspapers.com/image/749798323/.
49. "Video Clip: Martin Luther King, Jr.—'March on Ballot Boxes' Speech," C-SPAN, May 8, 2016, https://www.c-span.org/classroom/document/?18490.
50. "Video Clip: Martin Luther King, Jr.—'March on Ballot Boxes' Speech."
51. Author interview with Carl McNair, April 1, 2022.
52. Author interview with Carl McNair, April 1, 2022.
53. Nesbitt, "Astronaut Injured in Traffic Accident."
54. McNair, *In the Spirit*, 124.

Chapter 8: To Have and Have Not

1. "Opening the Space Frontier—The Next Giant Step," 16' x 72', acrylic on canvas, 1979; collection of NASA/Johnson Space Center, https://www.mccallstudios.com/a-new-dawn.
2. "A Hundred Years Ago: Birth of a Space Artist," NASA, December 20, 2019, https://www.nasa.gov/feature/100-years-ago-birth-of-space-artist-robert-mccall.
3. Douglas Yazell, "A Bright Future for People in Space," *Horizons*, May 2011, http://www.aiaahouston.org/Horizons/2011_05.pdf, pp. 8–14.

4. McCall, NASA Oral History Project, March 28, 2000.
5. Author interview with Sylvia Salinas Stottlemyer, May 6, 2022.
6. Sherr, *Sally Ride,* 109–10.
7. Sherr, *Sally Ride,* 125.
8. Author interview with Kathy Sullivan, October 14, 2020.
9. Sullivan, *Handprints on Hubble,* 31–32.
10. Sullivan, *Handprints on Hubble,* 31–32.
11. Sullivan, *Handprints on Hubble,* 32.
12. Sherr, *Sally Ride,* 112.
13. Sally Ride transcript, NASA Johnson Space Center Oral History Project, October 22, 2002, https://historycollection.jsc.nasa.gov/JSCHistoryPortal/history/oral_histories/RideSK/RideSK_10-22-02.htm.
14. *Who Was the First American Woman in Space?* Spark, 19:30, https://www.youtube.com/watch?v=HVcckmqxgFY.
15. Ride, NASA Oral History Project, October 22, 2002. Also, Shayler and Burgess, *NASA's First Space Shuttle Astronaut Selection,* 198–99.
16. Sullivan, *Handprints on Hubble,* 67.
17. Sherr, *Sally Ride,* 112–13.
18. Sherr, *Sally Ride,* 112–13.
19. Susan Okie, "Cool Hand Sally Showed the Right Stuff," *Washington Post,* May 10, 1983, https://www.washingtonpost.com/archive/politics/1983/05/10/cool-hand-sally-showed-the-right-stuff/a221e201-5388-468e-a22a-88344fb351eb.
20. Author interview with Kathy Sullivan, October 14, 2020.
21. Sullivan, *Handprints on Hubble,* 67.
22. Author interview with Kathy Sullivan, October 14, 2020.
23. Author interview with Kathy Sullivan, October 14, 2020.
24. Author interview with Kathy Sullivan, October 14, 2020.
25. Shayler and Burgess, *NASA's First Space Shuttle Astronaut Selection,* 199.
26. *Who Was the First American Woman in Space?* Spark, 19:22–19:50.
27. *Who Was the First American Woman in Space?* Spark, 19:22–19:50.
28. Spencer and Spolar, "The Epic Flight."
29. Spencer and Spolar, "The Epic Flight."
30. Spencer and Spolar, "The Epic Flight."
31. Spencer and Spolar, "The Epic Flight."
32. Dale Russakoff, "Space Shuttle Mission 51-L: The *Challenger* Seven: A Shared Romance with Space," *Washington Post,* February 2, 1986.
33. Spencer and Spolar, "The Epic Flight."
34. "Judith A. Resnik Biographical Data," NASA, https://www.nasa.gov/sites/default/files/atoms/files/resnik_judith_with_photo_0.pdf.
35. "Judith A. Resnik Biographical Data," NASA, https://www.nasa.gov/sites/default/files/atoms/files/resnik_judith_with_photo_0.pdf.
36. Russakoff, "Space Shuttle Mission 51-L."
37. Russakoff, "Space Shuttle Mission 51-L."
38. Shayler and Burgess, *NASA's First Space Shuttle Astronaut Selection,* 199.
39. Cassutt, *The Astronaut Maker,* 204.
40. Seddon, *Go for Orbit,* 87.
41. Author interview with Kathy Sullivan, April 14, 2022.

42. Sullivan, NASA Oral History Project, May 10, 2007.
43. Author interview with Kathy Sullivan, March 5, 2022.
44. Calla Cofield, "NASA's New Spacesuit Has a Built-In Toilet," Space.com, February 20, 2018, https://www.space.com/39710-orion-spacesuit-waste-disposal-system.html.
45. Shayler and Burgess, *NASA's First Space Shuttle Astronaut Selection,* 198–99.
46. Author interview with Guy Bluford, May and June 2021.
47. Shayler and Burgess, *NASA's First Space Shuttle Astronaut Selection*, 289.
48. Sherr, *Sally Ride,* 110.
49. Cassutt, *The Astronaut Maker,* 207.
50. Author interview with Anna Fisher, June 8, 2021.
51. Howard E. McCurdy, *The Space Station Decision: Incremental Politics and Technological Choice*. Baltimore, MD: John Hopkins University Press, 2007, p. 39.
52. Ken Nail Jr., *Chronology of KSC and KSC Related Events for 1982*, No. KHR-7, 1984, p. 87, https://docslib.org/doc/101220/chronology-of-ksc-and-ksc-related-events-for-1982.
53. Ken Nail Jr., *Chronology of KSC and KSC Related Events for 1982*, No. KHR-7, 1984, p. 87, https://docslib.org/doc/101220/chronology-of-ksc-and-ksc-related-events-for-1982.
54. Thomas J. Lewin and V. K. Narayanan, *Keeping the Dream Alive: Managing the Space Station Program, 1982–1986*. NASA History Office, 1990, pp. 7–17.
55. Joseph J. Trento, *Prescription for Disaster: From the Glory of Apollo to the Betrayal of the Shuttle*. New York: Random House, 1987, pp. 226–27.
56. Trento, *Prescription for Disaster*, 275.
57. Chris Gebhardt, "40 Years after STS-2: *Columbia*'s Second Flight and the Path to Reusability," NASASpaceflight.com, November 12, 2021, https://www.nasaspaceflight.com/2021/11/sts-2-40th-anniversary.
58. Cassutt, *The Astronaut Maker,* 231.
59. Author interview with Kathy Sullivan, June 14, 2021.
60. Author interview with Kathy Sullivan, June 14, 2021.
61. Author interview with Kathy Sullivan, June 14, 2021.
62. Gregory Cecil, "Our Spaceflight Heritage: The 33rd Anniversary of the Launch of STS-3," *Spaceflight Insider*, March 22, 2015, https://www.spaceflightinsider.com/space-flight-history/spaceflight-heritage-33rd-anniversary-launch-sts-3.
63. Michael Casutt, "The Secret Space Shuttles," *Air & Space Magazine*, August 2009, https://www.airspacemag.com/space/secret-space-shuttles-35318554/.
64. Cassutt, *The Astronaut Maker,* 243, 257–58.
65. *Report to the President by the Presidential Commission on the Space Shuttle Challenger Accident*, June 6, 1986, https://sma.nasa.gov/SignificantIncidents/assets/rogers_commission_report.pdf, p. 9.
66. *Report to the President by the Presidential Commission on the Space Shuttle Challenger Accident*, 182.
67. John M. Fabian transcript, NASA Johnson Space Center Oral History Project, February 10, 2006, https://historycollection.jsc.nasa.gov/JSCHistoryPortal/history/oral_histories/FabianJM/FabianJM_2-10-06.htm.
68. Author interview with Sylvia Salinas Stottlemyer, January 29, 2021.
69. Author interview with Sylvia Salinas Stottlemyer, January 29, 2021.

70. Mullane, *Riding Rockets,* 39.
71. Seddon, *Go for Orbit,* 74.
72. "The *Challenger* Disaster: NBC News Live Coverage 5:00 PM–6:00 PM," NBC News clip, 11:22–14:41, https://www.youtube.com/watch?v=hefu_qGLBcY&t=775s.
73. David J. Shayler and Ian A. Moule, *Women in Space—Following Valentina.* New York: Springer Praxis Books, 2005.
74. Barbara Galloway, "A Private Astronaut," *Akron Beacon Journal,* June 17, 1984, p. 15.
75. Sherr, *Sally Ride,* 109–10.
76. "May 1st 1981: The Day Billie Jean King Was Outed," Tennis Majors, https://www.tennismajors.com/our-features/on-this-day/may-1st-1981-the-day-billie-jean-king-was-outed-138210.html.
77. Sherr, *Sally Ride,* 121.
78. Sherr, *Sally Ride,* 121.
79. Sherr, *Sally Ride,* 121.
80. Sherr, *Sally Ride,* 121.
81. Sherr, *Sally Ride,* 121–22.
82. Sherr, *Sally Ride,* 124.
83. Matt Blitz, "How the NASA Wake-Up Call Went from an Inside Joke to a Beloved Tradition," *Popular Mechanics,* https://www.popularmechanics.com/space/a26229/nasa-wake-up-call.
84. Sherr, *Sally Ride,* 124–25.
85. John Uri, "100 Years Ago: Space Artist Robert McCall Born," NASA, January 2, 2020, https://roundupreads.jsc.nasa.gov/pages.ashx/1326/100.
86. Peter Larson, "Next Giant Leap for Womankind," *The Orlando Sentinel,* April 18, 1982.
87. *Who Was the First American Woman in Space?* Spark.
88. Cassutt, *The Astronaut Maker,* 241. Also, Mullane, *Riding Rockets,* 100.
89. Mullane, *Riding Rockets,* 100.
90. Mullane, *Riding Rockets,* 100.
91. Mullane, *Riding Rockets,* 100–01.
92. Mullane, *Riding Rockets,* 100–01.
93. Sherr, *Sally Ride,* 133.
94. Author interview with Anna Fisher, March 29, 2021.
95. Seddon, *Go for Orbit,* 134–35.
96. Seddon, *Go for Orbit,* 135.
97. Mullane, *Riding Rockets,* 101.
98. Okie, "Cool Hand Sally Showed the Right Stuff."
99. Okie, "Cool Hand Sally Showed the Right Stuff."
100. Sherr, *Sally Ride,* 133.
101. Cassutt, *The Astronaut Maker,* 241, and Sherr, *Sally Ride,* 133.
102. Cassutt, *The Astronaut Maker,* 241.
103. Cassutt, *The Astronaut Maker,* 241.
104. Elizabeth Kolbert, "Two Paths to the Stars: Turnings and Triumphs; Judith Resnik," *New York Times,* February 9, 1986, https://www.nytimes.com/1986/02/09/us/two-paths-to-the-stars-turnings-and-triumphs-judith-resnik.html.

105. Author interview with Sylvia Salinas Stottlemyer, July 2020.
106. Author interview with Michael Oldak, March 19, 2022.
107. McCall, NASA Oral History Project, March 28, 2000.

Chapter 9: A Feather in Her Cap

1. Bob Granath, "Kennedy's Beach House Reopens after Post-Hurricane Restoration," NASA, October 9, 2018, https://www.nasa.gov/feature/kennedys-beach-house-reopens-after-post-hurricane-restoration.
2. Jason Rhian, "Reflections: The Astronaut Beach House," *Spaceflight Insider*, January 25, 2015, https://www.spaceflightinsider.com/space-flight-history/reflections-astronaut-beach-house.
3. Sherr, *Sally Ride,* 153.
4. Author interview with George Abbey, November 1, 2019. Also, author interview with Sylvia Salinas Stottlemyer, July 20, 2020.
5. Sherr, *Sally Ride,* 153–54.
6. "Disneyland: A Day at Disneyland 1982 (direct capture)," https://www.youtube.com/watch?v=MOwPHB2mwWw.
7. Ride, NASA Oral History Project, October 22, 2002.
8. Sherr, *Sally Ride,* 128.
9. Sherr, *Sally Ride,* 128–29.
10. Sally said, "Dr. Kraft talked with me about the implications of being the first woman. He reminded me that I would get a lot of press attention and asked if I was ready for that. His message was just, 'Let us know when you need help; we're here to support you in any way and can offer whatever help you need.' It was a very reassuring message, coming from the head of the space center." Ride, NASA Oral History Project, October 22, 2002.
11. Sherr, *Sally Ride,* 129.
12. "Topic: *Challenger* STS-7—Sally's Ride," https://forum.nasaspaceflight.com/index.php?topic=35731.100.
13. *Who Was the First American Woman in Space?* Spark, 23:51–24:49.
14. Sherr, *Sally Ride,* 137.
15. Sherr, *Sally Ride,* 137.
16. The astronaut prayer was made famous by Alan Shepard, who said: "Dear Lord, please don't let me fuck up."
17. Sherr, *Sally Ride,* 134.
18. "Top 10 Must Sees & Hidden Gems of the Space Shuttle Endeavour," Discover Los Angeles, January 16, 2020, https://www.discoverlosangeles.com/things-to-do/top-10-must-sees-hidden-gems-of-the-space-shuttle-endeavour.
19. Sherr, *Sally Ride,* 144.
20. *Today* Staff and Wire Service Reports, "Ride a Big Draw for Shuttle Media," *Today,* June 19, 1983, https://www.newspapers.com/image/125244343/.
21. Sherr, *Sally Ride,* 144.
22. Kathryn D. Sullivan transcript, NASA Johnson Space Center Oral History Project, May 28, 2009, https://historycollection.jsc.nasa.gov/JSCHistoryPortal/history/oral_histories/SullivanKD/SullivanKD_5-28-09.htm.
23. Sullivan, NASA Oral History Project, May 28, 2009.
24. Ride, NASA Oral History Project, October 22, 2002.

25. Thomas O'Toole, "Sally Ride Soars at Her First News Session," *Washington Post*, May 25, 1983, https://www.washingtonpost.com/archive/politics/1983/05/25/sally-ride-soars-at-her-first-news-session/d6584bda-d927-4fd4-b2fb-2580143e51b0.
26. O'Toole, "Sally Ride Soars at Her First News Session."
27. O'Toole, "Sally Ride Soars at Her First News Session."
28. O'Toole, "Sally Ride Soars at Her First News Session."
29. Sherr, *Sally Ride*, 132.
30. *Who Was the First American Woman in Space?* Spark, 25:14–25:39; "Sally Ride on Dumb Questions," Blank on Blank, https://blankonblank.org/interviews/sally-ride-space-shuttle-first-woman-space-nasa.
31. Frederick H. Hauck transcript, NASA Johnson Space Center Oral History Project, November 20, 2003, https://historycollection.jsc.nasa.gov/JSCHistoryPortal/history/oral_histories/HauckFH/HauckFH_11-20-03.htm.
32. Michael Ryan, "A Ride in Space," *People*, June 20, 1983, https://people.com/archive/cover-story-a-ride-in-space-vol-19-no-24.
33. Carole Agus, "Sally Ride Has a Ticket to Ride Shuttle," *Miami Herald*, January 9, 1983.
34. Agus, "Sally Ride Has a Ticket to Ride Shuttle."
35. Sherr, *Sally Ride*, 149.
36. Agus, "Sally Ride Has a Ticket to Ride Shuttle."
37. Author interview with Susan Okie, April 8, 2022.
38. Author interview with Susan Okie, April 8, 2022.
39. Author interview with Susan Okie, April 8, 2022. Emphasis author's own.
40. Sherr, *Sally Ride*, 139.
41. Ryan, "A Ride in Space."
42. Sherr, *Sally Ride*, 138.
43. Sherr, *Sally Ride*, 138.
44. "Two Astronauts Tell Friends of Their Marriage Last Month," *New York Times*, August 15, 1982, https://timesmachine.nytimes.com/timesmachine/1982/08/15/085126.html?pageNumber=34.
45. Author interview with Steve Hawley, April 28, 2021.
46. Author interview with Susan Okie, April 8, 2022.
47. Author interview with Susan Okie, April 8, 2022.
48. Sherr, *Sally Ride*, 139.
49. Sherr, *Sally Ride*, 147.
50. Susan Okie, "Fame Finds Astronaut Determined to Ignore It," *Washington Post*, May 8, 1983, https://www.washingtonpost.com/archive/politics/1983/05/08/fame-finds-astronaut-determined-to-ignore-it/876c0c40-f205-4aaa-922d-e974240e8340.
51. Author interview with Susan Okie, April 8, 2022.
52. Author interview with Susan Okie, April 8, 2022.
53. Susan Okie, "Sally Ride Remains an Elusive Character—Even to a Close Friend," *The Baltimore Sun*, May 8, 1983, https://www.newspapers.com/image/377499519/.
54. Okie, "Sally Ride Remains an Elusive Character."
55. Okie, "Sally Ride Remains an Elusive Character."
56. Okie, "Sally Ride Remains an Elusive Character."

57. Okie, "Fame Finds Astronaut Determined to Ignore It."
58. Sherr, *Sally Ride,* 114.
59. Author interview with Sylvia Salinas Stottlemyer, July 20, 2020.
60. Author interview with Sylvia Salinas Stottlemyer, July 20, 2020.
61. Author interview with George Abbey, November 1, 2019. Also, author interview with Sylvia Salinas Stottlemyer, July 20, 2020.
62. Sherr, *Sally Ride,* 153–54.
63. Cheryl L. Mansfield, "If Walls Could Talk," NASA, June 28, 2005, https://www.nasa.gov/missions/shuttle/beach_house.html
64. Sherr, *Sally Ride,* 153–54.
65. Sherr, *Sally Ride,* 153–54.
66. "US Woman Flies into Space," UPI in Cape Canaveral, *San Francisco Examiner,* June 18, 1983.
67. "Lots of Souvenirs aboard *Challenger,*" *Reno Gazette-Journal,* June 20, 1983, p. 8, https://www.newspapers.com/image/149652639/.
68. "US Woman Flies Into Space."
69. Cassutt, *The Astronaut Maker,* 254.
70. Cassutt, *The Astronaut Maker,* 254.
71. Author interview with Anna Fisher, November 12, 2020.
72. "1983 STS 7 *Challenger* NASA," https://www.youtube.com/watch?v=x56omIeYXP4.
73. Sullivan, *Handprints on Hubble,* 50.
74. Sherr, *Sally Ride,* 155.
75. Sherr, *Sally Ride,* 155.
76. "STS-7 Launch and Land," NASA STI Program, https://www.youtube.com/watch?v=Vq8PAH0giKI.
77. Ride, NASA Oral History Project, October 22, 2002.
78. Sherr, *Sally Ride,* 156.
79. "Sally Ride on Dumb Questions." An "E ticket" was a special admission ticket used at Disneyland and Magic Kingdom from 1955 to 1982 that admitted holders to the newest, most thrilling rides and attractions at the theme parks.
80. "US Woman Flies Into Space." Also, Sherr, *Sally Ride,* 158.
81. Shayler and Burgess, *NASA's First Space Shuttle Astronaut Selection,* 309.
82. Robert Cooke, "From Headlines to T-Shirts, Sally-Mania Is Taking Hold," *Boston Globe,* June 20, 1983.
83. Associated Press, "Shuttle Ready, Astronauts Eager," *Democrat and Chronicle,* June 18, 1983.
84. "People Spotter," *Atlanta Constitution,* June 25, 1983, p. 19.
85. Amy Clark, "From Fonda to E.T.'s Pal, VIPs Cheered Sally's Ride," *Florida Today,* June 19, 1983.
86. "Ride Family Smiling with Pride," Associated Press report from Cape Canaveral, *Oshkosh Northwestern,* June 19, 1983. Also, Sherr, *Sally Ride,* 157–58.
87. Sherr, *Sally Ride,* 157–58.
88. Sherr, *Sally Ride,* 157–58.
89. Rony Laytner and Donald McLachlan, "Ride, Sally Ride: Her Place Is Space," *Chicago Tribune,* April 24, 1983.
90. Sherr, *Sally Ride,* 158.
91. Hanh Nguyen, "'The Vietnam War': How Jane Fonda Became One of the Most

Hated People Associated with the War," *IndieWire*, September 27, 2017, https://www.indiewire.com/2017/09/the-vietnam-war-jane-fonda-vietnam-photo-hanoi-jane-pbs-1201880919.

92. Clark, "From Fonda to E.T.'s Pal, VIPs Cheered Sally's Ride."
93. Trento, *Prescription for Disaster*, 247.
94. "Ride Family Smiling with Pride."
95. Associated Press, "Sally Rides into the Heavens," *Edmonton Journal*, January 19, 1983.
96. Sherr, *Sally Ride*, 163.
97. Ride, NASA Oral History Project, October 22, 2002.
98. Harry F. Rosenthal, "Sally Ride Oversees Ejection of TV Satellite Above Pacific," *Hartford Courant*, June 19, 1983. And "JSC 830—1983—We Deliver," NASA/JSC, https://archive.org/details/JSC_830_We_Deliver.wmv.
99. "JSC 830—1983—We Deliver."
100. John Noble Wilford, "*Challenger* Crew Snares Satellite," *New York Times*, June 23, 1983, https://www.nytimes.com/1983/06/23/us/challenger-crew-snares-satellite.html.
101. "JSC 830—1983—We Deliver."
102. Sally Ride, "Sally Ride Describes the Indescribable View from Orbit," *Air & Space Magazine*, July 2012, https://www.airspacemag.com/space/single-room-earth-view-5940961.
103. "Space Shuttle Flight 7 (STS-7) Post Flight Presentation, Narrated by the Astronauts," National Space Society, https://space.nss.org/space-shuttle-flight-7-video.
104. Sherr, *Sally Ride*, 165.
105. Sherr, *Sally Ride*, 168.
106. "1983: Sally Ride Returns from Space," *ABC News*, June 24, 1983, https://www.youtube.com/watch?v=nxo84aJJvWc.
107. "1983: Sally Ride Returns from Space."
108. Associated Press, "Astronaut Sally Ride Spurns Bouquet of Roses, Carnations," *Lancaster New Era*, June 25, 1983.
109. Amy E. Foster, *Integrating Women into the Astronaut Corps: Politics and Logistics at NASA, 1972–2004*, Baltimore, MD: Johns Hopkins University Press, 2011, and Associated Press, "Sally Ride Turns down Homecoming Bouquet," *El Paso Times*, June 26, 1983.
110. Sherr, *Sally Ride*, 169.
111. Sherr, *Sally Ride*, 169.
112. Mark Russell, "That's Politics," *Miami Herald*, July 13, 1983.
113. Associated Press photo and caption printed in *Los Angeles Times*, July 29, 1983.
114. Paul Beatty, "Sally Ride Upstages Jane Fonda," *Santa Cruz Sentinel*, August 16, 1983.
115. Sherr, *Sally Ride*, 171.
116. Sherr, *Sally Ride*, 171.
117. Shayler and Burgess, *NASA's First Space Shuttle Astronaut Selection*, 311.
118. Author interview with Steve Hawley, April 28, 2021.
119. Sherr, *Sally Ride*, 174.
120. John M. Fabian transcript, NASA Johnson Space Center Oral History Project,

February 10, 2006, https://historycollection.jsc.nasa.gov/JSCHistoryPortal/history/oral_histories/FabianJM/FabianJM_2-10-06.htm.

121. Sherr, *Sally Ride,* 174.
122. Ride, NASA Oral History Project, October 22, 2002.
123. Author interview with Susan Okie, April 8, 2022.
124. Sinéad Baker, "The Incredible Life of Sally Ride, Who Became the 1st American Woman in Space after Answering an Ad in Her College Paper," *Business Insider,* June 18, 2020, https://www.businessinsider.com/sally-ride-first-american-woman-in-space-life-legacy-2019-6.
125. Sherr, *Sally Ride,* 180.
126. Sherr, *Sally Ride,* 180.
127. Thom Patterson, "The Downing of Flight 007: 30 Years Later, a Cold War Tragedy Still Seems Surreal," CNN, August 31, 2013, https://www.cnn.com/2013/08/31/us/kal-fight-007-anniversary/index.html.
128. Sherr, *Sally Ride,* 181.
129. Sherr, *Sally Ride,* 181.
130. Author interview with Rick Hauck, May 2, 2022.
131. Sherr, *Sally Ride,* 183.
132. Sherr, *Sally Ride,* 183.
133. "Miss World Bob Hope Blooper 1970," https://www.youtube.com/watch?v=reCX3_OAkv8.
134. Sherr, *Sally Ride,* 176.
135. "Remembering Bob Hope, a Friend of NASA," NASA, July 28, 2003, https://www.nasa.gov/news/highlights/hope.html.
136. Sherr, *Sally Ride,* 178–79.
137. Author interview with John Fabian, January 21, 2021.
138. Sherr, *Sally Ride,* 177.
139. "Bob Hope's Salute to NASA: 25 Years of Reaching for the Stars," imdb.com, https://www.imdb.com/title/tt6104208/releaseinfo?ref_=ttfc_sa_1.
140. Sherr, *Sally Ride,* 177–78.
141. Tam O'Shaughnessy, *Sally Ride: A Photobiography of America's Pioneering Woman in Space.* New York: Roaring Brook Press, 2015, p. 112.

Chapter 10: Rocket Dawn

1. "STS-8 Press Kit," August 1983, https://spacepresskit.files.wordpress.com/2012/08/sts-8.pdf.
2. Author interview with Guy Bluford, March 9, 2021.
3. "ABC News Coverage of STS-8 Part 1," https://www.youtube.com/watch?v=H6AskWgEhm0.
4. Walter Leavy, "A Historic Step into Outer Space," *Ebony Magazine,* November 1983.
5. Leavy, "A Historic Step into Outer Space."
6. Leavy, "A Historic Step into Outer Space."
7. Author interview with Guy Bluford, March 8, 2021.
8. William Robbins, "Election of Black Mayor in Philadelphia Reflects a Decade of Change in City," *New York Times,* November 10, 1983, https://www.nytimes

.com/1983/11/10/us/election-of-black-mayor-in-philadelphia-reflects-a-decade-of-change-in-city.html. Also, Kevin Klose, "Washington Winner in Bitter Chicago Election for Mayor," *Washington Post*, April 13, 1983, https://www.washingtonpost.com/archive/politics/1983/04/13/washington-winner-in-bitter-chicago-election-for-mayor/af90b445-cf89-4ffd-8960-826c3e52abc1.

9. Milton Coleman, "Jackson Launches 1984 Candidacy," *Washington Post*, November 4, 1983, https://www.washingtonpost.com/archive/politics/1983/11/04/jackson-launches-1984-candidacy/3a977116-21c5-4516-9f9e-15bb5798173b.
10. Abigail Thernstrom and Stephan Thernstrom, "Black Progress: How Far We've Come, and How Far We Have to Go," *Brookings*, March 1, 1998, https://www.brookings.edu/articles/black-progress-how-far-weve-come-and-how-far-we-have-to-go.
11. Drew Desilver, "Black Unemployment Rate Is Consistently Twice That of Whites," Pew Research Center, August 21, 2013, https://www.pewresearch.org/fact-tank/2013/08/21/through-good-times-and-bad-black-unemployment-is-consistently-double-that-of-whites.
12. "Black Persons by Poverty Status in 1969, 1979, 1989, and 1999," US Census Bureau, https://www.census.gov/data/tables/time-series/dec/cph-series/cph-l/cph-l-166.html.
13. Spencer Rich, "Reagan Panel, Citing 'New Racism,' Urges Easing of EEOC Rules," *Washington Post*, January 30, 1981, https://www.washingtonpost.com/archive/politics/1981/01/30/reagan-panel-citing-new-racism-urges-easing-of-eeoc-rules/dfd7%E2%80%9DReagan 9721-7bbc-4ef0-91a3-ee5425904abe.
14. Lily Rothman, "How MLK Day Became a Holiday," *Time*, January 19, 2015, https://time.com/3661538/mlk-day-reagan-history.
15. Rothman, "How MLK Day Became a Holiday."
16. "Solid Rocket Boosters and Post-Launch Processing," Kennedy Space Center, https://www.nasa.gov/centers/kennedy/pdf/146685main_srb-et.pdf.
17. Phelps, *They Had a Dream*, 92.
18. Phelps, *They Had a Dream*, 92.
19. Phelps, *They Had a Dream*, 92.
20. Phelps, *They Had a Dream*, 92.
21. Phelps, *They Had a Dream*, 92.
22. Phelps, *They Had a Dream*, 92.
23. "STS-8 Post-Flight Crew Press Conference," NASA, September 13, 1983, p. 9, https://ia800607.us.archive.org/28/items/NasaAudioHighlightReels/Sts-8PressKitTranscript.pdf.
24. "Space Shuttle Propulsion Trivia," NASA, Marshall Space Flight Center, April 2005, https://www.nasa.gov/sites/default/files/113069main_shuttle_trivia.pdf.
25. Caitlin Shaw, *Gear Patrol*, October 4, 2017, https://www.gearpatrol.com/fitness/a393291/hook-maneuver-agsm-tutorial.
26. David Shayler, *Disasters and Accidents in Manned Spaceflight*. Germany: Springer, 2000, p. 136.
27. "Space Shuttle Solid Rocket Motor Plume Pressure and Heat Rate Measurements," https://ntrs.nasa.gov/citations/20120012051.
28. Kennedy Space Center, *KSC Science*, 1988, https://science.ksc.nasa.gov/shuttle/technology/sts-newsref/srb.html.

29. Bluford, NASA Oral History Project.
30. Bluford, NASA Oral History Project.
31. Author interview with Guy Bluford, March 8, 2021.
32. Author interview with Guy Bluford, March 8, 2021.
33. "Guy Bluford: Black Astonaut Makes First Space Mission," *Jet Magazine*, September 5, 1983, https://books.google.com/books?id=ILcDAAAAMBAJ&pg=PA20.
34. Bill Prochnau, "Guy Bluford: NASA's Reluctant Hero," *Washington Post*, August 21, 1983, https://www.washingtonpost.com/archive/politics/1983/08/21/guy-bluford-nasas-reluctant-hero/04a99e23-06df-499f-849a-0837848ae052/.
35. Prochnau, "Guy Bluford: NASA's Reluctant Hero."
36. Prochnau, "Guy Bluford: NASA's Reluctant Hero."
37. Prochnau, "Guy Bluford: NASA's Reluctant Hero."
38. Leavy, "A Historic Step into Outer Space."
39. James Haskins, *Black Eagles: African Americans in Aviation*. New York: Scholastic, 1995, p. 151.
40. Prochnau, "Guy Bluford: NASA's Reluctant Hero."
41. Author interview with Guy Bluford, December 5, 2020.
42. Prochnau, "Guy Bluford: NASA's Reluctant Hero."
43. Brian Anderson, "Innovator's Spotlight: Col. Guion S. Bluford," Air Force News Services, April 16, 2021, https://www.af.mil/News/Article-Display/Article/2575707/innovators-spotlight-col-guion-s-bluford/.
44. Prochnau, "Guy Bluford: NASA's Reluctant Hero."
45. Anderson, "Innovator's Spotlight: Col. Guion S. Bluford."
46. *Black in Space: Breaking the Color Barrier*, Smithsonian.
47. Phelps, *They Had a Dream*, 80–81. Also, Bluford, NASA Oral History Project.
48. *Black in Space: Breaking the Color Barrier*, Smithsonian.
49. Prochnau, "Guy Bluford: NASA's Reluctant Hero."
50. Prochnau, "Guy Bluford: NASA's Reluctant Hero."
51. Prochnau, "Guy Bluford: NASA's Reluctant Hero."
52. Prochnau, "Guy Bluford: NASA's Reluctant Hero."
53. "McDonnell Douglas F-4 Phantom II," Wikipedia, https://en.wikipedia.org/wiki/McDonnell_Douglas_F-4_Phantom_II.
54. Cooper Thomas, "Bombing Missions of the Vietnam War," https://storymaps.arcgis.com/stories/2eae918ca40a4bd7a55390bba4735cdb. Over the next decade, Allied pilots would drop over seven million pounds' worth of bombs over Vietnam, Laos, and Cambodia, more than tripling the total tonnage of explosives dropped over Germany and Japan during World War II.
55. Phelps, *They Had a Dream*, 90.
56. Author interview with Guy Bluford, December 5, 2020.
57. Bluford, NASA Oral History Project.
58. Bluford, NASA Oral History Project.
59. Author interview with Guy Bluford, December 5, 2020.
60. Bluford, NASA Oral History Project.
61. Charles B. Fancher Jr., "'Average Guy' Is Going into Orbit," *The Philadelphia Inquirer*, January 21, 1978.
62. "Astronaut Guion 'Guy' Bluford on What It Meant to Be the First Black Man to

Fly in Space," interview by De'Aundre Barnes, *OprahDaily.com*, June 15, 2021, https://www.oprahdaily.com/life/a36634334/guion-guy-bluford-interview.
63. Author interview with Guy Bluford, December 5, 2020.
64. Author interview with Guy Bluford, December 5, 2020.
65. Author interview with Guy Bluford, December 5, 2020.
66. Author interview with Guy Bluford, December 5, 2020.
67. Prochnau, "Guy Bluford: NASA's Reluctant Hero."
68. Brandenstein, NASA Oral History Project.
69. Bluford, NASA Oral History Project.
70. Author interview with Guy Bluford, December 5, 2020.
71. Gene Seymour, "'No Frills' Astronaut First American Black to Travel in Space," *The Akron Beacon Journal,* August 21, 1983.
72. "STS-8: The First Shuttle Night Launch," NASA, August 30, 2018.
73. "Satellite launch from Shuttle on the mark," *Florida Today,* September 1, 1983, p. 20A, https://www.newspapers.com/image/124978053.
74. "STS-8 Post-Flight Press Conference," https://space.nss.org/space-shuttle-flight-8-sts-8-post-flight-press-conference-video.
75. Author interview with Guy Bluford, March 8, 2021.
76. Bluford, NASA Oral History Project.
77. "STS-8 Post-Flight Press Conference."
78. Ian Haney-Lopez, "The Racism at the Heart of the Reagan Presidency," *Salon,* January 11, 2014, https://www.salon.com/2014/01/11/the_racism_at_the_heart_of_the_reagan_presidency/.
79. Robert C. Cowen, "Shuttle Landing Brings Successful Flight to Flawless End," *The Christian Science Monitor,* September 6, 1983. Also, Bluford, NASA Oral History Project.
80. Bluford, NASA Oral History Project.
81. Sharon Begley, "NASA's Nighttime Spectacular," *Newsweek,* September 5, 1983, p. 69. William J. Broad, "Man in the News; First US Black in Space," *New York Times,* August 31, 1983, https://www.nytimes.com/1983/08/31/us/man-in-the-news-first-us-black-in-space.html.
82. Sherr, *Sally Ride,* 130.
83. Prochnau, "Guy Bluford: NASA's Reluctant Hero."
84. "Survey Reveals Why Blacks Not Wanted on White Magazines' Covers," *Jet Magazine,* October 17, 1983, https://books.google.com/books?id=-rIDAAAAMBAJ&pg=PA30.
85. "Survey Reveals Why Blacks Not Wanted on White Magazines' Covers."
86. Bluford, NASA Oral History Project.
87. Bluford, NASA Oral History Project.
88. Bluford, NASA Oral History Project.
89. Bluford, NASA Oral History Project.
90. "Ed Dwight Talks about Astronauts HistoryMaker Guion Bluford and Frederick D. Gregory," The HistoryMakers, June 19, 2002, Tape 6, https://www.thehistorymakers.org/biography/ed-dwight-39.
91. Author interview with Guy Bluford, June 10, 2021. Also, James R. Hansen and Allan J. McDonald, *Truth, Lies, and O-Rings: Inside the Space Shuttle Challenger Disaster,* United States: University Press of Florida, 2018, p. 14 (Kindle ed.).
92. McDonald, *Truth, Lies, and O-Rings,* 11, 14–15.

93. Lewis, *The Voyages of Columbia,* 206–207.
94. Author interview with Guy Bluford, December 5, 2020.

Chapter 11: We Deliver

1. “Space Shuttle Flight 10 (STS-41B)—Post Flight Press Conference Video,” NASA, February 3, 1984, https://space.nss.org/space-shuttle-flight-10-post-flight-press-conference-video.
2. Heppenheimer, *The Space Shuttle Decision*, 245–290.
3. NASA handout film on the Space Shuttle, AP Archive, March 6, 1978, https://www.youtube.com/watch?v=I2p0e-qOtw4.
4. Greg Wayland, “Marking the End of a Chapter of Space Exploration,” *NECN*, March 25, 2014, https://www.necn.com/news/local/_necn__marking_the_end_of_a_chapter_of_space_exploration_necn/108509/.
5. Fisher, NASA Oral History Project, February 17, 2009.
6. NASA, “STS 41-B Post Flight Presentation,” National Space Society, 1984.
7. John Noble Wilford, “Missing Satellite Delays Launching by Space Shuttle,” *New York Times*, February 5, 1984.
8. “Space Shuttle Flight 10 (STS-41B)—Post Flight Press Conference Video.”
9. Olive Talley, “Baffling disappearance of Westar 6,” United Press International, February 4, 1984, https://www.upi.com/Archives/1984/02/04/Baffling-disappearance-of-Westar-6/3847444718800.
10. John Noble Wilford, “2d Satellite Lost as Failures Mar Shuttle Mission,” *New York Times*, February 7, 1984, https://www.nytimes.com/1984/02/07/us/2d-satellite-lost-as-failures-mar-shuttle-mission.html.
11. Wilford, “2d Satellite Lost as Failures Mar Shuttle Mission.”
12. Robert L. “Hoot” Gibson transcript, NASA Johnson Space Center Oral History Project, January 22, 2016, https://historycollection.jsc.nasa.gov/JSCHistoryPortal/history/oral_histories/GibsonRL/GibsonRL_1-22-16.htm.
13. “Space Shuttle Flight 10 (STS-41B)—Post Flight Press Conference Video.”
14. “Space Shuttle #535783,” *NBC Evening News,* February 7, 1984, https://tvnews.vanderbilt.edu/broadcasts/535783.
15. NASA handout film on the Space Shuttle, AP Archive, March 6, 1978, https://www.youtube.com/watch?v=I2p0e-qOtw4.
16. John W. Anderson, “Europe vs. America in Space,” *New York Times,* March 17, 1985, https://www.nytimes.com/1985/03/17/business/europe-vs-america-in-space.html.
17. Anderson, “Europe vs. America in Space.”
18. Anderson, “Europe vs. America in Space.”
19. President Ronald Reagan, State of the Union Address, January 25, 1984.
20. Joseph J. Trento and Susan B. Trento, “Why *Challenger* Was Doomed: The Story of the Ill-Fated Space Shuttle Goes Far Beyond O-Rings, Say the Officials Who Were Involved,” *Los Angeles Times*, January 19, 1987, https://www.latimes.com/archives/la-xpm-1987-01-18-tm-5326-story.html.
21. Trento and Trento, “Why *Challenger* Was Doomed.”
22. Trento and Trento, “Why *Challenger* Was Doomed.”
23. James M. Beggs transcript, NASA Johnson Space Center Oral History Project, March 7, 2002, https://historycollection.jsc.nasa.gov/JSCHistoryPortal

/history/oral_histories/NASA_HQ/Administrators/BeggsJM/BeggsJM_3-7-02.htm.

24. Beggs, NASA Oral History Project.
25. Beggs, NASA Oral History Project.
26. Trento, *Prescription for Disaster*, 246.
27. Anne Millbrooke, "Chapter 13 'More Favored than the Birds': The Manned Maneuvering Unit in Space," https://history.nasa.gov/SP-4219/Chapter13.html, and "Manned Maneuvering Unit: User's Guide," contractor report by J. A. Lenda, May 1, 1978, https://ntrs.nasa.gov/citations/19790008382.
28. Joseph P. Allen transcript, NASA Johnson Space Center Oral History Project, November 18, 2004, https://historycollection.jsc.nasa.gov/JSCHistoryPortal/history/oral_histories/AllenJP/AllenJP_11-18-04.htm.
29. "Bruce Mccandless, Who Made First Untethered Space Flight, Dies at 80," *BBC News*, December 23, 2017, https://www.bbc.com/news/world-us-canada-42465059.
30. McNair, *In the Spirit*, 132.
31. McNair, *In the Spirit*, 134.
32. John Kirby, "The Wonder of It All," *Oxford American*, November 19, 2019, https://main.oxfordamerican.org/magazine/item/1861-the-wonder-of-it-all.
33. "Space Station 20th: Music on ISS," May 12, 2020, https://www.nasa.gov/feature/space-station-20th-music-on-iss.
34. McNair, *In the Spirit*, 135.
35. Kurt Heisig, "Sax in Space," *North American Saxophone Association Magazine*, 1986, https://www.kurtheisigmusic.com/sax-in-space.
36. Kirby, "The Wonder of It All."
37. Kirby, "The Wonder of It All."
38. "Space Shuttle Flight 10 (STS-41B)—Post Flight Press Conference Video," 21:00.
39. "NASA Armstrong Fact Sheet: Sonic Booms," NASA, August 15, 2017, https://www.nasa.gov/centers/armstrong/news/FactSheets/FS-016-DFRC.html.
40. Mike Toner, "*Challenger*'s Cape Touchdown a 'Dream,'" *Miami Herald*, February 12, 1984.
41. Toner, "*Challenger*'s Cape Touchdown a 'Dream.'"
42. CBS, Internet Archive, 2011, https://archive.org/details/WUSA_20110708_110000_The_Early_Show/start/4200/end/4260.
43. Naden, *Ronald McNair*, 78.
44. Naden, *Ronald McNair*, 101–02.
45. Beggs, NASA Oral History Project.
46. Mark, *An Anxious Peace*, 577.
47. Mark, *An Anxious Peace*, 579–580.
48. Mark, *An Anxious Peace*, 579. Mark says he flew in the Air Force, but he was in the Navy and later the Utah Air National Guard.
49. Mark: "Jim Beggs would have to talk with Vice President Bush to see whether we could get him to talk with Senator Garn." Mark, *An Anxious Peace*, 580.

Chapter 12: Yellow Death

1. "Space Shuttle Solid Rocket Boosters," NASA, April 6, 2013, https://web.archive.org/web/20130406193019/http://www.nasa.gov/returntoflight/system/system

_SRB.html. Also, "Space Shuttle Main Engines," NASA, https://www.nasa.gov/returntoflight/system/system_SSME.html.

2. Mullane, *Riding Rockets,* 157.
3. Mike Leary, "Space Shot Aborted at Final Moment," *Philadelphia Inquirer,* June 27, 1984. James Fisher, "It May Be 2 Weeks before Nasa Tries Shuttle Launch," *Orlando Sentinel,* June 27, 1984.
4. Mullane, *Riding Rockets,* 158.
5. "Mission Workers Can't Hide Disappointment," United Press International, *Herald,* June 27, 1984.
6. John Uri, "35 Years Ago: STS-41D—First Flight of Space Shuttle Discovery," NASA, August 30, 2019, https://www.nasa.gov/feature/35-years-ago-sts-41d-first-flight-of-space-shuttle-discovery. Also, Mullane, *Riding Rockets,* p. 157.
7. Mullane, *Riding Rockets,* 158.
8. "STS-41D Pad Abort (6-26-84)," https://www.youtube.com/watch?v=-zVN9V5uBNc.
9. Rosemary Wittman Lamb, "Maiden Flight," *Guardian,* August 21, 1984.
10. Mullane, *Riding Rockets,* 159.
11. Henry W. "Hank" Hartsfield transcript, NASA Johnson Space Center Oral History Project, June 15, 2001, https://historycollection.jsc.nasa.gov/JSCHistoryPortal/history/oral_histories/HartsfieldHW/HartsfieldHW_6-15-01.htm.
12. William Harwood, "Launch Abort May Delay Discovery's Mission Indefinitely," UPI, *Herald,* June 27, 1984.
13. Thomas O'Toole, "Air Force Rocket Plan May Undermine Shuttle, NASA Chief Tells Hill," *Washington Post,* August 1, 1984, https://www.washingtonpost.com/archive/politics/1984/08/01/air-force-rocket-plan-may-undermine-shuttle-nasa-chief-tells-hill/f9a2dcdd-7f4e-4548-9667-a07af74a9449.
14. Mike Toner, "High Stakes Comeback Looms for Discovery," *Miami Herald,* August 27, 1984.
15. "Insulation Is Blamed for Discovery Abort; NASA Switches Engine," *The Burlington Free Press,* July 3, 1984.
16. Toner, "High Stakes Comeback Looms for Discovery."
17. Mullane, *Riding Rockets,* 171.
18. Michael L. Coats transcript, NASA Johnson Space Center Oral History Project, November 9, 2012, https://historycollection.jsc.nasa.gov/JSCHistoryPortal/history/oral_histories/CoatsML/CoatsML_11-9-12.htm.
19. Uri, "35 Years Ago: STS-41D—First Flight of Space Shuttle Discovery." Also, Mullane, *Riding Rockets,* 173.
20. "STS-41D Press Kit," August 1984, p. 15, https://historycollection.jsc.nasa.gov/JSCHistoryPortal/history/shuttle_pk/pk/Flight_012_STS-41D_Press_Kit.pdf.
21. "Space Shuttle Flight 12 (STS-41D) Post Flight Press Conference Video," National Space Society, https://space.nss.org/space-shuttle-flight-12-sts-41d-post-flight-press-conference-video.
22. "Space Shuttle Flight 12 (STS-41D) Post Flight Press Conference Video," National Space Society.
23. "STS-41D Press Kit."
24. Uri, "35 Years Ago: STS-41D—First Flight of Space Shuttle Discovery." Also, Mullane, *Riding Rockets,* 172.

25. "Photos: IMAX Cameras in Space," Space.com, April 5, 2012.
26. Mullane, *Riding Rockets,* 177.
27. Mullane, *Riding Rockets,* 177.
28. Associated Press, "Robot Arm Nudges Ice from Discovery's Side," *Daily Oklahoman,* September 4, 1984.
29. Associated Press, "Frizzies in Flight," *El Paso Times*, September 2, 1984.
30. Mullane, *Riding Rockets,* 178.
31. "Space Shuttle Flight 12 (STS-41D) Post Flight Press Conference Video," National Space Society.
32. "STS-41-D Background Briefing/5-21-84/Student Experiment," NASA, 1984, p. 56, https://ia600900.us.archive.org/19/items/STS-41D/Sts-41dPcTranscript.pdf.
33. Author interview with Steve Hawley, September 3, 2020.
34. Steven Hawley transcript, NASA Johnson Space Center Oral History Project, December 4, 2002, https://historycollection.jsc.nasa.gov/JSCHistoryPortal/history/oral_histories/HawleySA/HawleySA_12-4-02.htm.
35. Author interview with Steve Hawley, September 3, 2020.
36. Hawley, NASA Oral History Project, December 4, 2002.
37. Mullane, *Riding Rockets,* 184. Also, Richard M. Mullane transcript, NASA Johnson Space Center Oral History Project, January 24, 2003, https://historycollection.jsc.nasa.gov/JSCHistoryPortal/history/oral_histories/MullaneRM/MullaneRM_1-24-03.htm.
38. Associated Press, "Robot Arm Nudges Ice from Discovery's Side."
39. "STS-41D Space Shuttle Discovery Post-Flight Conference," https://www.youtube.com/watch?v=dRjm_NHEu1M.
40. John Noble Wilford, "Solar Power Mast Unfolded 73 Feet in Astronaut Test," *New York Times*, September 2, 1984, https://www.nytimes.com/1984/09/02/world/solar-power-mast-unfolded-73-feet-in-astronaut-test.html.
41. Mullane, *Riding Rockets,* 180–81.
42. "STS-41D," NASA, https://archive.org/details/STS-41D/624-AAE.wav
43. Brock R. "Randy" Stone transcript, NASA Johnson Space Center Oral History Project, November 14, 2006, https://historycollection.jsc.nasa.gov/JSCHistoryPortal/history/oral_histories/StoneBR/StoneBR_11-14-06.htm.
44. Author interview with Mike Coats, January 19, 2021.
45. *NASA Roundup*, September 21, 1984, vol. 23, no. 17, https://historycollection.jsc.nasa.gov/JSCHistoryPortal/history/roundups/issues/84-09-21.pdf.
46. Al Rossiter Jr., "Shuttle 'Ice Busters' Whip Problem," *The Sacramento Bee* (UPI), September 5, 1984.
47. Hawley, NASA Oral History Project, December 4, 2002.
48. "Message to the Congress Transmitting the Aeronautics and Space Report of the President," September 21, 1984, https://www.reaganlibrary.gov/archives/speech/message-congress-transmitting-aeronautics-and-space-report-president-0.

Chapter 13: Much Have I Traveled

1. Author interview with Kathy Sullivan, June 14, 2021.
2. Henry S. F. Cooper Jr., *Before Lift-Off: The Making of a Space Shuttle Crew*. Baltimore: Johns Hopkins University Press, 1987, pp. 40 and 84.

3. Author interview with Kathy Sullivan, March 10, 2021.
4. Author interview with Kathy Sullivan, October 14, 2020.
5. Sullivan, *Handprints on Hubble*, 53.
6. NASA was now using a new arcane number sequence to name its shuttle flights instead of 1,2,3,4 . . . The rumor was that a superstitious NASA administrator was trying to avoid an STS-13. 41-G was the thirteenth launch of the shuttle.
7. Sherr, *Sally Ride*, 188.
8. Author interview with Kathy Sullivan, June 14, 2021.
9. Author interview with Kathy Sullivan, June 14, 2021.
10. Author interview with Kathy Sullivan, June 14, 2021.
11. Sullivan, *Handprints on Hubble*, 63–64.
12. Sullivan, *Handprints on Hubble*, 64.
13. "Space Shuttle Flight 13 (STS-41G) Post Flight Presentation," National Space Society, https://www.youtube.com/watch?v=HHlgEY8fXpI.
14. "Space Shuttle Flight 13 (STS-41G) Post Flight Presentation."
15. "STS-41D Press Kit," NASA, August 1984, https://historycollection.jsc.nasa.gov/JSCHistoryPortal/history/shuttle_pk/pk/Flight_012_STS-41D_Press_Kit.pdf.
16. Sullivan, *Handprints on Hubble*, 66.
17. Sullivan, *Handprints on Hubble*, 66.
18. Sullivan, *Handprints on Hubble*, 66.
19. Sullivan, *Handprints on Hubble*, 68.
20. Sullivan, *Handprints on Hubble*, 68.
21. Sullivan, *Handprints on Hubble*, 68.
22. Sullivan, *Handprints on Hubble*, 69–70.
23. John Uri, "35 Years Ago: STS-41G—A Flight of Many Firsts," NASA, October 3, 2019, https://www.nasa.gov/feature/35-years-ago-sts-41g-a-flight-of-many-firsts. Also, Sullivan, *Handprints on Hubble*, 53.
24. Sullivan, *Handprints on Hubble*, 70.
25. Sullivan, *Handprints on Hubble*, 60.
26. Sullivan, *Handprints on Hubble*, 68.
27. Author interview with Kathy Sullivan, March 10, 2021. Also, Sullivan, *Handprints on Hubble*, 70.
28. Sullivan, *Handprints on Hubble*, 70.
29. Sullivan, *Handprints on Hubble*, 54.
30. Sullivan, *Handprints on Hubble*, 54. Also, "Space Shuttle Flight 13 (STS-41G) Post Flight Presentation."
31. "Space Shuttle Flight 13 (STS-41G) Post Flight Presentation."
32. Sullivan, *Handprints on Hubble*, 72.
33. Sullivan, *Handprints on Hubble*, 72.
34. Sullivan, *Handprints on Hubble*, 72.
35. Sullivan, *Handprints on Hubble*, 72–73.
36. Sullivan, *Handprints on Hubble*, 72–73.
37. Sullivan, *Handprints on Hubble*, 18.

Chapter 14: Send Me In, Coach

1. Author interview with Anna Fisher, November 12, 2020.
2. He was rejected once as a child and then again in 1978, when Anna was accepted.

3. Dick Rose, "Starry-Eyed, Astronaut's Ambitions Go to Mars and Beyond," *Arizona Republic*, May 16, 1981.
4. Rose, "Starry-Eyed, Astronaut's Ambitions Go to Mars and Beyond."
5. Author interview with Anna Fisher, November 12, 2020.
6. L. Tom Shaw Jr. and James R. Womack, "Launch Vehicle Integration Requirements for SP-100," Technical Information Report, SP-100 Program, March 1984, https://www.osti.gov/servlets/purl/10183807.
7. Author interview with Anna Fisher, November 12, 2020.
8. Megan A. Sholar, "The History of Family Leave Policies in the United States," *The American Historian*, https://www.oah.org/tah/issues/2016/november/the-history-of-family-leave-policies-in-the-united-states.
9. Jessica Contrera, "She Was Pregnant When NASA Offered to Send Her to Space. Anna Fisher Didn't Hesitate," *Washington Post*, May 11, 2019, https://www.washingtonpost.com/history/2019/05/11/she-was-pregnant-when-nasa-offered-send-her-space-anna-fisher-didnt-hesitate.
10. Fisher, NASA Oral History Project, February 17, 2009.
11. Contrera, "She Was Pregnant When NASA Offered to Send Her to Space."
12. Author interview with Anna Fisher, March 29, 2020.
13. Aimee Lee Ball, "When Mom Is an Astronaut," *Boston Globe,* October 28, 1984.
14. "STS 51-A Pre-Flight Crew Press Conference," NASA, November 12 and 14, 1984, p. 139, https://archive.org/stream/STS-51A/Sts-51aPcTranscript_djvu.txt.
15. Ball, "When Mom Is an Astronaut."
16. Ball, "When Mom Is an Astronaut."
17. Author interview with Anna Fisher, December 11, 2020.
18. *The Early Show,* CBS, July 9, 2011, https://archive.org/details/WJZ_20110709_120000_The_Early_Show/start/5640/end/5700.
19. John Noble Wilford, "Discovery Nears a Stray Satellite," *New York Times,* November 12, 1984, https://www.nytimes.com/1984/11/12/us/discovery-nears-a-stray-satellite.html.
20. Author interview with Anna Fisher, April 19, 2021.
21. Author interview with Anna Fisher, November 2, 2020.
22. Wilford, "Discovery Nears a Stray Satellite." Also, Ben Evans, "Fixing Solar Max: 30 Years Since Mission 41C (Part 2)," AmericaSpace, April 6, 2014, https://www.americaspace.com/2014/04/06/fixing-solar-max-30-years-since-mission-41c-part-2/.
23. Author interview with Anna Fisher, March 29, 2021.
24. Evans, "Fixing Solar Max: 30 Years Since Mission 41C (Part 2)."
25. Kenneth A. Young transcript, NASA Johnson Space Center Oral History Project, June 6, 2001, https://historycollection.jsc.nasa.gov/JSCHistoryPortal/history/oral_histories/YoungKA/YoungKA_6-6-01.htm.
26. Joseph P. Allen transcript, NASA Johnson Space Center Oral History Project, November 18, 2004, https://historycollection.jsc.nasa.gov/JSCHistoryPortal/history/oral_histories/AllenJP/AllenJP_11-18-04.htm.
27. Author interview with Anna Fisher, March 29, 2020.
28. Author interview with Anna Fisher, April 19, 2021.
29. Author interview with Anna Fisher, March 29, 2021.
30. Author interview with Anna Fisher, March 29, 2021.
31. Author interview with Anna Fisher, March 29, 2021.

32. Author interview with Anna Fisher, March 29, 2021.
33. "Space Shuttle Mission STS-51 A Press Kit," edited by Richard W. Orloff, NASA, April 1985, https://spacepresskit.files.wordpress.com/2012/08/sts-51a.pdf, pp. 5–6, 10. Also, Anna L. Fisher transcript, NASA Johnson Space Center Oral History Project, March 3, 2011, https://historycollection.jsc.nasa.gov/JSCHistoryPortal/history/oral_histories/FisherAL/FisherAL_3-3-11.htm.
34. John Noble Wilford, "Astronauts Deploy Canadian Satellite," *New York Times*, November 10, 1984.
35. Author interview with Anna Fisher, Rhea Seddon, and Kathy Sullivan, November 2, 2020.
36. Allen, NASA Oral History Project, November 18, 2004.
37. "Space Shuttle Mission STS-51 A Press Kit."
38. Allen, NASA Oral History Project, November 18, 2004.
39. Joseph Allen and Russell Martin, *Entering Space: An Astronaut's Odyssey*. New York: Stewart, Tabori & Chang, 1986, p. 113.
40. Allen and Russell, *Entering Space*, 224.
41. Allen, NASA Oral History Project, November 18, 2004.
42. Allen, NASA Oral History Project, November 18, 2004.
43. Author interview with Rick Hauck, May 2, 2022.
44. Allen and Russell, *Entering Space*, 233.
45. Fisher, NASA Oral History Project, March 3, 2011.
46. "Dale Gardner Atrapa El Satélite Westar 6 (NASA)," https://www.youtube.com/watch?v=VR6jlWBeHvY.
47. Allen, NASA Oral History Project, November 18, 2004.
48. Allen and Russell, *Entering Space*, 236.
49. "Shuttle Mission 51-A EVA Transcripts November 12 and 14, 1984," NASA, https://ia800607.us.archive.org/28/items/NasaAudioHighlightReels/Sts-51aTranscriptProfile.pdf, 92-93.
50. Evan Thomas, "Roaming the High Frontier," *Time*, November 26, 1984, https://time.com/vault/issue/1984-11-26/page/23.
51. Author interview with Anna Fisher, March 29, 2021.
52. Contrera, "She Was Pregnant When NASA Offered to Send Her to Space."
53. Author interview with Anna Fisher, April 19, 2021.
54. "Space Shuttle Flight 14 (STS-51A) Post Flight Presentation," National Space Society, https://www.youtube.com/watch?v=jSefxa9Ssl.
55. Author interview with Anna Fisher, April 19, 2021.
56. Thomas, "Roaming the High Frontier."
57. Thomas, "Roaming the High Frontier."
58. Author interview with Anna Fisher, March 1, 2022.

Chapter 15: Blood Moon

1. Jeffrey Richelson and William M. Arkin, "Spy Satellites: 'Secret,' But Much Is Known," *Washington Post*, January 6, 1985.
2. Thomas K. Mattingly II transcript, NASA Johnson Space Center Oral History Project, April 22, 2002, https://historycollection.jsc.nasa.gov/JSCHistoryPortal/history/oral_histories/MattinglyTK/MattinglyTK_4-22-02.htm Oral History.
3. Mattingly, NASA Oral History Project, April 22, 2002.

4. Pauline Yoshihashi, "3 Boys' Dreams of Space, 3 Deaths in the Sky; Ellison Onizuka," *New York Times*, February 11, 1986, https://www.nytimes.com/1986/02/11/us/3-boys-dreams-of-space-3-deaths-in-the-sky-ellison-onizuka.html.
5. Dennis Ogawa, "Scouting Helped Him to 'Be Prepared,'" *Hawaii Tribune-Herald*, February 9, 1986.
6. Ogawa, "Scouting Helped Him to 'Be Prepared,'" 17.
7. Ogawa, *Ellison S. Onizuka*, 29.
8. Jack Anderson and Dale Van Atta, "Hawaii's Samurai of Space," *Honolulu Star-Bulletin,* February 8, 1986.
9. Jack Anderson, "Hawaii's Son, Ellison Onizuka, Followed His Dream to Space," *Santa Cruz Sentinel,* February 9, 1986.
10. Ogawa, "Scouting Helped Him to 'Be Prepared,'" 56.
11. Dale Van Atta, "A High-Flying Dream Comes True," *Honolulu Star-Bulletin,* January 23, 1985.
12. Van Atta, "A High-Flying Dream Comes True."
13. Dennis M. Ogawa and Glen Grant, *Ellison S. Onizuka: A Remembrance.* Honolulu, HI: Onizuka Memorial Committee, 1987, p. 103.
14. Molly Solomon, "Once Lost, Internment Camp in Hawaii Now a National Monument," NPR, March 16, 2015, https://www.npr.org/sections/codeswitch/2015/03/16/393284680/in-hawaii-a-wwii-internment-camp-named-national-monument.
15. "Historical Overview—Hono'uli'uli National Historic Site (US National Park Service)," National Park Service, 2022, https://www.nps.gov/hono/learn/historical-overview.htm.
16. Ogawa and Grant, *Ellison S. Onizuka,* 32.
17. Ogawa and Grant, *Ellison S. Onizuka,* 95.
18. William J. Broad, "Pentagon Leaves the Shuttle Program," *New York Times,* August 7, 1989, https://www.nytimes.com/1989/08/07/us/pentagon-leaves-the-shuttle-program.html.
19. David, *Spies and Shuttles*, 209.
20. John M. Logsdon, ed., *Exploring the Unknown: Selected Documents in the History of the US Civilian Space Program,* Vol. 1. NASA: 1996, p. 455.
21. Sara Fritz, "House Chops Reagan Request, Votes $2.5 Billion for 'Star Wars' Spending," *Los Angeles Times,* June 21, 1985, https://www.latimes.com/archives/la-xpm-1985-06-21-mn-11520-story.html.
22. Cassutt, "The Secret Space Shuttles."
23. Fred Kaplan, "Military Has Influence on Program," *Odessa American* (reprinted from *Boston Globe*), April 20, 1986.
24. Victor K. McElheny, "Aerospace Contracts, Put at $1 Billion, Hinge on Debate Over Space Shuttles," *New York Times,* August 24, 1976, https://www.nytimes.com/1976/08/24/archives/aerospace-contracts-put-at-1-billion-hinge-on-debate-over-space.html.
25. "B-1813134 Construction of Space Shuttle Facilities at Vandenberg Air Force Base," June 2, 1977, https://www.gao.gov/assets/b-183134.pdf. Also, Robert Lindsey, "Getting Vandenberg off the Ground," *New York Times,* March 24, 1985, https://www.nytimes.com/1985/03/24/weekinreview/getting-vandenberg-off-the-ground.html.

26. Memorandum from Secretary of Defense to Secretaries of the Military Departments, et al., "Defense Space Launch Strategy," February 7, 1984, https://www.nap.edu/read/19350/chapter/14. Also, David, *Spies and Shuttles*, 227.
27. Thomas O'Toole, "Air Force Rocket Plan May Undermine Shuttle, NASA Chief Tells Hill," *Washington Post*, August 1, 1984, https://www.washingtonpost.com/archive/politics/1984/08/01/air-force-rocket-plan-may-undermine-shuttle-nasa-chief-tells-hill/f9a2dcdd-7f4e-4548-9667-a07af74a9449.
28. David, *Spies and Shuttles*, 226–29.
29. McDonald, *Truth, Lies, and O-Rings*, 47.
30. John Noble Wilford, "Shuttle Launched on Secret Mission," *New York Times*, January 25, 1985, https://www.nytimes.com/1985/01/25/us/shuttle-launched-on-secret-mission.html.
31. McDonald, *Truth, Lies, and O-Rings*, 47.
32. Croft and Youkauskas, *Come Fly with Us*, 185.
33. "Inertial Upper Stage," NASA STI Program, August 28, 2011, https://www.youtube.com/watch?v=d9JGJimUAAE.
34. Ben Evans, "Remembering Shuttle Discovery's Miracle Mission 51C, 35 Years On (Part 2)," *AmericaSpace*, January 26, 2020, https://www.americaspace.com/2020/01/26/remembering-shuttle-discoverys-miracle-mission-51c-35-years-on-part-2.
35. Don Oberdorfer, "US Sounds Out Soviets on SALT Flights," *Washington Post*, May 25, 1979, https://www.washingtonpost.com/archive/politics/1979/05/25/us-sounds-out-soviets-on-salt-flights/41c94e2a-1eb9-47a9-ac51-065d316987fb.
36. "Astronauts Onizuka and Shriver pose in middeck," STS51C, NASA, https://science.ksc.nasa.gov/mirrors/images/images/pao/STS51C/10062003.jpg.
37. "Japanese Internment Camps," History.com, October 29, 2021, https://www.history.com/topics/world-war-ii/japanese-american-relocation.
38. Author interview with June Scobee, June 10, 2021.
39. *NASA Facts: Space Shuttle Solid Rocket Booster Retrieval Ships*. Ebook. Merritt Island, FL: NASA, 2004, https://www.nasa.gov/centers/kennedy/pdf/167446main_SRBships06.pdf.
40. T.D. Wood, H.S. Kanner, D.M. Freeland, and D.T. Olson, "Solid Rocket Booster (SRB) Flight System Integration at Its Best," in *JANNAF 6th Liquid Propulsion Subcommittee Meeting* (No. M11-1379), December 2011, https://ntrs.nasa.gov/api/citations/20120003006/downloads/20120003006.pdf.
41. *SRB Retrieval Ships*, NASA, 2013, https://www.nasa.gov/content/srb-retrieval-ships-videos/.
42. "Solid Rocket Booster (SRB) Refurbishment Practices," https://llis.nasa.gov/lesson/836.
43. Each solid rocket booster has three segments, connected by three joints. The first is the forward segment, the piece closest to the top of the booster and the booster forward assembly. The forward segment is connected to the next segment, the forward center segment, by the forward field joint. The forward center segment is connected to the aft center segment, the piece closest to the base of the booster and the aft booster, by the center field joint. The aft field joint connects the aft center segment to the aft booster at the base of the booster. See S. Mohammad Gharouni, Hamid M. Panahiha, and Jafar Eskandari Jam, "Space Shuttle Solid Rocket Motor (SRM) Field Joint: Review Paper," *Metallurgical and Materials*

Engineering, September 30, 2014, vol. 20, no. 3, pp. 155–64, https://www.semanticscholar.org/paper/SPACE-SHUTTLE-SOLID-ROCKET-MOTOR-(SRM)-FIELD-JOINT%3A-Gharouni-Panahiha/6afbbbf74dfae5b11804eb7a6ec89b4a3a136339/figure/1.

44. "Volume 2: Appendix H—Flight Readiness Review Treatment of O-ring Problems," *Report of the Presidential Commission on the Space Shuttle Challenger Accident*, https://history.nasa.gov/rogersrep/v2apph.htm.
45. "Volume 2: Appendix H—Flight Readiness Review Treatment of O-ring Problems," 46.
46. McDonald, *Truth, Lies, and O-Rings*, 45.
47. McDonald, *Truth, Lies, and O-Rings*, 39.
48. McDonald, *Truth, Lies, and O-Rings*, 39.
49. McDonald, *Truth, Lies, and O-Rings*, 40.
50. "Chapter VI: An Accident Rooted in History," *Report of the Presidential Commission on the Space Shuttle Challenger Accident*, https://history.nasa.gov/rogersrep/v1ch6.htm.
51. McDonald, *Truth, Lies, and O-Rings*, 55.
52. Hearing of the Presidential Commission on the Space Shuttle *Challenger* Accident, February 25, 1986 session transcript, Section 1390–1391, *Report of the Presidential Commission on the Space Shuttle Challenger Accident,* Volume 4 Index, part 7, https://history.nasa.gov/rogersrep/v4part7.htm.
53. Hearing of the Presidential Commission on the Space Shuttle *Challenger* Accident, February 25, 1986 session transcript, Section 1394, *Report of the Presidential Commission on the Space Shuttle Challenger Accident,* Volume 4 Index, part 7, https://history.nasa.gov/rogersrep/v4part7.htm.
54. "Chapter VI: An Accident Rooted in History," *Report of the Presidential Commission on the Space Shuttle Challenger Accident*, https://history.nasa.gov/rogersrep/v1ch6.htm.
55. David, *Spies and Shuttles*, 229. Also, Edward C. "Pete" Aldridge transcript, NASA Oral History Project, May 29, 2009, https://historycollection.jsc.nasa.gov/JSCHistoryPortal/history/oral_histories/NASA_HQ/Administrators/AldridgeEC/AldridgeEC_5-29-09.htm.

Chapter 16: The Prince and the Politician

1. Seddon, *Go for Orbit*, 249.
2. Seddon, *Go for Orbit*, 244.
3. Sherr, *Sally Ride,* 191.
4. Author interview with Rhea Seddon and Hoot Gibson, October 5, 2020.
5. Author interview with Rhea Seddon and Hoot Gibson, October 5, 2020.
6. Seddon, *Go for Orbit*, 275.
7. Seddon, *Go for Orbit*, 274. Also, "Space Shuttle Fact Sheet," NASA, p. 1, https://www.nasa.gov/centers/johnson/pdf/167751main_FS_SpaceShuttle508c.pdf. And, "How Much Did the Space Shuttle Weigh?" https://coolcosmos.ipac.caltech.edu/ask/268-How-much-did-the-Space-Shuttle-weigh.
8. Seddon, *Go for Orbit*, 275.
9. Seddon, *Go for Orbit*, 274–79. Also, William Harwood, "Space Shuttle Brakes:

Growing Pains," UPI, May 25, 1985, https://www.upi.com/Archives/1985/05/25/Space-shuttle-brakes-growing-pains/8297485841600.

10. Bob Drogin, "Cuts in NASA Oversight of Shuttle Program Hit: Quality Control Allegedly Suffered; Workers Complain That Speed-Ups Have Led to Accidents," *Los Angeles Times*, February 5, 1986, https://www.latimes.com/archives/la-xpm-1986-02-05-mn-4352-story.html. Also, McConnell, *Challenger: A Major Malfunction*, 67.
11. McConnell, *Challenger: A Major Malfunction*, 66.
12. Drogin, "Cuts in NASA Oversight of Shuttle Program Hit."
13. "Chapter VIII: Pressures on the System," *Report of the Presidential Commission on the Space Shuttle Challenger Accident*, https://history.nasa.gov/rogersrep/v1ch8.htm. Program Manager Arnold Aldrich would explain the cuts by telling the Rogers Commission, ". . . intentional decisions were made to defer the heavy build-up of spare parts procurements in the program so that the funds could be devoted to other more pressing activities. It was a regular occurrence for several annual budget cycles."
14. McConnell, *Challenger: A Major Malfunction*, 25.
15. McConnell, *Challenger: A Major Malfunction*, 217–218.
16. Trento and Trento, "Why *Challenger* Was Doomed."
17. Mark, *The Space Station*, 215. Mark went on to become chancellor of the University of Texas System.
18. Kathleen Day, "Design Work Awarded for Space Station: Rockwell, McDonnell to Vie for Final Job; Lockheed Loses Out," *Los Angeles Times*, April 16, 1985, https://www.latimes.com/archives/la-xpm-1985-04-16-fi-23389-story.html.
19. Marcus Lindroos, "Power Tower Space Station," Astronautix, http://astronautix.com/p/powertowersestation-1984.html.
20. Robert C. Cohen, "Nations Join US Space Station Project for Political Reasons," *The Ottawa Citizen*, May 4, 1985, https://www.newspapers.com/image/464258589/.
21. Gregory, NASA Oral History Project. Also, "Space Shuttle Atlantis, Astronaut Fred Gregory at Kennedy Space Center," https://www.youtube.com/watch?v=iy5A33_tm2M.
22. McDonald, *Truth, Lies, and O-Rings*, 52.
23. McDonald, *Truth, Lies, and O-Rings*, 51–52. Also, author interview with Fred Gregory, June 21, 2019.
24. McConnell, *Challenger: A Major Malfunction*, 114.
25. Kriti Mehrotra, "Where Is Larry Mulloy Now?" *TheCinemaholic*, September 15, 2020, https://thecinemaholic.com/where-is-larry-mulloy-now.
26. McDonald, *Truth, Lies, and O-Rings*, 58–63.
27. McDonald, *Truth, Lies, and O-Rings*, 58.
28. McDonald, *Truth, Lies, and O-Rings*, 58.
29. McDonald, *Truth, Lies, and O-Rings*, 60.
30. "Memo from R. M. Boisjoly to R. K. Lund," National Archives Catalog, July 31, 1985, https://catalog.archives.gov/id/596263.
31. McConnell, *Challenger: A Major Malfunction*, 185–87.
32. McDonald, *Truth, Lies, and O-Rings*, 60.

33. McDonald, *Truth, Lies, and O-Rings*, 59.
34. McDonald, *Truth, Lies, and O-Rings*, 60.
35. Michael Lafferty, "Astronaut to signal end of holy month," *Florida Today*, May 5, 1985.
36. Fabian, NASA Oral History Project.
37. K.L. Bedingfield and R.D. Leach, "Spacecraft System Failures and Anomalies Attributed to the Natural Space Environment," NASA, August 1996, ed. M.B. Alexander, p. 32, https://ntrs.nasa.gov/api/citations/19960050463/downloads/19960050463.pdf.
38. Fabian, NASA Oral History Project.
39. *Space Shuttle Flight 18 (STS-51G) Post Flight Presentation*, NASA, 1985, 11:09–13:28, https://space.nss.org/space-shuttle-flight-18-post-flight-press-conference-video/.
40. "The Incredible Shannon Lucid," NASA, April 1, 2004. The SPARTAN-1 stood for the Shuttle Pointed Autonomous Research Tool for AstroNomy.
41. Ben Evans, "'A Job? But You're a Girl!': The Triumphant Career of Shannon Lucid," *AmericaSpace*, February 2, 2012, https://www.americaspace.com/2012/02/02/a-job-but-youre-a-girl-the-triumphant-career-of-shannon-lucid.
42. Evans, "'A Job? But You're a Girl!'"
43. Fabian, NASA Oral History Project.
44. Fabian, NASA Oral History Project.
45. Emily Carney, "A Deathblow to the Death Star: The Rise and Fall of NASA's Shuttle-Centaur," Ars Technica, October 8, 2015, https://arstechnica.com/science/2015/10/dispatches-from-the-death-star-the-rise-and-fall-of-nasas-shuttle-centaur.
46. McDonald, *Truth, Lies, and O-Rings*, 46.
47. Besides switching out the engines on a regular basis, NASA also awarded a contract worth hundreds of millions of dollars to the Pratt and Whitney division of United Technologies Corporation for the "development of improved hydrogen and oxygen turbo-pumps for the SSMEs." McDonald, *Truth, Lies, and O-Rings*, 45.
48. Stan M. Barauskas transcript, NASA Johnson Space Center Oral History Project, August 24, 2010, https://historycollection.jsc.nasa.gov/JSCHistoryPortal/history/oral_histories/STS-R/BarauskasSM/BarauskasSM_8-24-10.htm.
49. Lockheed, which had taken over shuttle refurbishment from Rockwell and Martin Marietta just two months prior, as a result of Beggs's Shuttle Sweepstakes had been responsible for checking out the units before launch. Barauskas, NASA Oral History Project.
50. "Volume 2: Appendix H—Flight Readiness Review Treatment of O-ring Problems."
51. McDonald, *Truth, Lies, and O-Rings*, 65.
52. McDonald, *Truth, Lies, and O-Rings*, 65–66.
53. McDonald, *Truth, Lies, and O-Rings*, 67.
54. McDonald, *Truth, Lies, and O-Rings*, 67.
55. McDonald, *Truth, Lies, and O-Rings*, 66–70.
56. Author interview with Anna Fisher, December 11, 2020.
57. Paul Dye, *Shuttle, Houston: My Life in the Center Seat of Mission Control*. New York: Hachette Books, 2020, p. 231.

Chapter 17: Beautiful, Like America

1. Sherr, *Sally Ride,* 191.
2. "Women's Caucus Looks to the Future," Bill Peterson, *Washington Post,* July 2, 1985, https://www.washingtonpost.com/archive/politics/1985/07/02/womens-caucus-looks-to-the-future/fab90ad2-d0d8-44c2-99b1-ce66f56227ad.
3. Sherr, *Sally Ride,* 194.
4. Sherr, *Sally Ride,* 194.
5. "Loving Sally Ride," Tam O'Shaughnessy interview with Madeline Sofia, June 24, 2021, *NPR,* https://www.npr.org/2021/06/22/1009098412/loving-sally-ride.
6. Sherr, *Sally Ride,* 195.
7. "Loving Sally Ride."
8. "Loving Sally Ride."
9. "Loving Sally Ride."
10. Sherr, *Sally Ride,* 151.
11. Chelsea Gohd, "This Pride, Be Inspired by Sally Ride's Legacy," Space.com, June 18, 2018, https://www.space.com/40916-sally-ride-pride-inspiration-legacy.html.
12. Sherr, *Sally Ride,* 197.
13. Sherr, *Sally Ride,* 142.
14. Sherr, *Sally Ride,* 199.
15. Gibson, NASA Oral History Project.
16. Cassutt, *The Astronaut Maker,* 288.
17. Charles F. Bolden transcript, NASA Johnson Space Center Oral History Project, January 6, 2004, https://historycollection.jsc.nasa.gov/JSCHistoryPortal/history/oral_histories/BoldenCF/BoldenCF_1-6-04.htm.
18. Nelson, NASA Oral History Project.
19. "NASA Nominee Bill Nelson Has Long History With Space Policy," March 25, 2021, https://www.aip.org/fyi/2021/nasa-nominee-bill-nelson-has-long-history-space-policy.
20. Bill Nelson and Jamie Buckingham, *Mission: An American Congressman's Voyage to Space.* United States: Harcourt Brace Jovanovich, 1988, pp. 35–36.
21. Rhea Seddon, "A Disorderly Christmas," https://astronautrheaseddon.com/a-disorderly-christmas.
22. Ben Evans, "The Real Mission Impossible: Remembering STS-61C's Quest for Space," AmericaSpace, January 12, 2020, https://www.americaspace.com/2020/01/12/the-real-mission-impossible-remembering-sts-61cs-quest-for-space.
23. Gibson, NASA Oral History Project, January 22, 2016.
24. Gibson, NASA Oral History Project, January 22, 2016.
25. Ben Evans, "The Real Mission Impossible: Remembering STS-61C's Quest for Space (Part 2)," AmericaSpace, January 12, 2020, https://www.americaspace.com/2020/01/12/the-real-mission-impossible-remembering-sts-61cs-quest-for-space, https://www.americaspace.com/2016/01/10/the-real-mission-impossible-30-years-since-mission-61c-part-2.
26. Bolden, NASA Oral History Project.
27. Robert D. Legler and Floyd V. Bennett, *Space Shuttle Missions Summary.*

Houston, TX: NASA, 2011, https://ntrs.nasa.gov/api/citations/20110001406/downloads/20110001406.pdf.

28. Steven A. Hawley transcript, NASA Johnson Space Center Oral History Project, December 17, 2002, https://historycollection.jsc.nasa.gov/JSCHistoryPortal/history/oral_histories/HawleySA/HawleySA_12-17-02.htm.
29. "STS-61C launch & landing (1-12-86)," https://www.youtube.com/watch?v=-l5UECsgzoU.
30. Evans, "The Real Mission Impossible: 30 Years Since Mission 61C (Part 2)."
31. Gibson, NASA Oral History Project, January 22, 2016.
32. Author interview with Steve Hawley, September 3, 2020.
33. Gibson, NASA Oral History Project, January 22, 2016.
34. David Middlecamp and Joe Johnston, "Vandenberg Was Poised to Be a Space Shuttle Launch Site—Then Tragedy Struck," *San Luis Obispo Tribune*, November 16, 2018.
35. "Space Shuttle Payload Flight Assignments. Technical Memorandum," NASA, June 1, 1985, https://ntrs.nasa.gov/citations/19850021678.
36. Ben Evans, "Up Against a Wall: What 1986 Might Have Been," *AmericaSpace*, February 8, 2015, https://www.americaspace.com/2015/02/08/up-against-a-wall-what-1986-might-have-been.
37. Carney, "A Deathblow to the Death Star."
38. Frederick H. Hauck transcript, NASA Johnson Space Center Oral History Project, November 20, 2003, https://historycollection.jsc.nasa.gov/JSCHistoryPortal/history/oral_histories/HauckFH/HauckFH_11-20-03.htm.
39. Hauck, NASA Oral History Project, November 20, 2003.
40. Author interview with Rick Hauck, May 2, 2022.
41. Trento and Trento, "Why *Challenger* Was Doomed."
42. "Apology for Ex-NASA Chief," *New York Times*, July 7, 1988, https://www.nytimes.com/1988/07/07/us/apology-for-ex-nasa-chief.html.
43. Cassutt, *The Astronaut Maker*, 267.
44. Cassutt, *The Astronaut Maker*, 267.
45. Associated Press, "Jesse Moore, A Top NASA Aide, Quits," *New York Times*, February 6, 1987, https://www.nytimes.com/1987/02/06/us/jesse-moore-a-top-nasa-aide-quits.html.
46. Associated Press, "Says Space Trip Safer Than Driving in Paris," *Vidette-Messenger of Porter County*, September 23, 1982, https://www.newspapers.com/image/333714908/.
47. Walter Pincus, "NASA's Push to Put Citizen in Space Overtook Fully 'Operational' Shuttle," *Washington Post*, March 5, 1986, https://www.washingtonpost.com/archive/politics/1986/03/05/nasas-push-to-put-citizen-in-space-overtook-fully-operational-shuttle/29fe2714-39b7-40dd-b15e-073441de636e.
48. Sarah Oates, "Applications for Ride on Shuttle Open to Journalists," *Los Angeles Times*, October 25, 1985, https://www.latimes.com/archives/la-xpm-1985-10-25-mn-14173-story.html.
49. Nathalia Holt, "Private Spaceflight and the Legacy of the *Challenger* Disaster," *Atlantic*, January 28, 2016, https://www.theatlantic.com/science/archive/2016/01/challenger-space-shuttle-anniversary-spacex/431685. Also, Jackie Wattles and Kerry Flynn, "There's a Long History of Failed Attempts to Put American Journalists in Space. Now, Michael Strahan is Going," CNN,

December 10, 2021, https://www.cnn.com/2021/12/10/media/blue-origin-strahan-history-journalists-in-space-scn/index.html.

50. Alan Boyle, "NASA Confirms Talks to Fly Big Bird on Doomed Shuttle *Challenger*," NBC News, May 4, 2015, https://www.nbcnews.com/science/weird-science/nasa-confirms-talks-fly-big-bird-doomed-Shuttle-challenger-n353521.
51. Philip M. Boffey, "First Shuttle Ride by Private Citizen to go to Teacher," *New York Times,* August 28, 1984, https://www.nytimes.com/1984/08/28/science/first-shuttle-ride-by-private-citizen-to-go-to-teacher.html.
52. "George Bush Announces Christa McAuliffe as the First Teacher to go to Space," The Space Archive, December 10, 2019, https://www.youtube.com/watch?v=7McTr25xJuA.
53. "Announcement by George Bush of the First Teacher to be on the Space Shuttle on July 19, 1985," Reagan Library, January 10, 2017, https://www.youtube.com/watch?v=04V8FNCJX0g.
54. "Teacher in Space," NASA STI Program, https://www.youtube.com/watch?v=0XqlOxvkB2o.
55. Rachel Chang, "How Teacher Christa McAuliffe Was Selected for the Disastrous *Challenger* Mission," Biography.com, September 15, 2020, https://www.biography.com/news/christa-mcauliffe-challenger-story.
56. Associated Press, "Future Astronaut Receives Shirt That Fits Her to a T," *Los Angeles Times,* September 5, 1985.
57. Beggs, NASA Oral History Project.
58. Mullane, *Riding Rockets,* 214. Also, author interview with Mary Cleave, December 9, 2020.
59. Yoshihashi, "3 Boys' Dreams of Space."
60. Author interview with June Scobee, June 10, 2021.
61. Cassutt, *The Astronaut Maker,* 277–78.
62. McConnell, *Challenger: A Major Malfunction,* 98.
63. John Noble Wilford, "A Teacher Trains for Outer Space," *New York Times,* January 5, 1986, https://www.nytimes.com/1986/01/05/magazine/a-teacher-trains-for-outer-space.html.
64. Author interview with Jim Bagian, April 7, 2021.
65. McConnell, *Challenger: A Major Malfunction,* 68–74.
66. McConnell, *Challenger: A Major Malfunction,* 74.
67. Cassutt, *The Astronaut Maker,* 290.
68. McConnell, *Challenger: A Major Malfunction,* 93.
69. McConnell, *Challenger: A Major Malfunction,* 94.
70. "*Challenger* Anniversary Resource Tape," NASA STI Program, https://www.youtube.com/watch?v=IviOm71Iml0&t=115s.
71. "*Challenger* Anniversary Resource Tape," NASA STI Program.
72. "*Challenger* Anniversary Resource Tape," NASA STI Program.
73. McConnell, *Challenger: A Major Malfunction,* 95.
74. McConnell, *Challenger: A Major Malfunction,* 104.

Chapter 18: Godspeed

1. Author interview with Sylvia Salinas Stottlemyer, July 20, 2020.
2. Author interview with Sylvia Salinas Stottlemyer, July 20, 2020.

3. Author interview with Sylvia Salinas Stottlemyer, July 20, 2020.
4. Author interview with Sylvia Salinas Stottlemyer, July 20, 2020.
5. Author interview with Sylvia Salinas Stottlemyer, July 20, 2020.
6. Author interview with Sylvia Salinas Stottlemyer, July 28, 2020.
7. Author interview with Sylvia Salinas Stottlemyer, July 28, 2020.
8. Author interview with Sylvia Salinas Stottlemyer, July 28, 2020.
9. Author interview with Sylvia Salinas Stottlemyer, January 29, 2021.
10. Author interview with Sylvia Salinas Stottlemyer, January 29, 2021.
11. Author interview with Sylvia Salinas Stottlemyer, January 29, 2021.
12. "'I'm Not Scared': Space Shuttle *Challenger* Astronaut Dr. Judith Resnik," Global News Archive, January 8, 1986, https://globalnews.ca/video/2481377/archive-im-not-scared-space-shuttle-challenger-astronaut-dr-judith-resnik.
13. "'I'm Not Scared,'" Global News Archive.
14. Author interview with Sylvia Salinas Stottlemyer, January 29, 2021.
15. Author interview with Sylvia Salinas Stottlemyer, January 29, 2021.
16. Author interview with Sylvia Salinas Stottlemyer, January 29, 2021.
17. Tonya Simpson, "NASA Astronaut Ellison Onizuka's Soccer Ball that Survived the *Challenger* Explosion," ESPN, January 28, 2021, https://www.espn.com/espn/story/_/id/30782213/nasa-astronaut-ellison-onizuka-soccer-ball-survived-challenger-explosion.
18. Simpson, "NASA Astronaut Ellison Onizuka's Soccer Ball that Survived the *Challenger* Explosion."
19. Simpson, "NASA Astronaut Ellison Onizuka's Soccer Ball that Survived the *Challenger* Explosion."
20. Jean-Michel Jarre, "Houston—A City in Concert," https://jeanmicheljarre.com/live/houston-a-city-in-concert.
21. Jonathan Hawkins, "The Tragic Triumph of the World's Largest Concert," CNN, April 5, 2016, https://www.cnn.com/2016/04/05/us/challenger-astronaut-and-saxophone/index.html.
22. Hawkins, "The Tragic Triumph of the World's Largest Concert."
23. Hawkins, "The Tragic Triumph of the World's Largest Concert."
24. Clendinen, "Two Pathes to the Stars."
25. Naden, *Ronald McNair*, 85.
26. Clendinen, "Two Pathes to the Stars."
27. Richard Lewis, *Challenger: The Final Voyage*, New York: Columbia University Press, 1988, p. 9.
28. Hugh Harris, *Challenger: An American Tragedy: The Inside Story from Launch Control*. New York: Open Road Media, 2022, p. 26.
29. Elizabeth Howell, "Halley's Comet: Facts about History's Most Famous Comet," Space.com, January 13, 2022, https://www.space.com/19878-halleys-comet.html.
30. John Uri, "35 Years Ago: Remembering *Challenger* and Her Crew," NASA, January 28, 2021, https://www.nasa.gov/feature/35-years-ago-remembering-challenger-and-her-crew.
31. John Noble Wilford, "Spacecraft Converge for First Look at Comet," *New York Times*, March 4, 1986.
32. "*Challenger* Disaster, Crew Spacewalk Training Session," Science Photo Library, https://www.sciencephoto.com/media/811159/view/challenger-disaster-crew-spacewalk-training-session.

33. Tyler David Peterson, "A Fire to Be Lighted: The Training of American Astronauts from 1959 to the Present," dissertation, May 2017, Texas A&M University Library, 2017, p. 422, https://oaktrust.library.tamu.edu/bitstream/handle/1969.1/161476/PETERSON-DISSERTATION-2017.pdf.
34. Peterson, dissertation, 422.
35. Kathryn Casey, "Remembering the *Challenger*," *Ladies Home Journal*, January 1996, https://archive.org/details/ladieshomejourna113janwye/page/n101/mode/2up.
36. The arctic ridge, which forecasters had been following, had stalled in the Florida Panhandle. "*Challenger*'s launch delayed 24 hours," Associated Press, January 26, 1986, *Dayton Daily News*.
37. Howard Benedict, "Shuttle Teacher Rides Bicycle Instead of a Rocket," Associated Press, January 26, 1986, https://apnews.com/article/e9317d1751ec3d749c65bb3c85a7ebe3.
38. Jay Hamburg, "*Challenger* Last Moments," *Orlando Sentinel*, January 28, 2006, https://www.orlandosentinel.com/space/orl-challenger-lastmoments-story.html.
39. David Ignatius, "Media May Have Played a Role in the Shuttle Disaster," *Minneapolis Star and Tribune*, April 1, 1986.
40. "Hatch Problems Delay Launch of the *Challenger*," United Press International, January 27, 1986, *Times-Mail*.
41. Hamburg, "*Challenger* Last Moments." Also, McConnell, *Challenger: A Major Malfunction*, 157.
42. McDonald, *Truth, Lies, and O-Rings*, 95.
43. Kevin Cook, *Burning Blue*. New York: Henry Holt and Co., 2021, pp. 140–141.
44. Before Lockheed had taken over the management of the orbiter, there had been a NASA technician on site that could make quick decisions. Now, the decision making had to work its way up a chain of command off site, which inevitably led to mind numbing delays. McConnell, *Challenger: A Major Malfunction*, 158.
45. Hamburg, "*Challenger* Last Moments."
46. David Ignatius, "Did the Media Goad NASA into the *Challenger* disaster?" *Washington Post*, March 30, 1986, https://www.washingtonpost.com/archive/opinions/1986/03/30/did-the-media-goad-nasa-into-the-challenger-disaster/e0c8669d-a809-4c8d-a4f8-50652b892274.
47. Cook, *Burning Blue*, 140–141.
48. June Scobee Rodgers, *Silver Linings: My Life Before and After Challenger 7*. Macon, GA: Smyth & Helwys Publishing, 2013, p. 146 (Kindle ed.).
49. The verbiage from a 1984 paper titled "Shuttle Centaur Project Perspective," written by Edwin T. Muckley of NASA's Lewis (now Glenn) Research Center, suggested that Jupiter would be the first of many deep-space destinations. Muckley optimistically announced the technology: "It's expected to meet the demands of a wide range of users including NASA, the DoD, private industry, and the European Space Agency (ESA)." Edwin T. Muckley, "Shuttle/Centaur Project Perspective," NASA, April 1, 1984, https://archive.org/details/nasa_techdoc_19850008682.
50. Carney, "A Deathblow to the Death Star."
51. Brian Lada, AccuWeather, https://www.accuweather.com/en/space-news/how-record-setting-cold-contributed-to-the-space-shuttle-challenger-disaster-on-jan-28-1986/337486.

52. *Report of the Presidential Commission on the Space Shuttle Challenger Accident*, https://history.nasa.gov/rogersrep/v1p83.htm.
53. Michael Isikoff, "Thiokol Was Seeking New Contract When Officials Approved Launch," *Washington Post*, February 27, 1986, https://www.washingtonpost.com/archive/politics/1986/02/27/thiokol-was-seeking-new-contract-when-officials-approved-launch/db2ab502-9723-4f90-afd0-8c75bf2237b4.
54. McDonald, *Truth, Lies, and O-Rings*, 95–96.
55. McDonald, *Truth, Lies, and O-Rings*, 97.
56. McDonald, *Truth, Lies, and O-Rings*, 97.
57. McDonald, *Truth, Lies, and O-Rings*, 97.
58. McDonald, *Truth, Lies, and O-Rings*, 103.
59. McDonald, *Truth, Lies, and O-Rings*, 101.
60. McDonald, *Truth, Lies, and O-Rings*, 104.
61. "Presidential Commission on the Space Shuttle *Challenger* Accident," C-SPAN, February 25, 1986, https://www.c-span.org/video/?126036-1/presidential-commission-space-shuttle-challenger-accident, 25:26–26:10.
62. "Presidential Commission on the Space Shuttle *Challenger* Accident," C-SPAN, February 25, 1986, https://www.c-span.org/video/?126036-1/presidential-commission-space-shuttle-challenger-accident, 25:26–26:10.
63. Rudy Abramson, "NASA Unveils Proposed $300-Million Redesign of Shuttle Rocket Joint," *Los Angeles Times*, August 13, 1986, https://www.latimes.com/archives/la-xpm-1986-08-13-mn-17744-story.html.
64. McDonald, *Truth, Lies, and O-Rings*, 104.
65. "Chapter V: The Contributing Cause of the Accident," *Report of the Presidential Commission on the Space Shuttle Challenger Accident*, https://history.nasa.gov/rogersrep/v1ch5.htm.
66. Hamburg, "*Challenger* Last Moments."
67. McDonald, *Truth, Lies, and O-Rings*, 104.
68. Hamburg, "*Challenger* Last Moments."
69. Hamburg, "*Challenger* Last Moments."
70. "Presidential Commission on the Space Shuttle *Challenger* Accident," C-SPAN, February 25, 1986, https://www.c-span.org/video/?126036-1/presidential-commission-space-shuttle-challenger-accident, 3:50:51-3:51:06.
71. McConnell, *Challenger: A Major Malfunction*, 198.
72. Boyce Rensberger, Kevin Klose, Walter Pincus, and Michael Isikoff, "Thiokol Engineers Tell of Being Overruled," *Washington Post*, February 2, 1986, https://www.washingtonpost.com/archive/politics/1986/02/26/thiokol-engineers-tell-of-being-overruled/3627b12c-e28f-4461-b20e-ca4e21be144c/.
73. McDonald, *Truth, Lies, and O-Rings*, 109–10.
74. Seddon, *Go for Orbit*, 298.
75. Scobee Rodgers, *Silver Linings*, 137–38.
76. Casey, "Remembering the *Challenger*."
77. Author interview with Sylvia Salinas Stottlemyer, January 29, 2021.
78. Phillip Booth, "Winter Wallop on Way," *Florida Today*, January 27, 1986, https://www.newspapers.com/image/125311886.
79. Hamburg, "*Challenger* Last Moments."
80. "Report—Investigation of the *Challenger* Accident," p. 232, https://www.govinfo.gov/content/pkg/GPO-CRPT-99hrpt1016/pdf/GPO-CRPT-99hrpt1016.pdf.

81. "Report—Investigation of the *Challenger* Accident," 232. Also, "Preventing Fires on the Launch Pad," NASA, August 12, 2004, https://www.nasa.gov/audience/forstudents/9-12/features/F_Preventing_Fires_on_the_Launchpad.html.
82. *Report of the Presidential Commission on the Space Shuttle Challenger Accident*, Volume 2: Appendix I—NASA Pre-Launch Activities Team Report, https://history.nasa.gov/rogersrep/v2appi.htm and Harris, *Challenger*, 21. Also, Jeff Stuckey and Anna Heiney, "Sound Suppression Test Unleashes a Flood," NASA, May 10, 2004, https://www.nasa.gov/missions/shuttle/f_watertest.html.
83. McConnell, *Challenger: A Major Malfunction*, 188.
84. McConnell, *Challenger: A Major Malfunction*, 188.
85. McConnell, *Challenger: A Major Malfunction*, 188.
86. McConnell, *Challenger: A Major Malfunction*, 189.
87. McConnell, *Challenger: A Major Malfunction*, 189.
88. Arnold D. Aldrich, "Challenger," NASA, August 27, 2008, p. 6, https://historycollection.jsc.nasa.gov/JSCHistoryPortal/history/oral_histories/SSP/AldrichAD_Challenger.pdf.
89. Aldrich, "Challenger," 6.
90. Harris, *Challenger*, location 99 (Kindle ed.).
91. Hamburg, "*Challenger* Last Moments." Also, Paul Dorian, "10:00 AM | Weather and the Shuttle *Challenger* Disaster Thirty Years Ago," Arcfield Weather, January 27, 2016, https://arcfieldweather.com/blog/2016/1/27/1000-am-weather-and-the-shuttle-challenger-disaster-thirty-years-ago.
92. McConnell, *Challenger: A Major Malfunction*, 219.
93. "Before Liftoff, Warnings of Ice," *New York Times*, June 5, 1986, https://www.nytimes.com/1986/06/05/us/before-liftoff-warnings-of-ice.html.
94. McConnell, *Challenger: A Major Malfunction*, 209–11.
95. Harris, *Challenger*, 29.
96. McConnell, *Challenger: A Major Malfunction*, 219.
97. Harris, *Challenger*, 21–22.
98. Author interview with Jim Bagian, April 2, 2021.
99. Author interview with Jim Bagian, April 2, 2021.
100. Scobee Rodgers, *Silver Linings*, 138.
101. Simpson, "NASA Astronaut Ellison Onizuka's Soccer Ball that Survived the *Challenger* Explosion."
102. Simpson, "NASA Astronaut Ellison Onizuka's Soccer Ball that Survived the *Challenger* Explosion."
103. William E. Schmidt, "The Shuttle Explosion; For the Families, the Moment of Doom," *New York Times*, January 29, 1986, https://www.nytimes.com/1986/01/29/us/the-shuttle-explosion-for-the-families-the-moment-of-doom.html. Also, Spencer and Spolar, "The Epic Flight."
104. Cassutt, *The Astronaut Maker*, 290–91
105. Harris, *Challenger*, 57. Also, "NASA's LCC: Building the Brains of Launch Operations," NASA's Kennedy Space Center, https://www.youtube.com/watch?v=R8aByl2fK5I.
106. Cassutt, *The Astronaut Maker*, 290–92.
107. Harry F. Rosenthal. "Shuttle Panel Disturbed by NASA's Rejection of Rockwell Warning," *AP News*, February 28, 1986, https://apnews.com/article/89632ce46072c5d8752949a749909310. Also, Cassutt, *The Astronaut Maker*, 291.

108. *Report of the Presidential Commission on the Space Shuttle Challenger Accident*, Volume 2: Appendix I—NASA Pre-Launch Activities Team Report, https://history.nasa.gov/rogersrep/v2appi.htm
109. McConnell, *Challenger: A Major Malfunction*, 228.
110. Aldrich, "Challenger," 9.
111. McConnell, *Challenger: A Major Malfunction*, 225.
112. Scobee Rodgers, *Silver Linings*, 139.
113. Author interview with Sylvia Salinas Stottlemyer, July 28, 2020.
114. Hamburg, "*Challenger* Last Moments." Also, David Hitt and Heather R. Smith. *Bold They Rise: The Space Shuttle Early Years, 1972–1986*. University of Nebraska Press, 2014, p. 297.
115. Author interview with Sylvia Salinas Stottlemyer, July 20, 2020.
116. Author interview with Fred Gregory, May 21, 2018.
117. Hitt and Smith, *Bold They Rise*, 287–88.
118. McConnell, *Challenger: A Major Malfunction*, 222.
119. "Countdown to Disaster: *Challenger*'s Last Flight: 9. ALL READY: 'What a Great Day for Flying,'" *Los Angeles Times*, February 9, 1986, https://www.latimes.com/archives/la-xpm-1986-02-09-ss-6012-story.html.
120. McConnell, *Challenger: A Major Malfunction*, 223. Also, "*Challenger* Accident—Part I Launch," https://www.youtube.com/watch?v=h7uAVs3tCL8.
121. Michael Isikoff, "Pathologists Work to Identify Remains of *Challenger* Crew," *Washington Post*, March 11, 1986, https://www.washingtonpost.com/archive/politics/1986/03/11/pathologists-work-to-identify-remains-of-challenger-crew/40663267-5569-42b5-a8e4-f45d31d95fed/.
122. "Countdown to Disaster: *Challenger*'s Last Flight: 9. ALL READY: 'What a Great Day for Flying,'" *Los Angeles Times*, February 9, 1986, https://www.latimes.com/archives/la-xpm-1986-02-09-ss-6012-story.html.
123. Lewis, *Challenger: The Final Voyage*, p. 5. Also, Hamburg, "*Challenger* Last Moments."
124. Hamburg, "*Challenger* Last Moments."
125. Hamburg, "*Challenger* Last Moments."
126. Cassutt, *The Astronaut Maker*, 290–92.
127. Trento and Trento, "Why *Challenger* Was Doomed."
128. Trento and Trento, "Why *Challenger* Was Doomed."
129. "Chapter III: The Accident," *Report of the Presidential Commission on the Space Shuttle Challenger Accident*, Volume 1, https://history.nasa.gov/rogersrep/v1ch3.htm.
130. "The *Challenger* Disaster: STS-51-L Countdown and Launch," January 28, 1986, https://www.youtube.com/watch?v=WqDxYFzETCk.
131. McConnell, *Challenger: A Major Malfunction*, 237.
132. Casey, "Remembering the *Challenger*."
133. Scobee Rodgers, *Silver Linings*, 140.
134. Cassutt, *The Astronaut Maker*, 291–292.
135. The Lox Arm is the liquid oxygen supply arm to the external tank. The beanie cap is the liquid oxygen vent cap.
136. Hamburg, "*Challenger* Last Moments."
137. "Chapter III: The Accident," *Report of the Presidential Commission on the Space Shuttle Challenger Accident*, https://history.nasa.gov/rogersrep/v1ch3.htm.

138. "The *Challenger* Disaster: STS-51-L Countdown and Launch." Also, https://science.ksc.nasa.gov/shuttle/technology/sts-newsref/stsover-launch.html.
139. Hamburg, "*Challenger* Last Moments."
140. "The *Challenger* Disaster: STS-51-L Countdown and Launch."
141. The "*Challenger* timeline" was assembled in the wake of the disaster by William Harwood, United Press International's Cape Canaveral bureau chief at the time of the accident, and Rob Navias, UPI radio's chief space correspondent. "*Challenger* Timeline," Spaceflight Now, https://spaceflightnow.com/challenger/timeline/.
142. House Committee. "Investigation of the *Challenger* Accident: Report of the Committee on Science and Technology, House of Representatives, Ninety-ninth Congress, Second Session," 1986, pp. 241–42, https://www.govinfo.gov/content/pkg/GPO-CRPT-99hrpt1016/pdf/GPO-CRPT-99hrpt1016.pdf.
143. "Chapter III: The Accident," *Report of the Presidential Commission on the Space Shuttle Challenger Accident*, https://history.nasa.gov/rogersrep/v1ch3.htm. See also, https://history.nasa.gov/rogersrep/v1p22.htm.
144. "Chapter III: The Accident," *Report of the Presidential Commission on the Space Shuttle Challenger Accident*, https://history.nasa.gov/rogersrep/v1ch3.htm.
145. "Chapter III: The Accident," *Report of the Presidential Commission on the Space Shuttle Challenger Accident*, https://history.nasa.gov/rogersrep/v1ch3.htm. Also, "NASA—STS-51L Mission Profile," December 5, 2005, NASA.
146. House Committee. "Investigation of the *Challenger* Accident," p. 188, https://www.govinfo.gov/content/pkg/GPO-CRPT-99hrpt1016/pdf/GPO-CRPT-99hrpt1016.pdf
147. "Chapter III: The Accident," *Report of the Presidential Commission on the Space Shuttle Challenger Accident*, https://history.nasa.gov/rogersrep/v1ch3.htm.
148. William Harwood, "*Challenger* Remembered: The Shuttle *Challenger*'s Final Voyage," CBS News, p. 10. http://studyphysics.ca/challenger.pdf
149. Hamburg, "*Challenger* Last Moments."
150. "STS-51L, *Challenger* (10), 25th Space Shuttle mission," Spacefacts, http://spacefacts.de/ballistic/english/sts-51l.htm.
151. Harwood, "*Challenger* Remembered," 11.
152. Hamburg, "*Challenger* Last Moments."
153. "*Challenger* Timeline," Spaceflight Now.
154. Harwood, "*Challenger* Remembered," 11.
155. Harwood, "*Challenger* Remembered," 8.
156. Harwood, "*Challenger* Remembered," 8.
157. Hamburg, "*Challenger* Last Moments."
158. "Report—Investigation of the *Challenger* Accident," p. 5, https://www.govinfo.gov/content/pkg/GPO-CHRG-99hhrg64294-vol1/pdf/GPO-CHRG-99hhrg64294-vol1.pdf.
159. "Report—Investigation of the *Challenger* Accident," pp. 78–79, https://www.govinfo.gov/content/pkg/GPO-CHRG-99hhrg64294-vol1/pdf/GPO-CHRG-99hhrg64294-vol1.pdf.
160. "Chapter III: The Accident," *Report of the Presidential Commission on the Space Shuttle Challenger Accident*, https://history.nasa.gov/rogersrep/v1ch3.htm.
161. Hamburg, "*Challenger* Last Moments."
162. "Chapter III: The Accident," *Report of the Presidential Commission on the Space Shuttle Challenger Accident*, https://history.nasa.gov/rogersrep/v1ch3.htm.

163. Hamburg, "*Challenger* Last Moments." Also, Harwood, "*Challenger* Remembered," 14.
164. "*Challenger* Timeline," Spaceflight Now.
165. Author interview with Fred Gregory, May 21, 2018.
166. Margaret Lazarus Dean, "The Oral History of the Space Shuttle *Challenger* Disaster," *Popular Mechanics*, January 28, 2021, https://www.popularmechanics.com/space/a18616/an-oral-history-of-the-space-shuttle-challenger-disaster.
167. "*Challenger* Timeline," Spaceflight Now.
168. Author interview with Fred Gregory, January 27, 2022.
169. Harwood, "*Challenger* Remembered," 16.
170. Michael Kranish, "And Then a Child Said "Where Is It?'" *Boston Globe*, January 29, 1986, http://archive.boston.com/news/nation/articles/1986/01/29/and_then_a_child_said_where_is_it.
171. David Tirrell-Wysocki, "Tears of Joy Turn into Tears of Terror for Shuttle Spectators," *AP News*, January 29, 1986, https://apnews.com/article/8c307dc37641bf7f10aa4fc435638545.
172. Kranish, "And Then a Child Said 'Where Is It?'"
173. Grace George Corrigan, *A Journal for Christa: Christa Mcauliffe, Teacher in Space*. Lincoln: University of Nebraska Press, 1993, p. 5.
174. Frederick D. Gregory transcript, NASA Johnson Oral History Project, March 14, 2005, https://historycollection.jsc.nasa.gov/JSCHistoryPortal/history/oral_histories/GregoryFD/GregoryFD_3-14-05.pdf.
175. Dean, "The Oral History of the Space Shuttle *Challenger* Disaster."
176. Dean, "The Oral History of the Space Shuttle *Challenger* Disaster."
177. Hamburg, "*Challenger* Last Moments."
178. McConnell, *Challenger: A Major Malfunction*, 246. Also, "Chapter IX: Other Safety Considerations," *Report of the Presidential Commission on the Space Shuttle Challenger Accident*, https://history.nasa.gov/rogersrep/v1ch9.htm.
179. Trento and Trento, "Why *Challenger* Was Doomed."
180. Trento and Trento, "Why *Challenger* Was Doomed."
181. Trento and Trento, "Why *Challenger* Was Doomed."
182. Trento and Trento, "Why *Challenger* Was Doomed."
183. Dean, "The Oral History of the Space Shuttle *Challenger* Disaster."
184. Scobee Rodgers, *Silver Linings*, 142.
185. Richard W. Nygren transcript, NASA Johnson Space Center Oral History Project, March 9, 2006, https://historycollection.jsc.nasa.gov/JSCHistoryPortal/history/oral_histories/NygrenRW/NygrenRW_3-9-06.htm.
186. *Challenger: The Final Flight*. Steven Leckart, Glen Zipper. Bad Robot, Sutter Road Picture Company, and Zipper Bros Films. *Netflix*.
187. *Challenger: The Final Flight*. Steven Leckart, Glen Zipper. Bad Robot, Sutter Road Picture Company, and Zipper Bros Films. *Netflix*.
188. Col. Frederick Drew Gregory, interviewed by Cheryl Butler, July 27, 2007, HistoryMakers Digital Archive, session 1, tape 5, story 4, https://da.thehistorymakers.org/story/629135.

Chapter 19: Speedbrake

1. Hamburg, "*Challenger* Last Moments."
2. Hitt and Smith, 289. (Per Dick Covey's memory.)

3. Author interview with Fred Gregory, May 21, 2018.
4. Harris, *Challenger*, 40.
5. Cassutt, *The Astronaut Maker*, 291–92.
6. Schmidt, "The Shuttle Explosion."
7. Harris, *Challenger*, 40 (Kindle ed.).
8. Jay Barbree, "Chapter 4: A Hellish Fireball," NBC, January 24, 2006, https://www.nbcnews.com/id/wbna3078064.
9. Cassutt, *The Astronaut Maker*, 291.
10. Scobee Rodgers, *Silver Linings*, 142.
11. Eric Braun, *Fatal Faults: The Story of the Challenger Explosion*. North Mankato, MN: Capstone Press, 2015, location 336 (Kindle ed.).
12. "*Challenger* Explosion, Live Audience Reaction, 25th Anniversary," *ABC News*, January 28, 2011, https://www.youtube.com/watch?v=pUALwYsXSm8.
13. Hamburg, "*Challenger* Last Moments."
14. Braun, *Fatal Faults*, location 336. And Bob Collins, "After Watching Her Daughter Die, Grace Corrigan Carried on Christa McAuliffe's Mission," Minnesota Public Radio, November 12, 2018, https://www.mprnews.org/story/2018/11/12/after-watching-her-daughter-die-grace-corrigan-carried-on-christa-mcauliffes-mission
15. Hamburg, "*Challenger* Last Moments."
16. Dan Zak, "We've Lost 'Em, God Bless 'Em: What It Was Like to Witness the *Challenger* Disaster," *Washington Post*, January 28, 2016, https://www.washingtonpost.com/news/arts-and-entertainment/wp/2016/01/28/weve-lost-em-god-bless-em-what-it-was-like-to-witness-the-challenger-disaster.
17. Hamburg, "*Challenger* Last Moments."
18. Author interview with Sylvia Salinas Stottlemeyer, July 28, 2020.
19. Hamburg, "*Challenger* Last Moments."
20. Fisher, NASA Oral History Project.
21. Author interview with Brewster Shaw, April 28, 2021.
22. Seddon, *Go for Orbit*, 304.
23. Author interview with Charlie Bolden, August 24, 2020.
24. Author interview with Anna Fisher, March 1, 2022.
25. Author interview with Brewster Shaw, April 28, 2021.
26. Sullivan, *Handprints on Hubble*, 133.
27. Author interview with Kathy Sullivan, November 2, 2020.
28. Author interview with Kathy Sullivan, November 2, 2020.
29. Sullivan, *Handprints on Hubble*, 134.
30. Sullivan, *Handprints on Hubble*, 134–35.
31. Scobee Rodgers, *Silver Linings*, 142–43.
32. Hamburg, "*Challenger* Last Moments."
33. Interview with Jim Bagian, May 2022.
34. Cassutt, *The Astronaut Maker*, 291–92.
35. "Episode 4: Nothing Ends Here," *Challenger: The Final Flight*. Directed by Daniel Junge and Steven Leckart, Santa Monica: Bad Robot Productions, 2020, https://www.netflix.com/title/81012137, 10:40.
36. Scobee Rodgers, *Silver Linings*, 143.
37. McNair, *In the Spirit*, 184.
38. Mullane, *Riding Rockets*, 226.

39. Hamburg, "*Challenger* Last Moments."
40. Hamburg, "*Challenger* Last Moments."
41. Scobee Rodgers, *Silver Linings,* 144.
42. Ben Bova, *Vision of the Future,* New York: Abrams, 1982.
43. Scobee Rodgers, *Silver Linings,* 48.
44. Author interview with Jim Bagian, April 2, 2021.
45. Author interview with Jim Bagian, April 2, 2021.
46. Author interview with Jim Bagian, April 2, 2021.
47. Author interview with Jim Bagian, April 2, 2021.
48. Cassutt, *The Astronaut Maker,* 292.
49. Hamburg, "*Challenger* Last Moments."
50. Scobee Rodgers, *Silver Linings,* 49.
51. Harris, *Challenger,* 16.
52. Harris, *Challenger,* location 447 (Kindle ed.). Also, "1986-*Challenger* Disaster—NASA Briefing," YouTube video uploaded by Cardinal de L'Étoile on July 17, 2014, https://www.youtube.com/watch?v=fEEqlcTFkWI.
53. Harris, *Challenger,* location 401 (Kindle ed.).
54. Statement by Jesse Moore, *New York Times,* January 29, 1986, https://www.nytimes.com/1986/01/29/us/the-shuttle-explosion-transcript-of-nasa-news-conference-on-the-shuttle-disaster.html.
55. Tricia Escobedo, "When a National Disaster Unfolded Live in 1986," CNN, March 31, 2016, https://www.cnn.com/2016/03/31/us/80s-cnn-challenger-coverage/index.html.
56. "CNN coverage of Space Shuttle *Challenger* explosion 1/28/86," YouTube video uploaded by Jason Hepner, 12:50–13:08, https://www.youtube.com/watch?v=Ync0f4e4tS4.
57. Statement by Jesse Moore.
58. "Explosion of the Space Shuttle *Challenger* Address to the Nation, January 28, 1986," President Ronald W. Reagan, NASA History Office, https://history.nasa.gov/reagan12886.html.
59. "Pilot Officer John Gillespie Magee: 'High Flight,'" National Museum of the United States Air Force, https://www.nationalmuseum.af.mil/Visit/Museum-Exhibits/Fact-Sheets/Display/Article/196844/pilot-officer-john-gillespie-magee-high-flight/.
60. "Explosion of the Space Shuttle *Challenger* Address to the Nation, January 28, 1986," Reagan. Also, John Gillespie Magee, "High Flight," https://nationalpoetryday.co.uk/poem/high-flight.
61. Becky Little, "Reagan Delayed the 1986 State of the Union to Mourn the *Challenger* Disaster," History.com, January 24, 2019, https://www.history.com/news/reagan-challenger-disaster-state-of-the-union-1986. Also, "Explosion of the Space Shuttle *Challenger* Address to the Nation, January 28, 1986," Reagan.
62. Scobee Rodgers, *Silver Linings,* 147.
63. Cassutt, *The Astronaut Maker,* 292.
64. Cassutt, *The Astronaut Maker,* 292.
65. Author interview with Fred Gregory, May 21, 2018.
66. Richard O. Covey transcript, NASA Johnson Space Center Oral History Project, November 15, 2006, https://historycollection.jsc.nasa.gov/JSCHistoryPortal/history/oral_histories/CoveyRO/CoveyRO_11-15-06.htm.

67. Jerry L. Ross and John Norberg, *Spacewalker: My Journey in Space and Faith as Nasa's Record-Setting Frequent Flyer.* West Lafayette, IN: Purdue University Press, 2013, location 2441 (Kindle ed.).
68. Gregory, NASA Oral History Project.
69. Nelson, NASA Oral History Project.
70. Cassutt, *The Astronaut Maker,* 292–93.
71. Author interview with Fred Gregory, May 21, 2018.
72. Seddon, *Go for Orbit,* 310.
73. Seddon, *Go for Orbit,* 311.
74. *Eternal Father, Strong to Save.* Words by William Whiting, 1860.
75. "Memorial Service for the Mission 51-L Crew," NASA STI Program, 3:54–4:23, https://www.youtube.com/watch?v=bJfoJFFQvyA.
76. "Transcript of the President's Eulogy for the Seven *Challenger* Astronauts," *New York Times,* February 1, 1986, https://www.nytimes.com/1986/02/01/us/transcript-of-the-president-s-eulogy-for-the-seven-challenger-astronauts.html.
77. Scobee Rodgers, *Silver Linings,* 149.
78. John A. Logsdon, "Chapter 15, Return to Flight: Richard H. Truly and the Recovery from the *Challenger* Accident," NASA History Office, https://history.nasa.gov/SP-4219/Chapter15.html.
79. Author interview with Mike Coats, January 19, 2021.
80. Author interview with Steve Hawley, April 28, 2021.
81. Author interview with Mike Coats, January 19, 2021.
82. Author interview with Brewster Shaw, April 28, 2021.
83. Volume 3: Appendix O—NASA Search, Recovery and Reconstruction Task Force Team Report, May 8, 1986, *Report of the Presidential Commission on the Space Shuttle Challenger Accident.*
84. Dennis E. Powell, "Obviously, a Major Malfunction," *Tropic* [*The Miami Herald Sunday Magazine*], November 13, 1988. Also, William Harwood, "Chapter Seven: The First Month," Voyage Into History, 1986, CBS News.
85. Volume 3: Appendix O—NASA Search, Recovery and Reconstruction Task Force Team Report, May 8, 1986, *Report of the Presidential Commission on the Space Shuttle Challenger Accident.*
86. Author interview with Mike Coats, January 19, 2021.
87. Harwood, "Chapter Seven: The First Month," Voyage Into History, 1986, CBS News.
88. Charles Fishman, "The Epic Search for the *Challenger,*" *Washington Post,* May 28, 1986, https://www.washingtonpost.com/archive/lifestyle/1986/05/28/the-epic-search-for-the-challenger/2c445a93-f39c-448b-b4c0-ca79efe3ad48/.
89. Harris, *Challenger,* 54–55.
90. William Harwood, "Reporters Recall *Challenger* Disaster 30 Years Later," CBS News, January 27, 2016, https://www.cbsnews.com/news/reporters-remember-challenger-disaster-30-years-later/.
91. Harris, *Challenger,* 54.
92. *New York Times Co. v. National Aeronautics & Space Administration,* 920 F.2d 1002 (DC Cir. 1990). https://casetext.com/case/new-york-times-co-v-nasa-3.
93. Harris, *Challenger,* 61, and Harwood, "Reporters Recall *Challenger* Disaster 30 Years Later."
94. Cassutt, *The Astronaut Maker,* 290.

95. Cassutt, *The Astronaut Maker,* 290–94.
96. Cassutt, *The Astronaut Maker,* 296.
97. Cassutt, *The Astronaut Maker,* 296.
98. Cassutt, *The Astronaut Maker,* 296.
99. Author interview with Jim Bagian, April 7, 2021.
100. "Manley Lanier 'Sonny' Carter Jr. Biographical Data," NASA, https://www.nasa.gov/sites/default/files/atoms/files/carter_manley_0.pdf.
101. Powell, "Obviously, a Major Malfunction."
102. Fishman, "The Epic Search for the *Challenger.*"
103. "The Space Shuttle *Challenger* Salvage Report," April 29, 1988, Department of the Navy Sea Systems Command. https://www.navsea.navy.mil/Portals/103/Documents/SUPSALV/SalvageReports/Space Shuttle Challenger.pdf.
104. Fishman, "The Epic Search for the *Challenger.*"
105. Fishman, "The Epic Search for the *Challenger.*"
106. Author interview with Jim Bagian, April 7, 2021.
107. Michael Lafferty, "Crews May Have Found Communications Satellite," *Florida Today,* February 13, 1986, p. 2A, https://www.newspapers.com/image/124839987/. Also, Michael Lafferty, "Engineer Joins Submarine Inspecting Sunken Right Booster," *Florida Today,* February 21, 1986, p. 2A, https://www.newspapers.com/image/125294894/. Also, Michael Lafferty, "Divers Retrieve Portion of Shuttle's 3 Engines," *Florida Today,* February 26, 1986, p. 2A, https://www.newspapers.com/image/125307718/. Also, William Harwood, "Engines, Booster Debris Found," UPI for *The Herald,* February 26, 1986, p. 24, https://www.newspapers.com/image/541450770/. And, William Harwood, "Salvage Crews Have Recovered Large Pieces of *Challenger*'s Rear . . .", UPI, March 8, 1986, https://www.upi.com/Archives/1986/03/08/Salvage-crews-have-recovered-large-pieces-of-Challengers-rear/9467510642000.
108. Fishman, "The Epic Search for the *Challenger.*" Also, Volume 3: Appendix O—NASA Search, Recovery and Reconstruction Task Force Team Report, Volume 2: Enclosure 9: CONTACT #131 AND #712 RECOVERED RH SRB AFT FIELD JOINT EVALUATION, May 8, 1986, *Report of the Presidential Commission on the Space Shuttle Challenger Accident,* https://history.nasa.gov/rogersrep/v3appoe9.htm.
109. Seddon, *Go for Orbit,* 315. Also, author interview with Rhea Seddon, August 27, 2022.
110. Seddon, *Go for Orbit,* 315.
111. Author interview with Jim Bagian, April 2, 2021.
112. Author interview with Jim Bagian, April 2, 2021.
113. Author interview with Jim Bagian, April 7, 2021.
114. Jim Leusner and Dan Tracy, "Standing Watch on the Jetties," *Orlando Sentinel,* March 14, 1986, https://www.newspapers.com/image/223369352/?terms=Jetty%20Park%20Challenger&match=1.
115. Leusner and Tracy, "Standing Watch on the Jetties."
116. Harris, *Challenger,* location 77–79.
117. Powell, "Obviously, a Major Malfunction."
118. Powell, "Obviously, a Major Malfunction."
119. Author interview with Jim Bagian, April 7, 2021.
120. Todd Halvorson, "They Found the Cabin and the Flag; They Lost Their Jobs,"

Florida Today, January 28, 1987, https://www.newspapers.com/image/177750150/?terms=%22Terry%20Bai-ley%22%20and%20McAllister&match=1.

121. "Divers Saw Flight Suit in Shuttle Wreck," *New York Times*, July 15, 1986, https://www.nytimes.com/1986/07/15/science/divers-saw-flight-suit-in-shuttle-wreck.html.
122. "Divers Mike McAllister and Terry Bailey Were 87 Feet . . . ," UPI Archives, July 13, 1986, https://www.upi.com/Archives/1986/07/13/Divers-%ADMike-%ADMcAllister-%ADand-%ADTerry-%ADBailey-%ADwere-87-feet/2254521611200.
123. Glenn Arnett, "Search for *Challenger*," *Air and Space Magazine*, vol. 9, no. 1, April-May 1994.
124. Powell, "Obviously, a Major Malfunction."
125. Powell, "Obviously, a Major Malfunction."
126. Powell, "Obviously, a Major Malfunction."
127. Powell, "Obviously, a Major Malfunction."

Chapter 20: All We Know of Heaven

1. Author Interview with Jim Bagian, April 7 and May 12, 2021.
2. Author interview with Jim Bagian, April 7, April 13, and May 12, 2021.
3. Author interview with Jim Bagian, April 2, 2021.
4. Author interview with Jim Bagian, April 2, 2021.
5. Author interview with Jim Bagian, April 2, 2021.
6. Author interview with Jim Bagian, April 2, 2021.
7. Author interview with Jim Bagian, April 7, 2021.
8. James Fisher, "Divers' Work Ends, Doctors' Begins," *Orlando Sentinel*, March 14, 1986, https://www.newspapers.com/image/223382597/.
9. Author interview with Jim Bagian, April 7, 2021.
10. Powell, "Obviously, a Major Malfunction."
11. Author interview with Jim Bagian, April 7, 2021.
12. Seddon, *Go for Orbit*, 317.
13. Powell, "Obviously, a Major Malfunction."
14. Fisher, "Divers' Work Ends, Doctors' Begins."
15. "*Challenger* Crew Cabin Discovered," William Harwood for UPI, March 10, 1986, *Lompoc Record*.
16. Mark Fisher, "Find Is a Sense of Relief for Kin, Doctor Says," *Miami Herald*, March 10, 1986, https://www.newspapers.com/image/632324964.
17. "Shuttle Crew Cabin, Remains Round," William Harwood for UPI, March 10, 1986, *Boston Globe*.
18. Scobee Rodgers, *Silver Linings*, 51.
19. Seddon, *Go for Orbit*, 316.
20. Hitt and Smith, *Bold They Rise*, 301–304.
21. Powell, "Obviously, a Major Malfunction."
22. Seddon, *Go for Orbit*, 318.
23. John Noble Wilford, "*Challenger* Blast Left Cabin Intact," *New York Times*, April 10, 1986, https://www.nytimes.com/1986/04/10/us/challenger-blast-left-cabin-intact.html.

24. Seddon, *Go for Orbit*, 320.
25. Author interview with Mike Coats, January 19, 2021.
26. Hitt and Smith, *Bold They Rise*, 303.
27. Seddon, *Go for Orbit*, 322–23.
28. Seddon, *Go for Orbit*, 323.
29. Seddon, *Go for Orbit*, 324.
30. Seddon, *Go for Orbit*, 324–25.
31. Seddon, *Go for Orbit*, 324–25.
32. William E. Schmidt, "All Shuttle Crew Remains Recovered, NASA Says," *New York Times*, April 20, 1986, https://www.nytimes.com/1986/04/20/us/all-shuttle-crew-remains-recovered-nasa-says.html.
33. William Harwood, "The Remains of the *Challenger* Seven Were Readied Monday . . ." UPI, April 28, 1986, https://www.upi.com/Archives/1986/04/28/The-remains-of-the-Challenger-seven-were-readied-Monday/4357515044800.
34. William E. Schmidt, "Bodies of Astronauts Flown to Delaware," April 30, 1986, *New York Times*, https://www.nytimes.com/1986/04/30/us/bodies-of-astronauts-flown-to-delaware.html.
35. Schmidt, "Bodies of Astronauts Flown to Delaware."
36. James Fisher and John J. Glisch, "*Challenger* Crew Remains Will Be Flown to Air Base," *Orlando Sentinel*, April 25, 1986, https://www.orlandosentinel.com/news/os-xpm-1986-04-25-0210460185-story.html.
37. Chris Polar and Charles Fishman, "Astronaut Remains Flown to Delaware," *Washington Post*, April 30, 1986, https://www.washingtonpost.com/archive/politics/1986/04/30/astronaut-remains-flown-to-delaware/ca99b32c-4bfc-4fa3-a537-d4df53059261/.
38. Author interview with Jim Bagian, April 2, 2021.
39. Author interview with Jim Bagian, April 2, 2021.
40. Author interview with Jim Bagian, April 2, 2021.
41. Author interview with Jim Bagian, April 2, 2021.
42. Author interview with Jim Bagian, April 2, 2021.
43. Seddon, *Go for Orbit*, 316.
44. Author interview with Jim Bagian, April 2, 2021.
45. Seddon, *Go for Orbit*, 322.
46. William G. Harwood, "The CBS News Space Reporter's Handbook STS-51L/107 Supplement," CBS News, https://www.cbsnews.com/network/news/space/SRH_Disasters.htm.
47. Author interview with Jim Bagian, April 2, 2021.
48. Author interview with Jim Bagian, April 2, 2021.
49. Author interview with Jim Bagian, April 2, 2021.
50. "Report to the President by the Presidential Commission on the Space Shuttle *Challenger* Accident," pp. 51–3.
51. "Report to the President by the Presidential Commission on the Space Shuttle *Challenger* Accident," pp. 19–39.
52. "*Challenger* Timeline," Spaceflight Now.
53. "*Challenger* Timeline," Spaceflight Now.
54. Karl Tate, "The Space Shuttle *Challenger* Disaster: What Happened? (Infographic)," Space.com, January 28, 2016, https://www.space.com/31732-space-shuttle-challenger-disaster-explained-infographic.html.

55. Jay Barbree, "Chapter 5: An Eternity of Descent," NBC News, January 25, 2004, https://www.nbcnews.com/id/wbna3078062.
56. Jan Ziegler, "The *Challenger*'s Crew Probably Survived at Least Several . . . ," UPI Archives, July 28, 1986, https://www.upi.com/Archives/1986/07/28/The-shuttle-Challengers-crew-probably-survived-at-least-several/9204522907200/.

Chapter 21: Nature Cannot Be Fooled

1. Ride, NASA Oral History Project, October 22, 2002.
2. Dean, "The Oral History of the Space Shuttle *Challenger* Disaster."
3. Ride, NASA Oral History Project, October 22, 2002.
4. Sherr, *Sally Ride,* 202.
5. Sherr, *Sally Ride,* 204.
6. Sherr, *Sally Ride,* 204.
7. Sherr, *Sally Ride,* 204.
8. "Two Small Notebooks Containing Ride's Notes from the Rogers Commission," Sally K. Ride Papers, page 1 of 17, image 1, National Air and Space Museum Archives, https://edan.si.edu/slideshow/viewer/?eadrefid=NASM.2014.0025_ref230. Also, Adam Bredenberg, "Sally Ride: A Retrospective," *Philadelphia Inquirer,* July 24, 2012, https://www.inquirer.com/archive/sally-ride-retrospective-20120724.html.
9. Richard P. Feynman, *"What Do You Care What Other People Think?" Further Adventures of a Curious Character.* New York: W. W. Norton, 2011, p. 125 (Kindle ed.).
10. Bredenberg, "Sally Ride: A Retrospective."
11. Dean, "The Oral History of the Space Shuttle *Challenger* Disaster."
12. "Reagan Panel to Probe Explosion of *Challenger,*" compiled from journal wires, February 4, 1986, *Albuquerque Journal,* https://www.newspapers.com/image/156829393/.
13. Dean, "The Oral History of the Space Shuttle *Challenger* Disaster."
14. "Two Small Notebooks Containing Ride's Notes from the Rogers Commission," Sally K. Ride Papers, page 1 of 17, image 2, National Air and Space Museum Archives, https://edan.si.edu/slideshow/viewer/?eadrefid=NASM.2014.0025_ref230.
15. Patrick Young, "Inquiry into Shuttle Disaster Is Raising Intriguing Questions," *Gazette* (Montreal, Canada), February 15, 1986, https://www.newspapers.com/image/422496792/.
16. "Yeager Breaks Record for Coast-to-Coast Flight," *AP News,* February 8, 1986, https://apnews.com/article/5d8242c20fbee57b40d4e89eb961fe5f.
17. Sherr, *Sally Ride,* 212.
18. "The *Challenger* Disaster 2-6-1986 CNN Coverage of 1st Rogers Commission Hearing," https://www.youtube.com/watch?v=-gw3kCUQZTg.
19. "Episode 4: Nothing Ends Here," *Challenger: The Final Flight.* Directed by Daniel Junge and Steven Leckart, Santa Monica: Bad Robot Productions, 2020, https://www.netflix.com/title/81012137, 16:45–17:29.
20. Hearing of the Presidential Commission on the Space Shuttle *Challenger* Accident, February 6, 1986 session transcript, Section 54, *Report of the Presidential Commission on the Space Shuttle Challenger Accident,* Volume 4 Index, part 1, https://history.nasa.gov/rogersrep/v4part1.htm.

21. Hearing of the Presidential Commission on the Space Shuttle *Challenger* Accident, February 6, 1986 session transcript, Section 55, *Report of the Presidential Commission on the Space Shuttle Challenger Accident*, Volume 4 Index, part 1, https://history.nasa.gov/rogersrep/v4part1.htm.
22. Hearing of the Presidential Commission on the Space Shuttle *Challenger* Accident, February 6, 1986 session transcript, Section 55, *Report of the Presidential Commission on the Space Shuttle Challenger Accident*, Volume 4 Index, part 1, https://history.nasa.gov/rogersrep/v4part1.htm.
23. Hearing of the Presidential Commission on the Space Shuttle *Challenger* Accident, February 6, 1986 session transcript, Sections 68–69, *Report of the Presidential Commission on the Space Shuttle Challenger Accident*, Volume 4 Index, part 1, https://history.nasa.gov/rogersrep/v4part1.htm.
24. "Presidential Commission on the Space Shuttle *Challenger* Accident," C-SPAN, February 6, 1986, https://www.c-span.org/video/?125982-1/presidential-commission-space-shuttle-challenger-accident.
25. McConnell, *Challenger: A Major Malfunction*, 228.
26. Aldrich, "Challenger," 9–10. According to Aldrich's own account, he asked two Rockwell managers for a specific recommendation on whether to launch or not, but the managers "seemed to back off somewhat on the degree of additional risk they felt we were being exposed to and they would not offer a firm recommendation." As a result, Aldrich said, "I could not feel that the RI [Rockwell] concerns should override the positive go that I had received from each of the other organizations represented . . ."
27. Hearing of the Presidential Commission on the Space Shuttle *Challenger* Accident, February 6, 1986 session transcript, Section 55, *Report of the Presidential Commission on the Space Shuttle Challenger Accident*, Volume 4 Index, part 1, https://history.nasa.gov/rogersrep/v4part1.htm.
28. Hearing of the Presidential Commission on the Space Shuttle *Challenger* Accident, February 6, 1986 session transcript, Section 167, *Report of the Presidential Commission on the Space Shuttle Challenger Accident*, Volume 4 Index, part 1, https://history.nasa.gov/rogersrep/v4part1.htm.
29. Hearing of the Presidential Commission on the Space Shuttle *Challenger* Accident, February 6, 1986 session transcript, Sections 160–161, *Report of the Presidential Commission on the Space Shuttle Challenger Accident*, Volume 4 Index, part 1, https://history.nasa.gov/rogersrep/v4part1.htm.
30. Richard Cook, *Challenger Revealed: An Insider's Account of How the Reagan Administration Caused the Greatest Tragedy of the Space Age*. New York: Basic Books, 2006, location 2713 (Kindle ed.)
31. McDonald, *Truth, Lies, and O-Rings*, 63.
32. McDonald, *Truth, Lies, and O-Rings*, 146. When McDonald watched this part of Dr. Lovingood's testimony on TV, he thought to himself, "Boy, did [Lovingood] just dodge a bullet!"
33. "Episode 4: Nothing Ends Here," *Challenger: The Final Flight*, Junge and Leckart, 21:15–21:25. Quote from David Sanger, *New York Times*.
34. Sherr, *Sally Ride*, 206–07. Also, Dean, "The Oral History of the Space Shuttle *Challenger* Disaster."
35. Author interview with Steve Hawley, April 28, 2021.
36. Author interview with Steve Hawley, April 28, 2021.

37. Dean, "The Oral History of the Space Shuttle *Challenger* Disaster."
38. Sherr, *Sally Ride,* 205.
39. Dean, "The Oral History of the Space Shuttle *Challenger* Disaster." Also, "Donald J. Kutyna—Reaching for the Stars," interview by Aleksandra Ziolkowska-Boehm, January 19, 2018, https://aleksandraziolkowskaboehm.blogspot.com/2018/01/donald-jkutyna-reaching-for-stars.html.
40. Sherr, *Sally Ride,* 205.
41. "Two Small Notebooks Containing Ride's Notes from the Rogers Commission," Sally K. Ride Papers, page 1 of 17, images 8 and 9, National Air and Space Museum Archives, https://edan.si.edu/slideshow/viewer/?eadrefid=NASM.2014.0025_ref230.
42. "Two Small Notebooks Containing Ride's Notes from the Rogers Commission," Sally K. Ride Papers, page 1 of 17, image 8, National Air and Space Museum Archives, https://edan.si.edu/slideshow/viewer/?eadrefid=NASM.2014.0025_ref230.
43. Dean, "The Oral History of the Space Shuttle *Challenger* Disaster."
44. "General Donald J. Kutyna," Air Force, January 1991, https://www.af.mil/About-Us/Biographies/Display/Article/106524/general-donald-j-kutyna/general-donald-j-kutyna/.
45. Dean, "The Oral History of the Space Shuttle *Challenger* Disaster."
46. Feynman, *"What Do You Care What Other People Think?,"* 130.
47. Phil Boffey, "NASA Had Warning of a Disaster Risk Posed By Booster," *New York Times,* February 9, 1986, https://www.nytimes.com/1986/02/09/us/nasa-had-warning-of-a-disaster-risk-posed-by-booster.html.
48. Boffey, "NASA Had Warning of a Disaster Risk Posed By Booster."
49. Boffey, "NASA Had Warning of a Disaster Risk Posed By Booster." Also, "Memorandum: Problem with SRB Seals. [From: BRC/Richard Cook; To: BRC/Michael Mann Subject: Problem with SRB seals; Date: 7/23/85].," Volume 4 Index, Hearings of the Presidential Commission on the Space Shuttle *Challenger* Accident: February 6, 1986 to February 25, 1986, February 10, 1986 Session, https://history.nasa.gov/rogersrep/v4p273.htm.
50. Cook, *Challenger Revealed,* 2748–54.
51. Boffey, "NASA Had Warning of a Disaster Risk Posed By Booster."
52. McDonald, *Truth, Lies, and O-Rings,* 148.
53. "Episode 4: Nothing Ends Here," *Challenger: The Final Flight,* Junge and Leckart, 26:33–27:44.
54. McDonald, *Truth, Lies, and O-Rings,* 157.
55. "Two Small Notebooks Containing Ride's Notes from the Rogers Commission," Sally K. Ride Papers, page 2 of 17, images 7-8, National Air and Space Museum Archives, https://edan.si.edu/slideshow/viewer/?eadrefid=NASM.2014.0025_ref230.
56. McDonald, *Truth, Lies, and O-Rings,* 158. McDonald does not mention the name of the spokesmen and this was a closed-door session, so a public record was not readily available. See also Sally Ride's notes of the day: "Two Small Notebooks Containing Ride's Notes from the Rogers Commission," Sally K. Ride Papers, page 2 of 17, image 7 through page 3 of 17, image 1, National Air and Space Museum Archives, https://edan.si.edu/slideshow/viewer/?eadrefid=NASM.2014.0025_ref230.

57. McDonald, *Truth, Lies, and O-Rings*, 159.
58. McDonald, *Truth, Lies, and O-Rings*, 159.
59. McDonald, *Truth, Lies, and O-Rings*, 159.
60. "Episode 4: Nothing Ends Here," *Challenger: The Final Flight,* Junge and Leckart, 26:33–27:44.
61. McDonald, *Truth, Lies, and O-Rings*, 160.
62. McDonald, *Truth, Lies, and O-Rings*, 47. Even though the ambient temperature of that launch was 62°, the temperature of the O-rings in the SRBs was just 53°F.
63. McDonald, *Truth, Lies, and O-Rings*, 160.
64. McDonald, *Truth, Lies, and O-Rings*, 159–161.
65. McDonald, *Truth, Lies, and O-Rings*, 161.
66. McDonald, *Truth, Lies, and O-Rings*, 161.
67. McDonald, *Truth, Lies, and O-Rings*, 162.
68. Dean, "The Oral History of the Space Shuttle *Challenger* Disaster."
69. McDonald, *Truth, Lies, and O-Rings*, 162.
70. Dean, "The Oral History of the Space Shuttle *Challenger* Disaster."
71. McDonald, *Truth, Lies, and O-Rings*, 162.
72. "Chapter III: The Accident," *Report of the Presidential Commission on the Space Shuttle Challenger Accident*, Volume 1, https://history.nasa.gov/rogersrep/v1ch3.htm.
73. McDonald, *Truth, Lies, and O-Rings*, 165.
74. McDonald, *Truth, Lies, and O-Rings*, 75, 165.
75. McDonald, *Truth, Lies, and O-Rings*, 166.
76. McDonald, *Truth, Lies, and O-Rings*, 166.
77. McDonald, *Truth, Lies, and O-Rings*, 167.
78. Sherr, *Sally Ride,* 208–09.
79. McDonald, *Truth, Lies, and O-Rings*, 187. In Sally's notes from that day, she sketched a neat black star: "leak port in 'same area' as black puff + flame (another trail)," "crack in case = another trail." She also wrote, "O-ring people expressed concern. Telecon (MSFC, Thiokol, KSC) Thiokol rec. not launch below 53° Then changed their minds." "Two Small Notebooks Containing Ride's Notes from the Rogers Commission," Sally K. Ride Papers, page 2 of 17, image 9, National Air and Space Museum Archives, https://edan.si.edu/slideshow/viewer/?eadrefid=NASM.2014.0025_ref230.
80. McDonald, *Truth, Lies, and O-Rings*, 171.
81. "Episode 4: Nothing Ends Here," *Challenger: The Final Flight,* Junge and Leckart, 27:44–27:54.
82. "Two Small Notebooks Containing Ride's Notes from the Rogers Commission," Sally K. Ride Papers, page 2 of 17, image 9, National Air and Space Museum Archives, https://edan.si.edu/slideshow/viewer/?eadrefid=NASM.2014.0025_ref230.
83. "Two Small Notebooks Containing Ride's Notes from the Rogers Commission," Sally K. Ride Papers, page 2 of 17, image 11, National Air and Space Museum Archives, https://edan.si.edu/slideshow/viewer/?eadrefid=NASM.2014.0025_ref230.
84. Sherr, *Sally Ride,* 204.
85. Author interview with Steve Hawley, April 28, 2021.
86. Sherr, *Sally Ride,* 210.
87. Sherr, *Sally Ride,* 211.

88. "Weather History Results for Washington, DC (20418) February 11th, 1986," Farmers' Almanac, https://www.farmersalmanac.com/weather-history-results/zipcode-20418/1986/02/11.
89. Volume 4 Index, Hearing of the Presidential Commission on the Space Shuttle *Challenger* Accident, part 4.
90. Volume 4 Index, Hearing of the Presidential Commission on the Space Shuttle *Challenger* Accident, part 4.
91. Volume 4 Index, Hearing of the Presidential Commission on the Space Shuttle *Challenger* Accident, part 4.
92. "Presidential Commission on the Space Shuttle *Challenger* Accident," February 11, 1986, C-SPAN, 29:16, https://www.c-span.org/video/?125993-1/presidential-commission-space-shuttle-challenger-accident.
93. "Presidential Commission on the Space Shuttle *Challenger* Accident," February 11, 1986, C-SPAN, https://www.c-span.org/video/?125993-1/presidential-commission-space-shuttle-challenger-accident.
94. Volume 4 Index, Hearing of the Presidential Commission on the Space Shuttle *Challenger* Accident, part 4.
95. Volume 4 Index, Hearing of the Presidential Commission on the Space Shuttle *Challenger* Accident, part 4.
96. Feynman, *"What Do You Care What Other People Think?,"* 150 (Kindle ed.).
97. Dean, "The Oral History of the Space Shuttle *Challenger* Disaster."
98. Dean, "The Oral History of the Space Shuttle *Challenger* Disaster."
99. Volume 4 Index, Hearing of the Presidential Commission on the Space Shuttle *Challenger* Accident, part 4.
100. Volume 4 Index, Hearing of the Presidential Commission on the Space Shuttle *Challenger* Accident, part 4.
101. Volume 4 Index, Hearing of the Presidential Commission on the Space Shuttle *Challenger* Accident, part 4.
102. Volume 4 Index, Hearing of the Presidential Commission on the Space Shuttle *Challenger* Accident, part 6.
103. Cassandra is a figure from Greek mythology. She was the daughter of the king of Troy, gifted with the ability to predict the future and cursed that no one would ever believe her warnings.
104. McDonald, *Truth, Lies, and O-Rings*, 47, 111.
105. Volume 4 Index, Hearing of the Presidential Commission on the Space Shuttle *Challenger* Accident, part 6.
106. Volume 4 Index, Hearing of the Presidential Commission on the Space Shuttle *Challenger* Accident, part 6.
107. Volume 4 Index, Hearing of the Presidential Commission on the Space Shuttle *Challenger* Accident, part 6.
108. "Two Small Notebooks Containing Ride's Notes from the Rogers Commission," Sally K. Ride Papers, page 4 of 17, image 3, National Air and Space Museum Archives, https://edan.si.edu/slideshow/viewer/?eadrefid=NASM.2014.0025_ref230.
109. "Two Small Notebooks Containing Ride's Notes from the Rogers Commission," Sally K. Ride Papers, page 4 of 17, image 4, National Air and Space Museum Archives, https://edan.si.edu/slideshow/viewer/?eadrefid=NASM.2014.0025_ref230.

110. Michael Tackett and Ann Marie Lipinski, "NASA Upset over Probe Exile," *Chicago Tribune,* February, 17, 1986.
111. "Presidential Commission on the Space Shuttle *Challenger* Accident," C-SPAN, February 26, 1986, 00:12–04:15, https://www.c-span.org/video/?126041-1/presidential-commission-space-shuttle-challenger-accident.
112. "Presidential Commission on the Space Shuttle *Challenger* Accident," C-SPAN, February 26, 1986, 04:10–05:11, https://www.c-span.org/video/?126041-1/presidential-commission-space-shuttle-challenger-accident.
113. "Presidential Commission on the Space Shuttle *Challenger* Accident," C-SPAN, February 26, 1986, 29:56–30:14, https://www.c-span.org/video/?126041-1/presidential-commission-space-shuttle-challenger-accident.
114. Volume 5 Index, February 26, 1986, session transcript, section 1514, *Report of the Presidential Commission on the Space Shuttle Challenger Accident*, https://history.nasa.gov/rogersrep/v5part1a.htm.
115. Volume 5 Index, February 26, 1986 Session transcript, section 1549, *Report of the Presidential Commission on the Space Shuttle Challenger Accident*, https://history.nasa.gov/rogersrep/v5part1a.htm.
116. McDonald, *Truth, Lies, and O-Rings,* 251.
117. McDonald, who watched the hearing on television, said, "Apparently, Mulloy was not aware of the controversial 31° launch criteria when we discussed it the evening before the launch because it was never brought up; furthermore, he and his associates, by the end of that meeting, had supported a launch at 26°!" McDonald, *Truth, Lies, and O-Rings,* 251.
118. The solid rocket motors, or SRMS, inside the solid rocket boosters are the first of their kind to be built for a manned spacecraft and the largest solid propellant motor ever developed for space flight. "Space Shuttle—Solid Rocket Boosters," NASA, https://www.nasa.gov/returntoflight/system/system_SRB.html.
119. McDonald, *Truth, Lies, and O-Rings,* 251.
120. McDonald, *Truth, Lies, and O-Rings,* 55.
121. McDonald, *Truth, Lies, and O-Rings,* 255–56.
122. McDonald, *Truth, Lies, and O-Rings,* 257.
123. McDonald, *Truth, Lies, and O-Rings,* 251, 255, 257.
124. Sherr, *Sally Ride,* 209–10.
125. Sherr, *Sally Ride,* 209–10.
126. David E. Sanger, "Engineers Tell of Punishment for Shuttle Testimony," *New York Times,* May 11, 1986, https://www.nytimes.com/1986/05/11/us/engineers-tell-of-punishment-for-shuttle-testimony.html.
127. Maura Dolan, "NASA Rocket Manager Named in Wrongful Death Suit to Retire," *Los Angeles Times,* July 17, 1986, https://www.latimes.com/archives/la-xpm-1986-07-17-mn-21673-story.html.
128. John Noble Wilford, "Key NASA Rocket Official Quits," *New York Times,* June 5, 1986, https://www.nytimes.com/1986/06/05/us/key-nasa-rocket-official-quits.html.
129. Arnold D. Aldrich transcript, NASA Johnson Space Center Oral History Project, April 28, 2008, https://historycollection.jsc.nasa.gov/JSCHistoryPortal/history/oral_histories/AldrichAD/AldrichAD_4-28-08.htm.
130. "Jesse Moore, A Top NASA Aide, Quits," Associated Press, *New York Times,* February 6, 1987, https://www.nytimes.com/1987/02/06/us/jesse-moore-a-top-nasa-aide-quits.html.

131. Sanger, "Engineers Tell of Punishment for Shuttle Testimony."
132. Clay Risen, "Allan McDonald Dies at 83; Tried to Stop the *Challenger* Launch," *New York Times*, March 9, 2021, https://www.nytimes.com/2021/03/09/us/allan-mcdonald-dead.html.
133. Douglas Martin, "Roger Boisjoly, 73, Dies; Warned of Shuttle Danger," *New York Times*, February 3, 2012, https://www.nytimes.com/2012/02/04/us/roger-boisjoly-73-dies-warned-of-shuttle-danger.html.
134. "Chapter V: The Contributing Cause of the Accident," *Report of the Presidential Commission on the Space Shuttle Challenger Accident*, https://history.nasa.gov/rogersrep/v1ch5.htm.
135. "Chapter V: The Contributing Cause of the Accident," *Report of the Presidential Commission on the Space Shuttle Challenger Accident*, https://history.nasa.gov/rogersrep/v1ch5.htm.
136. "Chapter VI: An Accident Rooted in History," *Report of the Presidential Commission on the Space Shuttle Challenger Accident*, https://history.nasa.gov/rogersrep/v1ch6.htm.
137. Volume 2: Appendix F—Personal Observations on Reliability of Shuttle, R. P. Feynman, *Report of the Presidential Commission on the Space Shuttle Challenger Accident*, https://history.nasa.gov/rogersrep/v2appf.htm.
138. Volume 2: Appendix F—Personal Observations on Reliability of Shuttle.
139. Volume 2: Appendix F—Personal Observations on Reliability of Shuttle.
140. Letter, *Report of the Presidential Commission on the Space Shuttle Challenger Accident*, https://history.nasa.gov/rogersrep/letter.htm.
141. Ride, NASA Oral History Project, October 22, 2002.
142. Author interview with George Abbey, December 4, 2020.
143. Cassutt, *The Astronaut Maker*, 298.
144. Volume 5 Index, April 3, 1986, session transcript, *Report of the Presidential Commission on the Space Shuttle Challenger Accident*, https://history.nasa.gov/rogersrep/v5part5.htm.
145. Volume 5 Index, April 3, 1986, session transcript, section 2382–83.
146. Volume 5 Index, April 3, 1986, session transcript, section 2394.
147. Young, *Forever Young*, 247, 257.
148. Author interview with George Abbey, December 4, 2020.
149. Walter Cunningham, *The All-American Boys*, United Kingdom: iBooks, 2003, location 7055–103 (Kindle ed.).
150. Cunningham, *The All-American Boys*, location 7055–103 (Kindle ed.).
151. Author interview with Bob Crippen, January 28, 2022.
152. Author interview with Kathy Sullivan, March 10, 2021.
153. Author interview with Kathy Sullivan, March 10, 2021.
154. Author interview with Steve Hawley, April 28, 2021.
155. Randy Avera, *The Truth About Challenger*. Good Hope, GA: Randolph Pub., 2003, p. 95.
156. Author interview with Steve Hawley, April 28, 2021.
157. Author interview with Steve Hawley, April 28, 2021.
158. Author interview with Steve Hawley, April 28, 2021.
159. Author interview with Steve Hawley, April 28, 2021.
160. Sherr, *Sally Ride*, 223–24.
161. Myrna Oliver, "Dale Ride; Father of First US Woman Astronaut," *Los Angeles*

Times, September 11, 1989, https://www.latimes.com/archives/la-xpm-1989-09-11-mn-1330-story.html.

162. Sherr, *Sally Ride,* 237.
163. Sherr, *Sally Ride,* 237.
164. Kathy Sawyer, "*Challenger* Pilot Widow Files $1.5 Billion Suit," *Washington Post,* May 7, 1987, https://www.washingtonpost.com/archive/politics/1987/05/07/challenger-pilot-widow-files-15-billion-suit/5f51e5b9-ad80-46d0-add0-18fdfa4cc7e1/.
165. Carolyn Click, "Lawyer for Astronaut's Widow Hails 'Accountability,'" UPI Archives, August 23, 1988, https://www.upi.com/Archives/1988/08/23/Lawyer-for-astronauts-widow-hails-accountability/9193588312000/.
166. Associated Press, "Some Shuttle Heirs Got Funds, Others Didn't," *Albuquerque Journal,* March 14, 1988, https://www.newspapers.com/image/157945227/.
167. "Some Shuttle Heirs Got Funds, Others Didn't." Despite the malfunction on *Challenger,* Morton Thiokol continued to be a subcontractor for NASA, dedicated to re-designing the SRBs. The SRB re-design, as well as the work to replace the reusable hardware lost on *Challenger,* was worth about $409 million—which Thiokol performed "at no profit." Also, Harry F. Rosenthal, "Morton Thiokol Gives Up $10 Million in Profits in Face of NASA Fine," *AP News,* February 24, 1987, https://apnews.com/article/27ffadba339e45ff1ec674e0daa30a9e.
168. The Resniks filed suit, with Judy's ex-husband Michael Oldak representing them. Judy's relatives had been offered less because Judy had no spouse or children to support. Oldak reached a separate settlement with Morton Thiokol that was rumored to be similar to what Jarvis's family had received. Ironically, Judy's mother, Sarah, sued separately from the rest of the family and was awarded an undisclosed amount. Ron McNair's family hired the well-known attorney who had represented Gus Grissom's widow in the Apollo One fire, Ronald Krist. Krist's filings listed Morton Thiokol and five of the company's executives as co-defendants. Krist told the press that Morton Thiokol had played "Russian roulette with the astronauts' lives." It took two years, but the case finally settled with Thiokol for an undisclosed amount. Mary Thornton, "US Settles with Kin of Astronauts," *Washington Post,* December 30, 1986, https://www.washingtonpost.com/archive/politics/1986/12/30/us-settles-with-kin-of-astronauts/9d6e3062-8223-4e68-9e59-d9909d81ed56/. Also, UPI, "The Father of *Challenger* Astronaut Judith Resnik accepted a . . . ," February 18, 1988, https://www.upi.com/Archives/1988/02/19/The-father-of-Challenger-astronaut-Judith-Resnik-accepted-a/1171572245200/.
169. Avera, *The Truth About Challenger,* 155–156.
170. Robert Z. Pearlman, "NASA Exhibits Space Shuttles *Challenger, Columbia* Debris for First Time," Space.com, June 29, 2015, https://www.space.com/29794-space-shuttles-challenger-columbia-debris-exhibit.html.

Chapter 22: God Help You If You Screw This Up

1. "Max Qs First Gig," video courtesy of Hoot Gibson, *Smithsonian Magazine,* https://www.airspacemag.com/videos/category/space-exploration/max-qs-first-gig.
2. Robert L. "Hoot" Gibson transcript, NASA Johnson Oral History Project,

April 10, 2018, https://historycollection.jsc.nasa.gov/JSCHistoryPortal/history/oral_histories/GibsonRL/GibsonRL_4-10-18.htm.

3. Author interview with Estella Gillete, July 1, 2020.
4. John A. Logsdon, "Chapter 15 Return to Flight: Richard H. Truly and the Recovery from the *Challenger* Accident," NASA History Office, https://history.nasa.gov/SP-4219/Chapter15.html.
5. Charles Fishman, "Astronauts' Portrait of Work at Odds with Public Image," *Washington Post*, April 1, 1986, https://www.washingtonpost.com/archive/politics/1986/04/01/astronauts-portrait-of-work-at-odds-with-public-image/4004e705-d070-420f-b986-8e86daa1958b/.
6. Gregory, NASA Oral History Project, March 14, 2005.
7. Shayler and Burgess, *NASA's First Space Shuttle Astronaut Selection*, location 10676–709 (Kindle ed.).
8. Kathy Sawyer, "NASA Is Developing Shuttle Escape System," *Washington Post*, December 25, 1986.
9. Mark Betancourt, "They Said It Wasn't Possible to Escape the Space Shuttle. These Guys Showed It Was," *Air and Space Magazine,* September 2020, https://www.airspacemag.com/airspacemag/escape-speeding-shuttle-180975606/. Also, William B. Scott, "NASA Selects Telescoping Pole for Shuttle Crew Escape System," *Aviation Week and Space Technology,* April 11, 1988, https://archive.aviationweek.com/issue/19880411.
10. Steven R. Nagel transcript, NASA Johnson Space Center Oral History Project, December 20, 2002, https://historycollection.jsc.nasa.gov/JSCHistoryPortal/history/oral_histories/NagelSR/NagelSR_12-20-02.htm.
11. Thomas H. Maugh II, "Escape System May Be Added to Next Shuttle," *Los Angeles Times,* January 21, 1987.
12. Paul Recer, "New Space Shuttle Brake Design Now Ready for Use: NASA Engineer," Associated Press, June 12, 1986.
13. John Noble Wilford, "American Astronauts Roar Back to Space, Renewing the Nation's Hope for Shuttle," *New York Times*, September 30, 1988, https://www.nytimes.com/1988/09/30/us/american-astronauts-roar-back-space-renewing-nation-s-hopes-for-shuttle-long.html.
14. Author interview with Guy Bluford, December 5, 2020.
15. Shayler and Burgess, *NASA's First Space Shuttle Astronaut Selection*, 368.
16. Sullivan, *Handprints on Hubble*, 154, 226.
17. "J. C. Fletcher; Headed NASA Before, After Space Disaster," *Times* Staff and Wire Reports, *Los Angeles Times,* December 23, 1991, https://www.latimes.com/archives/la-xpm-1991-12-23-mn-596-story.html.
18. Fishman, "Astronauts' Portrait of Work at Odds with Public Image."
19. Cunningham, *The All-American Boys*, 361–62.
20. Mullane, *Riding Rockets,* 215.
21. Cassutt, *Astronaut Maker,* 311–12
22. Author interview with Fred Gregory, May 17, 2021.
23. Mullane, *Riding Rockets,* 264.
24. Mullane, *Riding Rockets*, 264.
25. Author interview with Jim Bagian, February 24, 2022.
26. McDonald, *Truth, Lies, and O-Rings*, 451.
27. "Report to the President, Implementation of the Recommendations of the

Presidential Commission on the Space Shuttle *Challenger* Accident," NASA, June 1987, https://ntrs.nasa.gov/api/citations/19930010801/downloads/19930010801.pdf, 13. Also, McDonald, *Truth, Lies, and O-Rings*, 428.

28. McDonald, *Truth, Lies, and O-Rings*, 499.
29. Logsdon, "Chapter 15 Return to Flight: Richard H. Truly and the Recovery from the *Challenger* Accident." Also, McDonald, *Truth, Lies, and O-Rings*, 499–500.
30. Logsdon, "Chapter 15 Return to Flight: Richard H. Truly and the Recovery from the *Challenger* Accident."
31. James Fisher and John J. Glisch, "Is the Shuttle Ready to Fly?" *Orlando Sentinel*, September 11, 1988, https://www.newspapers.com/image/230000075.
32. Fisher and Glisch, "Is the Shuttle Ready to Fly?"
33. John J. Glisch, "Shuttle Commander: The Bird Is Ready, We're Ready," *Orlando Sentinel*, September 27, 1988, https://www.newspapers.com/image/229969793/.
34. John M. Lounge transcript, NASA Johnson Space Center Oral History Project, February 7, 2008, https://historycollection.jsc.nasa.gov/JSCHistoryPortal/history/oral_histories/LoungeJM/LoungeJM_2-7-08.htm.
35. John Noble Wilford, "Astronaut on the Ground Making Final Decision," *New York Times*, September 29, 1988, https://www.nytimes.com/1988/09/29/us/astronaut-on-the-ground-making-final-decision.html.
36. "Robert L. Crippen Biography," https://www.nasa.gov/centers/kennedy/about/biographies/crippen.html.
37. Leon Jaroff, "The Magic Is Back!" *Time*, October 10, 1988, http://content.time.com/time/magazine/article/0,9171,968628,00.html.
38. Richard Tribou, "September 29 in Space History: Shuttle Program Rebounds," *Orlando Sentinel*, September 29, 2016, https://www.orlandosentinel.com/space/os-september-29-in-space-history-shuttle-program-gets-back-in-action-20160928-story.html.
39. Author interview with Bob Crippen, Octboer 22, 2020.
40. Nelson, NASA Oral History Project.
41. John Noble Wilford, "American Astronauts Roar Back to Space, Renewing the Nation's Hopes for Shuttle; Long Hiatus Ends," *New York Times*, September 30, 1988, https://www.nytimes.com/1988/09/30/us/american-astronauts-roar-back-space-renewing-nation-s-hopes-for-shuttle-long.html.
42. Sullivan, *Handprints on Hubble*, 166–67, 171–73.
43. "'We Have Resumed the Journey . . . for You,'" *Washington Post*, October 3, 1988, https://www.washingtonpost.com/archive/politics/1988/10/03/we-have-resumed-the-journey-for-you/073a59a2-8e3b-4a09-ac26-9ccaeda55a5f.
44. Cassutt, *The Astronaut Maker*, 308.
45. Jason Treat, Jay Bennett, and Christopher Turner, "How 'The Right Stuff' Has Changed," *National Geographic*, November 6, 2020, https://www.nationalgeographic.com/science/graphics/charting-how-nasa-astronaut-demographics-have-changed-over-time.
46. Treat, Bennett, and Turner, "How 'The Right Stuff' Has Changed."
47. Logsdon, "Chapter 15 Return to Flight: Richard H. Truly and the Recovery from the *Challenger* Accident."
48. Harry F. Rosenthal, "Reagan Orders NASA to Halt Launch of Commercial Payloads," Associated Press, August 15, 1986, https://apnews.com/article/7e6b76c27ec65f93f14fd7913cf95c48.

49. Craig Covault, "New Manifest for Space Shuttle Generates Payload Sponsor Debate," *Aviation Week & Space Technology,* October 13, 1986, https://archive.aviationweek.com/issue/19861013. Also, John H. Cushman Jr., "Air Force to Build New Rocket Fleet to Loft Satellites," *New York Times,* August 1, 1986, https://www.nytimes.com/1986/08/01/us/air-force-to-build-new-rocket-fleet-to-loft-satellites.html?searchResultPosition=1.
50. Associated Press, "US Space Plan Takes New Direction," *The Daily Journal,* February 12, 1988, https://www.newspapers.com/image/155130011/.
51. Robert A. Jones, "It's 'All or Nothing': US Space Program at Crossroads," *Los Angeles Times,* September 6, 1988, https://www.latimes.com/archives/la-xpm-1988-09-06-mn-1742-story.html.
52. "12 Nations Back Space Station," Associated Press, *Los Angeles Times,* September 29, 1988, https://www.latimes.com/archives/la-xpm-1988-09-29-mn-6218-story.html.
53. Agreement among the United States of America, governments of Member States of the European Space Agency, the government of Japan, and the government of Canada on Cooperation in the Detailed Design, Development, Operation, and Utilization of the Permanently Manned Civil Space Station, Sept. 29, 1988.

Chapter 23: Through a Glass Darkly

1. "Chamber B," NASA, https://www.nasa.gov/centers/johnson/engineering/integrated_environments/human_space_environment_testing/chamber_B/index.html.
2. Sullivan, *Handprints on Hubble,* 190.
3. Sullivan, *Handprints on Hubble,* 85.
4. Sullivan, *Handprints on Hubble,* 153, 173–74.
5. Author interview with Kathy Sullivan, March 10, 2021.
6. Sullivan, *Handprints on Hubble,* 186.
7. "Neutral Buoyancy Space Simulator," National Park Service, https://www.nps.gov/articles/neutral-buoyancy-space-simulator.htm.
8. "Chamber B," NASA, June 27, 2017, https://www.nasa.gov/feature/chamber-b. Also, "Chamber B Thermal/Vacuum Chamber," NASA, https://www.nasa.gov/centers/johnson/pdf/638559main_ChamberB-User_Test_Planning_Guide.pdf.
9. Sullivan, *Handprints on Hubble,* 188.
10. Cooper, *Before Lift-Off,* 182.
11. Sullivan, *Handprints on Hubble,* 189.
12. Sullivan, *Handprints on Hubble,* 190.
13. Sullivan, *Handprints on Hubble,* 190–91.
14. Sullivan, *Handprints on Hubble,* 104–05.
15. Author interview with Kathy Sullivan, March 10, 2021.
16. Author interview with Kathy Sullivan, March 10, 2021.
17. Sullivan, *Handprints on Hubble,* 112.
18. Sullivan, *Handprints on Hubble,* 112.
19. "About the Hubble Space Telescope," NASA, https://www.nasa.gov/mission_pages/hubble/story/index.html.
20. Sullivan, *Handprints on Hubble,* 180–81.
21. Sullivan, *Handprints on Hubble,* 114.

22. Sullivan, *Handprints on Hubble*, 107 and 143.
23. Sullivan, *Handprints on Hubble*, 114.
24. Sullivan, *Handprints on Hubble*, 200.
25. Sullivan, *Handprints on Hubble*, 200–01.
26. Sullivan, NASA Oral History Project. Also, Sarah Loff, "Deployment of the Hubble Space Telescope," NASA, April 24, 2015, https://www.nasa.gov/image-feature/april-25-1990-deployment-of-the-hubble-space-telescope.
27. Sullivan, NASA Oral History Project, March 12, 2008.
28. Sullivan, *Handprints on Hubble*, 210.
29. Steven A. Hawley transcript, NASA Johnson Space Center Oral History Project, December 17, 2002, https://historycollection.jsc.nasa.gov/JSCHistoryPortal/history/oral_histories/HawleySA/HawleySA_12-17-02.htm.
30. Sullivan, *Handprints on Hubble*, 210–11.
31. Sullivan, *Handprints on Hubble*, 212.
32. Author interview with Kathy Sullivan, March 10, 2021.
33. Sullivan, NASA Oral History Project, March 12, 2008.
34. Sullivan, *Handprints on Hubble*, 219.
35. Sullivan, *Handprints on Hubble*, 224–25.
36. Sullivan, *Handprints on Hubble*, 222.
37. Jon Van, "Scientists on Hubble: 'Any One of Us Should Have . . . ,'" *Chicago Tribune*, June 28, 1990, https://www.chicagotribune.com/news/ct-xpm-1990-06-29-9002220516-story.html.
38. "Almanac: Hubble Space Telescope," CBS News, April 24, 2016, https://www.cbsnews.com/news/almanac-hubble-space-telescope/.
39. "Topic: *Columbia* STS-35," NASA Spaceflight Forum, https://forum.nasaspaceflight.com/index.php?topic=42754.100.
40. "Hubble's Mirror Flaw," NASA, https://www.nasa.gov/content/hubbles-mirror-flaw. Also, Sarah Scoles, "Hubble's Blurry Years," *Physics Today*, April 1, 2020, https://physicstoday.scitation.org/do/10.1063/pt.6.4.20200401d/full.
41. Thomas H. Maugh II, "NASA Grounds Space Shuttles; Woes Pile Up," *Los Angeles Times*, June 30, 1990, https://www.latimes.com/archives/la-xpm-1990-06-30-mn-609-story.html.
42. "Hubble Space Telescope Problems," C-SPAN, June 29, 1990, https://www.c-span.org/video/?12954-1/hubble-space-telescope-problems.
43. Thomas H. Maugh II, "NASA Grounds Space Shuttles; Woes Pile Up," *Los Angeles Times*, June 30, 1990, https://www.latimes.com/archives/la-xpm-1990-06-30-mn-609-story.html.
44. Sullivan, *Handprints on Hubble*, 226.
45. Sullivan, *Handprints on Hubble*, 227–28.
46. Author interview with Kathy Sullivan, March 10, 2021.
47. Sullivan, *Handprints on Hubble*, 229–30.
48. In total, there would be five servicing and repair missions for Hubble. Besides STS-61 in 1993, they were: STS-82 in February 1997; STS-103 in December 1999; STS-109 in March 2002; and STS-125 in May 2009. "About—Hubble Servicing Missions," NASA, https://www.nasa.gov/mission_pages/hubble/servicing/index.html.
49. Sullivan, *Handprints on Hubble*, 236–37.
50. "About—Hubble Servicing Missions | SM1," NASA, https://www.nasa.gov

/content/about-hubble-servicing-missions-sm1. Also, “STS-61,” Space Shuttle Mission Archives, NASA, https://www.nasa.gov/mission_pages/shuttle/shuttlemissions/archives/sts-61.html.

51. William Harwood, “How NASA Fixed Hubble’s Flawed Vision—and Reputation,” CBS News, April 22, 2015, https://www.cbsnews.com/news/an-ingenius-fix-for-hubbles-famously-flawed-vision.
52. Sullivan, *Handprints on Hubble*, 237.
53. Sullivan, *Handprints on Hubble*, 238.
54. Kathy Sawyer, “Given New Focus, Hubble Can Almost See Forever,” *Washington Post,* January 14, 1994, https://www.washingtonpost.com/archive/politics/1994/01/14/given-new-focus-hubble-can-almost-see-forever/d87e01da-720f-4382-8eee-e43105e22e97.
55. Sullivan, *Handprints on Hubble*, 244–45.
56. J. D. Harrington, “Hubble Finds First Organic Molecule on an Exoplanet,” NASA, March 19, 2008, https://www.nasa.gov/mission_pages/hubble/science/hst_img_20080319.html.
57. “Hubble Discovers a Fifth Moon Orbiting Pluto,” NASA, July 11, 2012, https://www.nasa.gov/mission_pages/hubble/science/new-pluto-moon.html.
58. "Black Holes Dwell in Most Galaxies, According to Hubble Census,” NASA press release, January 13, 1997, https://hubblesite.org/contents/news-releases/1997/news-1997-01.html.
59. Sullivan, *Handprints on Hubble*, 3.
60. “Hubble Accomplishments,” NASA, https://www.nasa.gov/content/hubble-accomplishments.
61. Sullivan, *Handprints on Hubble*, 238.

Chapter 24: Closer to God

1. Shannon Lucid. *Tumbleweed: Six Months Living on Mir.* MkEk, 2020, p. 15 (Kindle ed.).
2. “US-Russian Summits, 1992–2000,” Prepared by the Office of the Historian, Bureau of Public Affairs, US Department of State, July 2000, https://1997-2001.state.gov/regions/nis/chron_summits_russia_us.html.
3. Bryan Burrough, *Dragonfly: An Epic Adventure of Survival in Outer Space.* New York: HarperCollins, 2000, p. 245.
4. Gardiner Morse, “Leading Ferociously,” *Harvard Business Review,* May 2002, https://hbr.org/2002/05/leading-ferociously.
5. Kathy Sawyer, “The Man on the Moon,” *Washington Post,* July 20, 1994, https://www.washingtonpost.com/archive/lifestyle/1994/07/20/the-man-on-the-moon/0fc3cc3d-ea96-401d-9947-302ad1edad5a.
6. Cassutt, *The Astronaut Maker,* 347.
7. Cassutt, *The Astronaut Maker,* 347–48.
8. Cassutt, *The Astronaut Maker,* 363.
9. Clay Morgan, *Shuttle-Mir.* Vol. 4225, National Aeronautics and Space Administration, Lyndon B. Johnson Space Center, 2001, 6–7.
10. Cassutt, *The Astronaut Maker,* 367.
11. Cassutt, *The Astronaut Maker,* 373.
12. Author interview with Shannon Lucid, September 17, 2020.

13. Shannon W. Lucid transcript, NASA Johnson Space Center Oral History Project, June 17, 1998, https://historycollection.jsc.nasa.gov/JSCHistoryPortal/history/oral_histories/Shuttle-Mir/LucidSW/LucidSW_6-17-98.htm.
14. Lucid, *Tumbleweed,* 12.
15. *People* Magazine Staff, "NASA Picks Six Women Astronauts with the Message: You're Going a Long Way, Lady," *People,* February 6, 1978, https://investing.suaramasa.com/host-https-people.com/archive/nasa-picks-six-women-astronauts-with-the-message-youre-going-a-long-way-baby-vol-9-no-5/.
16. "Out of This World," Oklahoma Medical Research Foundation, https://omrf.org/findings/out-of-this-world/.
17. Author interview with Shannon Lucid, September 17, 2020.
18. Author interview with Shannon Lucid, September 17, 2020.
19. Author interview with Shannon Lucid, September 17, 2020.
20. Author interview with Shannon Lucid, September 17, 2020.
21. Author interview with Shannon Lucid, September 17, 2020.
22. Author interview with Shannon Lucid, September 17, 2020.
23. Author interview with Shannon Lucid, September 17, 2020.
24. Sharon Begley, "Down to Earth," *Newsweek,* October 6, 1996, https://www.newsweek.com/down-earth-179194.
25. Lucid, *Tumbleweed,* 14–15.
26. Polina Ivanova, "The Woman Who Fell from the Sky," *Reuters,* November 13, 2020, https://www.reuters.com/investigates/special-report/health-coronavirus-russia-starcity.
27. Burrough, *Dragonfly,* 98.
28. Lucid, *Tumbleweed,* 52–53.
29. Lucid, *Tumbleweed,* 52–53.
30. Lucid, *Tumbleweed,* 50–52.
31. Lucid, *Tumbleweed,* 33.
32. Lucid, *Tumbleweed,* 49.
33. Lucid, *Tumbleweed,* 73.
34. "JAXA Astronaut Activity Report, June, 2014," JAXA, June 2014, https://iss.jaxa.jp/en/astro/report/2014/1406.html.
35. Lucid, *Tumbleweed,* 75.
36. Lucid, *Tumbleweed,* 75.
37. "US astronaut Shannon Lucid (C) w. Russian cosmonauts Yuri Usachev (L) and Yuri Onufriyenko aboard their 10-yr-old space station *Mir*; she joined them from the space shuttle Atlantis for five months of research," *LIFE Photo Collection,* March 1996, https://artsandculture.google.com/asset/PAGWDcBB94gJJA?hl=en March 1996.
38. Lucid, *Tumbleweed,* 79.
39. Lucid, *Tumbleweed,* 85.
40. Lucid, *Tumbleweed,* 85.
41. Lucid, *Tumbleweed,* 62.
42. "NASA and NACA Center Directors," *NASA History,* https://history.nasa.gov/director.html.
43. "Huntoon to Lead Planning Effort for Life Sciences Institute," *NASA News Release,* August 4, 1995, https://www.nasa.gov/home/hqnews/1995/95-132.txt.
44. Cassutt, *The Astronaut Maker,* 379.

45. Burrough, *Dragonfly,* 331.
46. Lucid, *Tumbleweed,* 62.
47. Burrough, *Dragonfly,* 332–33.
48. Burrough, *Dragonfly,* 332–33.
49. Lucid, *Tumbleweed,* 62–65.
50. Burrough, *Dragonfly,* 332–33.
51. Sharon Begley, "Lucid's Long Road Home," *Newsweek,* September 29, 1996, https://www.newsweek.com/lucids-long-road-home-178032.
52. William H. Gerstenmaier transcript, NASA Shuttle-*Mir* Oral History Project, September 22, 1998, https://historycollection.jsc.nasa.gov/JSCHistoryPortal/history/oral_histories/Shuttle-Mir/GerstenmaierWH/GerstenmaierWH_9-22-98.htm.
53. NASA Shuttle-*Mir* Audio/Video Gallery, https://historycollection.jsc.nasa.gov/history/shuttle-mir/multimedia/m-av.htm.
54. Lucid, *Tumbleweed,* 170. "The number is not a constant because radiation comes from more than one source. There are galactic cosmic rays (GCR) and solar cosmic rays (SCR). There is also radiation trapped within the magnetic field, auroral precipitation, and albedo radiation." "How Much Radiation Are ISS Astronauts Exposed To?" *Forbes,* November 13, 2018, https://www.forbes.com/sites/quora/2018/11/13/how-much-radiation-are-iss-astronauts-exposed-to/?sh=4ecbb30518a9.
55. "Lucid on Treadmill in Russian *Mir* Space Station," NASA, Archive.org, March 28, 1996, https://archive.org/details/GPN-2000-001034.
56. Lucid, *Tumbleweed,* 142.
57. Lucid, *Tumbleweed,* 197.
58. Lucid, *Tumbleweed,* 105.
59. Lucid, *Tumbleweed,* 133.
60. "Gelatin? It's Sunday in Space," Reuters, *Oklahoman,* June 17, 1996, https://www.oklahoman.com/article/2542447/gelatin-its-sunday-in-space.
61. Lucid, *Tumbleweed,* 125.
62. Lucid, *Tumbleweed,* 151.
63. *Mir*-21 Mission Interviews, NASA, https://history.nasa.gov/SP-4225/documentation/mir-summaries/mir21/interviews.htm.
64. Anatoly Zak, "*Mir*'s *Priroda* module," *Russian Space Web,* http://www.russianspaceweb.com/mir_priroda.html. Also, Shannon Lucid, "Shuttle-*Mir* Stories: Lucid on Pink Socks and Jello," NASA, May 19, 1996, https://historycollection.jsc.nasa.gov/history/shuttle-mir/history/to-h-f-lucid-socks.htm.
65. Lucid, *Tumbleweed,* 161–62.
66. Lucid, *Tumbleweed,* 161–62.
67. Lucid, "Shuttle-*Mir* Stories: Lucid on Pink Socks and Jello."
68. Lucid, *Tumbleweed,* 165.
69. Lucid, "Shuttle-*Mir* Stories: Lucid on Pink Socks and Jello."
70. Lucid, *Tumbleweed,* 145.
71. "Shuttle's Delay Strands Astronaut," *Washington Post,* July 13, 1996, https://www.washingtonpost.com/archive/politics/1996/07/13/shuttles-delay-strands-astronaut/9d569d71-a401-4982-bdca-6e69234d9544.
72. Lucid, *Tumbleweed,* 187.

73. Shayler and Burgess, *NASA's First Space Shuttle Astronaut Selection,* 551.
74. Lucid, *Tumbleweed,* 156.
75. Lucid, *Tumbleweed,* 198.
76. "NASA-2 Shannon Lucid: Enduring Qualities, March 22-September 26, 1996," *NASA History,* https://history.nasa.gov/SP-4225/nasa2/nasa2.htm.
77. "NASA-2 Shannon Lucid: Enduring Qualities, March 22-September 26, 1996."
78. "NASA-2 Shannon Lucid: Enduring Qualities, March 22-September 26, 1996."
79. Begley, "Lucid's Long Road Home."
80. Begley, "Lucid's Long Road Home."
81. Begley, "Down to Earth."
82. Marcia Dunn, "Down to Earth Shannon Lucid Looks to the Future, Not Just to Her High-Flying Record," *Chicago Tribune,* December 1, 1996.
83. "USA: Astronaut Shannon Lucid Returns after 188 Days in Space," AP Archive.
84. "USA: Astronaut Shannon Lucid Returns after 188 Days in Space," AP Archive.

Chapter 25: Everything That Rises Must Converge

1. Author interview with Anna Fisher, December 11, 2020.
2. Author interview with Anna Fisher, May 4, 2021.
3. Author interview with Anna Fisher, December 11, 2020.
4. "Astronaut Bio: William F. Fisher," NASA, https://www.nasa.gov/sites/default/files/atoms/files/fisher-wf.pdf.
5. Paul Recer, "Study Concludes Space Station May Require Up To 10 Space Walks A Week," *AP News,* July 20, 1990, https://apnews.com/article/dd9a9ef9711aaf0d3f9df4907c0d3f07.
6. William J. Broad, "Astronaut, Quitting NASA, Urges Overhaul of Space Station," *New York Times,* January 9, 1991, https://www.nytimes.com/1991/01/09/us/astronaut-quitting-nasa-urges-overhaul-of-space-station.html.
7. Author interview with Anna Fisher, December 11, 2020.
8. Author interview with Anna Fisher, December 11, 2020.
9. Fisher, NASA Oral History Project, May 3, 2011.
10. Fisher, NASA Oral History Project, May 3, 2011.
11. Fisher, NASA Oral History Project, May 3, 2011.
12. Author interview with Anna Fisher, December 11, 2020.
13. Fisher, NASA Oral History Project, May 3, 2011.
14. Author interview with Anna Fisher, December 11, 2020.
15. Kristin Fisher's Instagram, https://www.instagram.com/p/lz-TgKjlbg/.
16. Author interview with Anna Fisher, December 11, 2020.
17. Author interview with Anna Fisher, December 11, 2020.
18. Author interview with Anna Fisher, December 11, 2020.
19. Author interview with George Abbey, December 4, 2020.
20. "Women in the American Workforce," US Equal Employment Opportunity Commission, https://www.eeoc.gov/special-report/women-american-workforce.
21. Author interview with Anna Fisher, April 19, 2021.
22. Author interview with Anna Fisher, April 19, 2021.
23. "International Space Station Status Report #98-3," 5:30 AM EST, Friday, November 20, 1998, Mission Control Center, Korolev, Russia, NASA, https://www.nasa.gov/centers/johnson/news/station/1998/iss98-03.html.

24. “Zarya Launch Footage,” NASA Johnson Space Center, 2018, https://www.youtube.com/watch?v=6qGgtpuTtoo.
25. “Baikonur Cosmodrome,” *Travels in Orbit,* 2015, http://www.travelsinorbit.com/wp-content/uploads/2015/12/Baikonur_Cosmodrome_Map_2015-1.jpg.
26. “A New Era of Endeavour: STS-88 Space Shuttle Press Kit,” NASA, November 20, 1998, https://historycollection.jsc.nasa.gov/JSCHistoryPortal/history/shuttle_pk/pk/Flight_093_STS-088_Press_Kit.pdf.
27. “Zarya Cargo Module,” NASA, https://www.nasa.gov/mission_pages/station/structure/elements/zarya-cargo-module.
28. By October 1998, one month before *Zarya*’s launch, NASA was asking Congress to approve an additional $660 million in spending for the ISS over the next five years. Independent aerospace experts estimated that the project would require anywhere from $600 million to $3 billion above that to complete by 2004. “International Space Station Is Worth It,” *North Jersey Herald & News,* October 19, 1998, https://www.newspapers.com/image/529651455/. Also, Ronald Kotulak and Colin McMahon, “Politics, Not Science, Propels Costly Space Station, Critics Say,” *Austin American-Statesman,* July 5, 1998, https://www.newspapers.com/image/357229561.
29. “Zarya Launch Footage,” NASA Johnson Space Center.
30. “International Space Station Status Report #98-3,” 5:30 AM EST, Friday, November 20, 1998, Mission Control Center, Korolev, Russia, NASA, https://www.nasa.gov/centers/johnson/news/station/1998/iss98-03.html.
31. “The International Space Station Begins: Part 2,” *Houston, We Have A Podcast,* NASA, https://soundcloud.com/nasa/the-international-space-station-begins-part-2. Also, “Space Shuttle Flight 93 (STS-88) — Post Flight Presentation Video,” *National Space Society,* https://space.nss.org/space-shuttle-flight-93-sts-88-post-flight-presentation-video.
32. Richard C. Paddock, “Space Module Blazes a New Trail in Orbit,” *Los Angeles Times,* November 21, 1998, https://www.latimes.com/archives/la-xpm-1998-nov-21-mn-46133-story.html.
33. Cliff Lethbridge, “STS-51A Fact Sheet,” Spaceline.org, https://www.spaceline.org/united-states-manned-space-flight/space-shuttle-mission-program-fact-sheets/sts-51a. Also, “STS-88,” Space Shuttle Mission Archives, NASA, https://www.nasa.gov/mission_pages/shuttle/shuttlemissions/archives/sts-88.html.
34. Kelli Mars, “20 Years Ago, Space Station Construction Begins,” *NASA History,* November 20, 2018, https://www.nasa.gov/feature/20-years-ago-iss-construction-begins.
35. “STS-88 Day Five Highlights,” Mission Control Center, *KSC Science,* https://science.ksc.nasa.gov/shuttle/missions/sts-88/sts-88-day-05-highlights.html.
36. “STS-88 Day Five Highlights,” Mission Control Center, *KSC Science,* https://science.ksc.nasa.gov/shuttle/missions/sts-88/sts-88-day-05-highlights.html.
37. Author interview with Anna Fisher, July 30, 2020.
38. Author interview with Anna Fisher, July 30, 2020.
39. “NASA Astronaut Group 8.”
40. Author interview with Anna Fisher, December 11, 2020.
41. Frank Morring, “Goldin Changed NASA Forever, But Successor Must Pay Costs,” *Aviation Week,* November 12, 2001, https://archive.aviationweek.com/issue/20011112.

42. Andrew Chaikin, "Aiming High in Hard Times," *Popular Science*, February 1996, vol. 248, issue 2.
43. Cassutt, *The Astronaut Maker,* 410.
44. Morring, "Goldin Changed NASA Forever, But Successor Must Pay Costs."
45. Cassutt, *The Astronaut Maker,* 411.
46. "NASA Administrator Appoints Johnson Space Center Director to Senior Assistant Position," NASA press release, February 23, 2001, https://www.nasa.gov/home/hqnews/2001/01-027.txt.
47. Cassutt, *The Astronaut Maker,* 413–15.
48. The first American to *Mir* was TFNG Norm Thagard, who launched aboard a Soyuz TM-21 in 1995, a year before Shannon began her 188-day stay on the Russian station.
49. Author interview with Anna Fisher, April 27, 2022.
50. "Frank Culbertson Letter," NASA Human Spaceflight, September 12, 2001, https://web.archive.org/web/20011018060839/http://spaceflight.nasa.gov/station/crew/exp3/culbertsonletter.html.
51. Warren E. Leary, "Shuttle to Lift Off Tonight for Space Station," *New York Times,* November 30, 2000, https://www.nytimes.com/2000/11/30/us/shuttle-to-lift-off-tonight-for-space-station.html.

Chapter 26: Yesterday in Texas

1. Michael Leinbach and Jonathan H. Ward, *Bringing Columbia Home: The Untold Story of a Lost Space Shuttle and Her Crew*. New York: Arcade, 2020, p. 42.
2. Leinbach and Ward, *Bringing Columbia Home,* 45.
3. Leinbach and Ward, *Bringing Columbia Home,* 43–44.
4. Byron Starr, *Finding Heroes: The Search for Columbia's Astronauts*. Vancouver, BC: Liaison Press, 2006, pp. 9–10 (Kindle ed.).
5. Leinbach and Ward, *Bringing Columbia Home,* 43–44.
6. Leinbach and Ward, *Bringing Columbia Home,* 43–44.
7. William Langewiesche, "*Columbia*'s Last Flight," *Atlantic*, November 2003, https://www.theatlantic.com/magazine/archive/2003/11/columbias-last-flight/304204.
8. Leinbach and Ward, *Bringing Columbia Home,* 44.
9. Author interview with Fred Gregory, May 21, 2018.
10. Author interview with Fred Gregory, October 23, 2020, and March 3, 2021.
11. "Frederick D. Gregory," *NASA History,* https://history.nasa.gov/gregory.htm.
12. Author interview with Fred Gregory, March 3, 2021.
13. "STS-107 Mission Archives," *NASA Space Shuttle Missions,* https://www.nasa.gov/mission_pages/shuttle/shuttlemissions/archives/sts-107.html.
14. Harold W. Gehman, *Columbia Accident Investigation Board Report,* vol. 1, United States, *Columbia* Accident Investigation Board, 2003, 27, http://s3.amazonaws.com/akamai.netstorage/anon.nasa-global/CAIB/CAIB_lowres_chapter2.pdf.
15. "Kalpana Chawla Biographical Data," NASA, https://www.nasa.gov/sites/default/files/atoms/files/chawla_kalpana.pdf. Also, "NASA's African-American Astronauts," https://www.nasa.gov/sites/default/files/atoms/files/african_american_astronauts_fs.pdf.

16. "Laurel Blair Salton Clark Biographical Data," NASA, https://history.nasa.gov/columbia/Troxell/Columbia%20Web%20Site/Biographies/Crew%20Profile%20.
17. Leinbach and Ward, *Bringing Columbia Home,* 23.
18. Leinbach and Ward, *Bringing Columbia Home,* 26.
19. Leinbach and Ward, *Bringing Columbia Home,* 28.
20. Leinbach and Ward, *Bringing Columbia Home,* 25–26.
21. Leinbach and Ward, *Bringing Columbia Home,* 27.
22. R. Jeffrey Smith, "Mistakes of NASA Toted Up," *Washington Post,* July 13, 2003, https://www.washingtonpost.com/archive/politics/2003/07/13/mistakes-of-nasa-toted-up/8676a83f-c795-468f-8e6b-f65cb8cc906d.
23. Leinbach and Ward, *Bringing Columbia Home,* 27.
24. Leinbach and Ward, *Bringing Columbia Home,* 29.
25. Remarks by the President on the Loss of Space Shuttle *Columbia*, NASA, https://history.nasa.gov/columbia/Troxell/Columbia Web Site/Documents/Executive Branch/President Bush/president1.html.
26. Author interview with Fred Gregory, May 6, 2021.
27. Langewiesche, "*Columbia*'s Last Flight." Also, Sean O'Keefe transcript, NASA Headquarters Oral History Project, September 23, 2014, https://historycollection.jsc.nasa.gov/JSCHistoryPortal/history/oral_histories/NASA_HQ/Administrators/OKeefeS/OKeefeS_9-23-14.htm.
28. Author interview with Fred Gregory, May 6, 2021.
29. *Columbia Accident Investigation Board Report*, vol. 1, United States, *Columbia* Accident Investigation Board, 2003, pp. 44–45, http://s3.amazonaws.com/akamai.netstorage/anon.nasa-global/CAIB/CAIB_lowres_full.pdf.
30. Leinbach and Ward, *Bringing Columbia Home,* 79.
31. Author interview with Fred Gregory, May 6, 2021.
32. Leinbach and Ward, *Bringing Columbia Home,* 67–69.
33. Leinbach and Ward, *Bringing Columbia Home,* 67.
34. Leinbach and Ward, *Bringing Columbia Home,* 67.
35. Leinbach and Ward, *Bringing Columbia Home,* 68.
36. Langewiesche, "*Columbia*'s Last Flight."
37. Leinbach and Ward, *Bringing Columbia Home,* 73.
38. According to Leinbach and Ward, "Out of respect for the crew's families, NASA has never released details about the identity, location, or condition of any crew member's remains during the recovery." Leinbach and Ward, *Bringing Columbia Home,* 73.
39. Leinbach and Ward, *Bringing Columbia Home,* 72–73.
40. Leinbach and Ward, *Bringing Columbia Home,* 98–99.
41. Leinbach and Ward, *Bringing Columbia Home,* 185.
42. Leinbach and Ward, *Bringing Columbia Home,* 72. Also, Paul Keller, "Searching for and Recovering the Space Shuttle *Columbia*: Documenting the USDA Forest Service Role In This Unprecedented 'All-Risk' Incident, February 1 through May 10, 2003," National Wildfire Coordinating Group, p. 13, https://www.nwcg.gov/sites/default/files/wfldp/docs/searching-and-recovering-space-shuttle-columbia.pdf.
43. Leinbach and Ward, *Bringing Columbia Home,* 73–75.

44. Leinbach and Ward, *Bringing Columbia Home,* 75.
45. Remarks by the President at the Memorial Service in Honor of the STS-107 Crew, Space Shuttle *Columbia,* February 4, 2003, The White House, https://georgewbush-whitehouse.archives.gov/news/releases/2003/02/20030204-1.html.
46. Leinbach and Ward, *Bringing Columbia Home,* 154.
47. John Schwartz, "Video Shows Astronauts Before Shuttle Disintegrated," *New York Times,* March 1, 2003, https://www.nytimes.com/2003/03/01/us/video-shows-astronauts-before-shuttle-disintegrated.html.
48. Schwartz, "Video Shows Astronauts Before Shuttle Disintegrated."
49. Leinbach and Ward, *Bringing Columbia Home,* 178.
50. Leinbach and Ward, *Bringing Columbia Home,* 196.
51. Leinbach and Ward, *Bringing Columbia Home,* 172, 191. Also, "American Indian Fire Crews Help with the Recovery Effort of Space Shuttle *Columbia,*" February 7, 2003, Office of the Assistant Secretary—Indian Affairs, US Department of the Interior, https://www.doi.gov/sites/default/files/archive/news/archive/03_News_Releases/030207b.htm. Also, Keller, "Searching For and Recovering the Space Shuttle *Columbia,*" 27.
52. Author interview with Fred Gregory, May 6, 2021.
53. Leinbach and Ward, *Bringing Columbia Home,* 171.
54. Leinbach and Ward, *Bringing Columbia Home,* 171.
55. Langewiesche, "*Columbia*'s Last Flight."
56. Author interview with Fred Gregory, March 3, 2021.
57. Leinbach and Ward, *Bringing Columbia Home,* 225.
58. Author interview with Fred Gregory, March 3, 2021.
59. Keller, "Searching For and Recovering the Space Shuttle *Columbia,*" 36–39.
60. Leinbach and Ward, *Bringing Columbia Home,* 105.
61. Leinbach and Ward, *Bringing Columbia Home,* 258.
62. "Appendix A: The Investigation," *Columbia Accident Investigation Board Report,* vol. 1, p. 232, https://s3.amazonaws.com/akamai.netstorage/anon.nasa-global/CAIB/CAIB_lowres_full.pdf.
63. Langewiesche, "*Columbia*'s Last Flight." Also, "Appendix A: The Investigation," *Columbia Accident Investigation Board Report,* vol. 1, p. 232, https://s3.amazonaws.com/akamai.netstorage/anon.nasa-global/CAIB/CAIB_lowres_full.pdf.
64. "The Task Force," *Washington Post,* March 14, 2003, https://www.washingtonpost.com/archive/politics/2003/03/14/the-task-force/c32c680e-bf83-476a-9f29-5518f2f66769.
65. Sherr, *Sally Ride,* 279.
66. Sherr, *Sally Ride,* 279.
67. Melissa Harris, "Science Festival to Go On," *Orlando Sentinel,* February 2, 2003, https://www.newspapers.com/image/269637744/. Also, "America Begins Moving On," Associated Press, *South Florida Sun Sentinel,* February 3, 2003, https://www.newspapers.com/image/284711259/.
68. Harris, "Science Festival to Go On."
69. Harris, "Science Festival to Go On."
70. Sherr, *Sally Ride,* 280.
71. Sherr, *Sally Ride,* 280.

72. Sherr, *Sally Ride,* 280.
73. Sherr, *Sally Ride,* 281.
74. Leinbach and Ward, *Bringing Columbia Home,* 63.
75. Sally K. Ride Papers, "Ride's *Columbia* Investigation Notes," p. 1 of 12, image 7, https://edan.si.edu/slideshow/viewer/?eadrefid=NASM.2014.0025_ref259.
76. Langewiesche, "*Columbia*'s Last Flight."
77. "STS-27R OV-104 Orbiter TPS Damage Review Team," vol. 1, John W. Thomas, NASA, February 1, 1989, https://ntrs.nasa.gov/citations/19890010807.
78. Michael Cabbage and William Harwood, *Comm Check . . . : The Final Flight of Shuttle Columbia*. Free Press, 2004, 58.
79. *Columbia Accident Investigation Board Report,* vol. 1, 53.
80. *Columbia Accident Investigation Board Report,* vol. 1, 122.
81. *Columbia Accident Investigation Board Report,* vol. 1, 102, 122.
82. Cabbage and Harwood, *Comm Check,* 59.
83. Leinback and Ward, *Bringing Columbia Home,* 265.
84. Leinback and Ward, *Bringing Columbia Home,* 28.
85. Ham's quote is taken from a meeting five days after *Columbia*'s launch. Eric Pianin and R. Jeffrey Smith, "Shuttle Flight's Chief Takes Responsibility on Decision," *Washington Post,* July 23, 2003, https://www.washingtonpost.com/archive/politics/2003/07/23/shuttle-flights-chief-takes-responsibility-on-decision/e79d0b3d-60ce-42b4-9021-ee23a2e9a142.
86. Langewiesche, "*Columbia*'s Last Flight."
87. *Columbia Accident Investigation Board Report,* vol. 1, 57.
88. Langewiesche, "*Columbia*'s Last Flight."
89. *Columbia Accident Investigation Board Report,* vol. 1, 140, 150–53.
90. *Columbia Accident Investigation Board Report,* vol. 1, 143.
91. *Columbia Accident Investigation Board Report,* vol. 1, 143.
92. Langewiesche, "*Columbia*'s Last Flight."
93. Langewiesche, "*Columbia*'s Last Flight."
94. Langewiesche, "*Columbia*'s Last Flight."
95. Sherr, *Sally Ride,* 281.
96. John Schwartz with Matthew L. Wald, "Echoes of *Challenger*: Shuttle Panel Considers Longstanding Flaws in NASA's System," *New York Times,* April 13, 2003, https://www.nytimes.com/2003/04/13/us/echoes-of-challenger-shuttle-panel-considers-longstanding-flaws-in-nasa-s-system.html.
97. Sherr, *Sally Ride,* 282.
98. Langewiesche, "*Columbia*'s Last Flight."
99. Langewiesche, "*Columbia*'s Last Flight."
100. Author interview with Anna Fisher, May 4, 2021.
101. Author interview with Anna Fisher, May 4, 2021.
102. Author interview with Anna Fisher, July 30, 2020.
103. Author interview with Anna Fisher, January 21, 2022.
104. Langewiesche, "*Columbia*'s Last Flight."
105. Langewiesche, "*Columbia*'s Last Flight."
106. Author interview with Anna Fisher, July 30, 2020.
107. Sherr, *Sally Ride,* 282–83.
108. Sherr, *Sally Ride,* 283.
109. Sherr, *Sally Ride,* 283.

110. Todd Halvorson, "Daring Rescue May Have Saved *Columbia* Crew," *Florida Today,* May 21, 2003, http://www.collectspace.com/ubb/Forum23/HTML/000560.html.
111. Halvorson, "Daring Rescue May Have Saved *Columbia* Crew."
112. Sherr, *Sally Ride,* 284.
113. Sherr, *Sally Ride,* 284.
114. Gail Gibson, "A Highly Successful Mission Until the Last Moments," *Baltimore Sun* published in *South Florida Sun Sentinel,* February 2, 2003, https://www.sun-sentinel.com/news/bal-te.flight02feb02-story.html.
115. Marcia Dunn, "NASA Remembers its Losses and Lessons," NBC News, January 28, 2004, https://www.nbcnews.com/id/wbna4088037.
116. "Excerpts from Report of the *Columbia* Accident Investigation Board," August 27, 2003, https://www.nytimes.com/2003/08/27/national/nationalspecial/excerpts-from-report-of-the-columbia-accident.html. Also, *Columbia Accident Investigation Board Report,* vol. 1, 125, 138.
117. Ralph Vartabedian, "3 Shuttle Managers Reassigned," *Los Angeles Times,* July 3, 2003, https://www.latimes.com/archives/la-xpm-2003-jul-03-na-shuttle3-story.html.
118. Claudia Dreifus, "A Conversation with Sally Ride; Painful Questions from an Ex-Astronaut," *New York Times,* August 26, 2003, https://www.nytimes.com/2003/08/26/science/a-conversation-with-sally-ride-painful-questions-from-an-ex-astronaut.html.
119. *Columbia Accident Investigation Board Report,* vol. 1, 115.
120. Author interview with Fred Gregory, May 6, 2021.
121. *Columbia Accident Investigation Board Report,* vol. 1, 209–211.
122. "President Bush Delivers Remarks on US Space Policy," *NASA Facts,* January 14, 2004, https://www.nasa.gov/pdf/54868main_bush_trans.pdf.
123. "President Bush Delivers Remarks on US Space Policy."
124. Leinbach and Ward, *Bringing Columbia Home,* 202. Also, Keller, "Searching for and Recovering the Space Shuttle Columbia," 40.
125. Leinbach and Ward, *Bringing Columbia Home,* 202.
126. Leinbach and Ward, *Bringing Columbia Home,* 201.
127. Leinbach and Ward, *Bringing Columbia Home,* 200.
128. Leinbach and Ward, *Bringing Columbia Home,* 200.
129. Leinbach and Ward, *Bringing Columbia Home,* 203–04.
130. Leinbach and Ward, *Bringing Columbia Home,* 200.
131. William Harwood, "STS-114 Mission Archive (Final)," CBS News/Kennedy Space Center, August 11, 2005, https://www.cbsnews.com/network/news/space/STS-114_Archive.html.
132. Leinbach and Ward, *Bringing Columbia Home,* 204.
133. "SPACE FLIGHT 2005—International Space Station," NASA, https://www.nasa.gov/directorates/heo/reports/2005/iss.html.
134. "International Space Station," NASA, p. EC 5-2, https://www.nasa.gov/pdf/55411main_28 ISS.pdf.

Chapter 27: God Bless

1. Author interview with Anna Fisher, June 8, 2021.
2. Author interview with Fred Gregory, May 6, 2021.

3. Author interview with Rhea Seddon, September 3, 2020.
4. "Margaret Rhea Seddon," 100 Distinguished Alumni, *Tennessee Alumnus,* https://our.tennessee.edu/100-distinguished-alumni/margaret-rhea-seddon/.
5. "Fair Treatment of Experienced Pilots Act," *FAA,* May 9, 2019, https://www.faa.gov/other_visit/aviation_industry/airline_operators/airline_safety/info/all_infos/media/age65_qa.pdf.
6. Robin White, "The Man Who's Flown Everything," *Air & Space Magazine,* April 30, 2009, https://www.airspacemag.com/flight-today/the-man-whos-flown-everything-57719824/.
7. Author interview with Anna Fisher, July 30, 2020.
8. "NASA Selects New Astronauts for Future Space Exploration," NASA, https://www.nasa.gov/astronauts/ascans2009.html.
9. Author interview with Fred Gregory, June 21, 2019.
10. "Frederick D. Gregory Biographical Data," NASA, https://www.nasa.gov/sites/default/files/atoms/files/gregory_frederick.pdf.
11. Author interview with Fred Gregory, May 6, 2021.
12. "STS-135: The Final Voyage," NASA, July 27, 2011, https://www.nasa.gov/mission_pages/shuttle/shuttlemissions/sts135/launch/sts-135_mission-overview.html.
13. Kathryn Sullivan, "Why the Hubble Is Unlike Any Other Satellite in History," *MIT Technology Review,* November 19, 2019, https://www.technologyreview.com/2019/11/19/131905/why-the-hubble-is-unlike-any-other-satellite-in-history/.
14. "Galileo Mission Overview," *NASA Science Solar System Exploration,* https://solarsystem.nasa.gov/missions/galileo/overview.
15. "Building the International Space Station," European Space Agency, https://www.esa.int/Science_Exploration/Human_and_Robotic_Exploration/International_Space_Station/Building_the_International_Space_Station3. Also, Mark Garcia, "Space Station Spacewalks," NASA, July 21, 2022, https://www.nasa.gov/mission_pages/station/spacewalks/. Also, Robert Dempsey, *The International Space Station: Operating an Outpost in the New Frontier,* NASA, p. 10, https://www.nasa.gov/sites/default/files/atoms/files/iss-operating_an_outpost-tagged.pdf.
16. "Reference Guide to the International Space Station," NASA, September 2015, https://www.nasa.gov/sites/default/files/atoms/files/np-2015-05-022-jsc-iss-guide-2015-update-111015-508c.pdf.
17. "Astronaut Bob Crippen at STS-135 Final Shuttle Launch," https://www.youtube.com/watch?v=t5-NYHKqZOg.
18. "Astronaut Bob Crippen at STS-135 Final Shuttle Launch."
19. J.D. Gallop, "Climate Official Discusses Weather During Melbourne Visit," *Florida Today,* July 8, 2021, https://www.newspapers.com/image/360541507/.
20. Author interview with Kathy Sullivan, December 12, 2020.
21. Sullivan, NASA Oral History Project.
22. Sherr, *Sally Ride,* 264.
23. Sherr, *Sally Ride,* 295.
24. "Loving Sally Ride."
25. Sherr, *Sally Ride,* 298.
26. Sherr, *Sally Ride,* 299.

27. Sherr, *Sally Ride,* 286.
28. "Summary Report of the Review of US Human Space Flight Plans Committee," September 8, 2009, SpaceRef, http://www.spaceref.com/news/viewsr.html?pid=32327. Also, "Augustine Commission—DC—Sally Ride—3 of 6," https://www.youtube.com/watch?v=6RwqYIw_C-4.
29. Dennis Overbye, "NASA Panel Grapples with Cost of Space Plans," *New York Times,* August 12, 2009, https://www.nytimes.com/2009/08/13/science/space/13nasa.html?searchResultPosition=1.
30. Sherr, *Sally Ride,* 286. Also, "Summary Report of the Review of US Human Space Flight Plans Committee."
31. Sherr, *Sally Ride,* 287.
32. Sherr, *Sally Ride,* 301.
33. According to Sherr, *Sally Ride,* 302, the tumor had "involvement with the surrounding blood vessels" which suggests the "regional" classification on "Survival Rates for Pancreatic Cancer," American Cancer Society, https://www.cancer.org/cancer/pancreatic-cancer/detection-diagnosis-staging/survival-rates.html.
34. Sherr, *Sally Ride,* 309.
35. Sherr, *Sally Ride,* 310.
36. Sherr, *Sally Ride,* 313.
37. "Space Shuttle Era Facts," NASA, https://www.nasa.gov/pdf/566250main_2011.07.05 SHUTTLE ERA FACTS.pdf.
38. Houston, *Wheels Stop,* location 4506 (Kindle ed.).
39. Houston, *Wheels Stop,* 312.
40. Christ Gebhardt, "STS-135/ULF-7—The Final Flight's Timeline Takes Shape," NASASpaceflight.com, June 17, 2011, https://www.nasaspaceflight.com/2011/06/sts-135-final-flights-timeline-takes-shape.
41. Houston, *Wheels Stop,* location 4589 (Kindle ed.).
42. Cary Funk and Kim Parker, "1. Diversity in the STEM Workforce Varies Widely Across Jobs," Pew Research Center, January 9, 2018, https://www.pewresearch.org/social-trends/2018/01/09/diversity-in-the-stem-workforce-varies-widely-across-jobs.
43. *Black in Space: Breaking the Color Barrier,* Smithsonian, 50:11–50:17, https://www.youtube.com/watch?v=I7jJ8jEh608.
44. Author interview with Shannon Lucid, September 17, 2020.
45. "Record-Setting Female Astronaut Shannon Lucid Leaving NASA," Space.com, February 3, 2012, https://www.space.com/14457-nasa-astronaut-shannon-lucid-retires.html.
46. Author interview with Guy Bluford, December 5, 2020.
47. "Guion 'Guy' S. Bluford Jr.," National Aviation Hall of Fame, https://www.nationalaviation.org/enshrinee/guion- guy-s-bluford-jr/.
48. Author interview with George Abbey, August 28, 2018, and Cassutt, *The Astronaut Maker,* 417.
49. "Joyce Abbey," *SAIC,* https://www.saic.com/blogs/authors/joyce-abbey.
50. Cassutt, *The Astronaut Maker,* 418–19.
51. Author interview with George Abbey, November 1, 2019.
52. Cassutt, *The Astronaut Maker,* xix.
53. Author interview with Fred Gregory, June 21, 2019.

54. Kristin Fisher, *CBS,* July 9, 2011, https://archive.org/details/WUSA_20110709_120000_The_Early_Show/start/5760/end/5820.
55. Houston, *Wheels Stop,* location 4695 (Kindle ed.).
56. Cheryl L. Mansfield, "STS-135: The Final Voyage," NASA, July 27, 2011, https://www.nasa.gov/mission_pages/shuttle/shuttlemissions/sts135/launch/sts-135_mission-overview.html.
57. Sharon Gaudin, "Atlantis Blasts Off on Historic Last Mission," *Computerworld,* July 8, 2011, https://www.computerworld.com/article/2510071/atlantis-blasts-off-on-historic-last-mission.html. Also, Matthew Calamia, "Android Joins IPhone in Space," *Mobiledia,* July 8, 2011, https://web.archive.org/web/20110927173200/http://www.mobiledia.com/news/97297.html.
58. Author interview with Anna Fisher, April 19, 2021.
59. Fisher, NASA Oral History Project.
60. Houston, *Wheels Stop,* location 4718.
61. Lucid, "Shuttle-*Mir* Stories: Lucid on Pink Socks and Jello."
62. "STS-135 Mission Highlights," NASA, 38:35, https://www.youtube.com/watch?v=e7dISGGxLRk&t=2315s. Also, "Magnus Configures Raffaello for Ingress," NASA, July 11, 2011, https://images.nasa.gov/details-s135e007401.html.
63. "STS-135 Flight Day 4 Crew Wake Up Call," NASA, https://www.youtube.com/watch?v=ur-fPpy8sdQ."
64. "STS-135 Mission Highlights," NASA.
65. Robert X. Pearlman, "President Obama Reveals Astronauts' Secret Souvenir on Final Shuttle Mission," Space.com, July 15, 2011, https://www.space.com/12309-obama-space-astronauts-secret-american-flag.html.
66. "Flag Day 2020—One Small American Flag's Incredible Journey," *NASA History,* June 14, 2020, https://www.nasa.gov/feature/flag-day-2020-one-small-american-flag-s-incredible-journey.
67. "STS-135 Space Shuttle Atlantis—Flight Day 14 Wake Up—God Bless America," https://www.youtube.com/watch?v=zfGLEjsWZd0.
68. Cheryl L. Mansfield, "STS-135: The Final Voyage," NASA, July 27, 2011, https://www.nasa.gov/mission_pages/shuttle/shuttlemissions/sts135/launch/sts-135_mission-overview.html.
69. "15 Ways the International Space Station is Benefiting Earth," NASA, October 30, 2015, https://www.nasa.gov/mission_pages/station/research/news/15_ways_iss_benefits_earth.
70. Erin Winick, "20 Breakthroughs from 20 Years of Science aboard the International Space Station," NASA, October 26, 2020, https://www.nasa.gov/mission_pages/station/research/news/iss-20-years-20-breakthroughs.
71. Houston, *Wheels Stop,* location 4807. Also, author interview with Mike Fossum, July 27, 2018.
72. "Atlantis Makes Shuttle Program's Final Landing," NASA, July 21, 2011, https://www.nasa.gov/multimedia/podcasting/sts135landing.html.
73. Houston, *Wheels Stop,* location 4811.
74. Miriam Kramer, "NASA's Space Shuttles: Where Are They Now?" Space.com, July 4, 2013, www.space.com/21804-nasa-space-shuttles-where-are-they.html.
75. "NASA's Space Shuttle Museum Flights: Complete Coverage," Space.com,

September 20, 2012, https://www.space.com/15268-space-shuttle-museum-flights-coverage.html.

76. Tariq Malik, "Shuttle Endeavour Takes Off on Historic California Sightseeing Flight," Space.com, September 21, 2012, https://www.space.com/17713-shuttle-endeavour-begins-california-sightseeing-flight.html.
77. Merryl Azriel, "Endeavour Squeaks Down Los Angeles Streets, Reaches New Home," *Space Safety Magazine,* October 15, 2012, http://www.spacesafetymagazine.com/space-on-earth/everyday-life/endeavour-squeaks-los-angeles-streets-reaches-home.
78. "Space Shuttle Discovery Buzzes D.C. Monuments (Pictures)," *National Geographic,* April 18, 2012, https://www.nationalgeographic.com/science/article/120417-space-shuttle-discovery-washington-dc-science-nation-final-flight-smithsonian.
79. "Space Shuttle Discovery Flies to the Smithsonian," *Smithsonian,* April 17, 2012, https://www.si.edu/newsdesk/releases/space-shuttle-discovery-flies-smithsonian.
80. Author interview with Shannon Lucid, September 17, 2020.
81. Author interview with Shannon Lucid, September 17, 2020.
82. Brooke Boen, "NASA Space Launch System (SLS) Rocket," NASA, August 13, 2014, https://www.nasa.gov/sls/multimedia/gallery/sls-infographic3.html.
83. "Loving Sally Ride."
84. Sherr, *Sally Ride,* 306.
85. "Loving Sally Ride."
86. Sherr, *Sally Ride,* 311.
87. Sherr, *Sally Ride,* 313.
88. Sherr, *Sally Ride,* 311.
89. Tracy Baim, "The Late Sally Ride Also Gay Trailblazer," *Chicago Tribune,* July 26, 2012, https://www.newspapers.com/image/240636995/. Also, Associated Press, "Death of Sally Ride Sparks Debate About 'Coming Out' Posthumously," *Journal Gazette* (Mattoon, Illinois), July 26, 2012, https://www.newspapers.com/image/84593791/.
90. Marian Ely, "Letters to the Editor: Who Cares if Sally Ride Was Gay? She Was an Excellent Role Model," *Corvallis Gazette-Times,* July 27, 2012, https://www.newspapers.com/image/587882033/.
91. Seth Borenstein, "Sally Ride, First US Woman in Space, Dies," *Tribune* (Coshocton, Ohio), July 24, 2012, https://www.newspapers.com/image/360878943/.
92. Rich Abdill, "Sally Ride's Sister, on the Wikipedia Debate over Homosexuality: 'Sally Hated Labels of Every Kind,'" *Broward Palm Beach New Times,* July 25, 2012, https://www.browardpalmbeach.com/news/sally-rides-sister-on-the-wikipedia-debate-over-homosexuality-sally-hated-labels-of-every-kind-6473119.
93. Seddon, *Go for Orbit,* 311.
94. "Grove of Oaks Honors NASA's Fallen Astronauts / 7 more to be planted at Johnson Space Center memorial," *SF Gate,* March 16, 2003, https://www.sfgate.com/news/article/Grove-of-oaks-honors-NASA-s-fallen-astronauts-7-2628783.php.
95. "NASA Family Honors Dr. Sally Ride on September 18 with Tree-Planting Ceremony," NASA, https://www.youtube.com/watch?v=67E3kxAA4k4.

96. Cassutt, *The Astronaut Maker,* 418–20.
97. "Memorial Trees," NASA, https://www.nasa.gov/Starport/ProgramsAndEvents/MemorialTrees.
98. Author interview with June Scobee, June 10, 2021.
99. Email to the author from Lorna Onizuka.
100. "Guion S. Bluford Jr. Biographical Data," NASA, https://www.nasa.gov/sites/default/files/atoms/files/bluford_guion.pdf.
101. Heather Murphy, "First American Woman to Walk in Space Reaches Deepest Spot in the Ocean," *New York Times,* June 8, 2020, https://www.nytimes.com/2020/06/08/science/challenger-deep-kathy-sullivan-astronaut.html.
102. Kelly-Leigh Cooper, "Kathy Sullivan: The Woman Who's Made History in Sea and Space," *BBC News,* June 14, 2020, https://www.bbc.com/news/world-us-canada-53008948.
103. Author interview with Rhea Seddon, November 20,2020.
104. "Astronaut Profile Dr. Anna L. Fisher," *Space Center Houston,* https://spacecenter.org/wp-content/uploads/2019/04/Anna-Fisher-Bio-Card-web.pdf.
105. Author interview with Anna Fisher, June 22, 2020.
106. "Jacket, In-Flight Suit, Shuttle, Sally Ride, STS-7," Smithsonian Air and Space Museum, https://airandspace.si.edu/collection-objects/jacket-flight-suit-shuttle-sally-ride-sts-7/nasm_A19830241000.
107. Author interview with Susan Okie, April 8, 2022.
108. Author interview with Fred Gregory, May 6, 2021.
109. Author interview with Fred Gregory, May 6, 2021.

Photo Credits

Grateful acknowledgment is made to the following for the use of the photographs that appear in the art insert and appendix:

Art Insert

p. 1, top and middle: NASA
p. 1, bottom: Anna Fisher
p. 2, top: Getty Images
p. 2, middle and bottom: NASA
p. 3, all: NASA
p. 4, top: The National Archives
p. 4, bottom: NASA
p. 5, top: Rhea Seddon
p. 5, bottom: NASA
p. 6, top: Image used by permission of *Ms. Magazine*, © 1983
p. 6, middle: NASA
p. 6, bottom: Getty Images
p. 7, all: NASA
p. 8, all: NASA
p. 9, all: NASA
p. 10, top: Getty Images
p. 10, middle and bottom: NASA
p. 11, top: Anna Fisher
p. 11, bottom: NASA
p. 12, all: NASA
p. 13, top: The Ronald Reagan Presidential Library
p. 13, bottom: The National Archives
p. 14, top: AP Images
p. 14, middle: The National Archives
p. 14, bottom: NASA

p. 15, top and middle: NASA
p. 15, bottom: Getty Images
p. 16, top: NASA
p. 16, bottom: Anna Fisher

Appendix

p. 407: NASA

Index

About the Author

Meredith Bagby is a nonfiction writer and a partner at Big Swing Productions, a film and TV production company. Her previous books include *We've Got Issues, Rational Exuberance,* and *The Annual Report of the United States of America*. Bagby was a senior film development executive at DreamWorks SKG, a political reporter and producer for CNN, and a teaching fellow at Harvard's Institute of Politics. Her education includes Columbia Law School and Harvard College.